Ecology and Biodiversity of Benthos

Ecology and Biodiversity of Benthos

Edited by

PRINCE S. GODSON
Assistant Professor, Centre for Marine Science and Technology, Manonmaniam Sundaranar University, Marina Campus, Rajakkamangalam, Tamil Nadu, India

SALOM GNANA THANGA VINCENT
Professor, Department of Environmental Sciences, University of Kerala, Kariavattom Campus, Thiruvananthapuram, Kerala, India

S. KRISHNAKUMAR
Assistant Professor, Department of Geology, Malankara Catholic College, Mariagiri, Kaliakkavilai, Tamil Nadu, India

ELSEVIER

Elsevier
Radarweg 29, PO Box 211, 1000 AE Amsterdam, Netherlands
The Boulevard, Langford Lane, Kidlington, Oxford OX5 1GB, United Kingdom
50 Hampshire Street, 5th Floor, Cambridge, MA 02139, United States

ISBN: 978-0-12-821161-8

For Information on all Elsevier publications
visit our website at https://www.elsevier.com/books-and-journals

Publisher: Candice Janco
Acquisitions Editor: Louisa Munro
Editorial Project Manager: Pat Gonzalez
Production Project Manager: Kumar Anbazhagan
Cover Designer: Christian J. Bilbow

Typeset by MPS Limited, Chennai, India

Contents

List of contributors

K.U. Abdul Jaleel
CSIR-National Institute of Oceanography, Regional Centre, Kochi, Kerala, India

D.B. Arya
Department of Environmental Sciences, University of Kerala, Thiruvananthapuram, Kerala, India

S. Bijoy Nandan
Department of Marine Biology, Microbiology and Biochemistry, School of Marine Sciences, Cochin University of Science and Technology, Kochi, India

A. Biju Kumar
Department of Aquatic Biology & Fisheries, University of Kerala, Kariyavattom, Thiruvananthapuram, Kerala, India

N. Chandrasekar
Department of Geology, Francis Xavier Engineering College, Tirunelveli, Tamil Nadu, India

M.M. El-Sherbiny
Department of Marine Biology, Faculty of Marine Sciences, King Abdulaziz University, Jeddah, Saudi Arabia; Department of Marine Sciences, Faculty of Science, Suez Canal University, Ismailia, Egypt

Prince S. Godson
Centre for Marine Science and Technology, Manonmaniam Sundaranar University, Tamil Nadu, India

Aiswarya Gopal
Centre for Marine Living Resources & Ecology, Ministry of Earth Sciences, Kochi, Kerala, India

S.M. Hussain
Department of Geology, School of Earth and Atmospheric Sciences, University of Madras, Chennai, India

P.R. Jayachandran
Department of Marine Biology, Microbiology and Biochemistry, School of Marine Sciences, Cochin University of Science and Technology, Kochi, India

M. Jima
Department of Marine Biology, Microbiology and Biochemistry, School of Marine Sciences, Cochin University of Science and Technology, Kochi, India

Thadickal V. Joydas
Center for Environment and Marine Studies, Research Institute, King Fahd University of Petroleum and Minerals (KFUPM), Dhahran, Saudi Arabia

Ajmal Khan
Center for Environment and Marine Studies, Research Institute, King Fahd University of Petroleum and Minerals (KFUPM), Dhahran, Saudi Arabia; Centre of Advanced Study in Marine Biology, Annamalai University, Parangipettai, Tamil Nadu, India

S. Krishnakumar
Department of Geology, Malankara Catholic College, Mariyagiri, Tamil Nadu, India

Seerangan Manokaran
Center for Environment and Marine Studies, Research Institute, King Fahd University of Petroleum and Minerals (KFUPM), Dhahran, Saudi Arabia

Usha V. Parameswaran
Centre for Marine Living Resources & Ecology, Ministry of Earth Sciences, Kochi, Kerala, India

J. Philomina
Department of Marine Biology, Microbiology and Biochemistry, School of Marine Sciences, Cochin University of Science and Technology, Kochi, India

K. Radhakrishnan
Department of Geology, School of Earth and Atmospheric Sciences, University of Madras, Chennai, India

R. Ravinesh
Coastal and Marine Ecology Division, Gujarat Institute of Desert Ecology, Bhuj (Kachchh), Gujarat, India; Department of Aquatic Biology & Fisheries, University of Kerala, Kariyavattom, Thiruvananthapuram, Kerala, India

S. Satheesh
Department of Marine Biology, Faculty of Marine Sciences, King Abdulaziz University, Jeddah, Saudi Arabia

J.V. Sinu
P.G. Department of Zoology, Christian College, Chengannur, Kerala, India

V. Sivapriya
Department of Geology, School of Earth and Atmospheric Sciences, University of Madras, Chennai, India

Salom Gnana Thanga Vincent
Department of Environmental Sciences, University of Kerala, Thiruvananthapuram, Kerala, India

N.K. Vishnudattan
Department of Marine Biology, Microbiology and Biochemistry, School of Marine Sciences, Cochin University of Science and Technology, Kochi, India

Angelo Mark P. Walag
Department of Science Education, College of Science and Technology Education, University of Science and Technology of Southern Philippines, Cagayan de Oro, Philippines; Center for Advanced Plant Science and Phytotechnologies, University of Science and Technology of Southern Philippines, Cagayan de Oro City, Philippines

Preface

Ecology and Biodiversity of Benthos creates an understanding of the general characteristic features of marine benthos, which play a significant role in the coastal ecosystem functioning, as communities in the base level of the food chain. Ecosystem services develop shared approaches and scientific methods to achieve knowledge-based sustainable planning and management of the benthic ecosystem. This book gives hands-on multidisciplinary reflections by various researchers involved in the coastal ecosystem service community. The book would be a helpful tool for those working on marine and coastal environments as well as for any person who requires basic knowledge of benthic ecology. The book targets newcomers to the field, isolated workers who have no ready access to the literature, and those outside the topic whose work led them into benthic studies.

As editors, we are indebted to all the authors involved in the chapters for their tremendous cooperation and enthusiasm showed toward the successful completion of the work that began in mid-2019. Each author, an authority in their speciality, has explicitly been invited to write a review chapter. Therefore the challenges are much more significant than in the case of typical contributed articles. Despite all the hurdles caused by the *Covid-19* pandemic, the authors have finished their tasks efficaciously. The authors have also given us thought-provoking ideas that have helped in the furtherance of the general structure of the book. We thank *Elsevier Science* for accepting to publish this book. We also extend our thanks to the publishing editors—Louisa Munro, who first believed that our proposal is realizable, and Ali Afzal-Khan, Pat Gonzalez, and Kumar for their direction and encouragement throughout the course of the work and for handling the publishing processes meticulously.

Our hope is that the book will provide instructional material for students and researchers in benthic ecology and serve as a general reference volume for those directly involved in marine benthic research. The book has widespread scientific coverage providing to the requirements of marine benthologists, marine ecologists, aquatic researchers, scientists, teachers, and research scholars. In addition, it also serves as a reference material for postgraduate/undergraduate students. Policymakers can make use of the book for making laws and legislations. Focusing on various beach types of temperate as well as tropical countries, and addressing issues from the behavioural and physiological adaptations of the benthic environment to study the impact of pollution and anthropogenic activities, this book would aid as a standard reference for researchers in the field of marine biology.

Prince S. Godson
Salom Gnana Thanga Vincent
S. Krishnakumar

CHAPTER 1

Understanding the World of benthos: an introduction to benthology

Angelo Mark P. Walag[1,2]
[1]Department of Science Education, College of Science and Technology Education, University of Science and Technology of Southern Philippines, Cagayan de Oro, Philippines
[2]Center for Advanced Plant Science and Phytotechnologies, University of Science and Technology of Southern Philippines, Cagayan de Oro City, Philippines

1.1 Introduction

The term benthos is derived from a Greek word that means depth. This term is used to classify organisms, both flora and fauna, occurring on, in, or close to this region of the aquatic ecosystem, including lakes, streams, rivers, and the sea (Reynolds, 2013). The benthic zone presents a greater variety of living organisms and physical conditions, which differ in depth, light, temperature, degree of water immersion, and substrate (Lalli & Parsons, 1997). This region is a rich source of both marine ecological productivity and economic potential, as derived by the products discovered from organisms inhabiting this region (Walag, 2017). More so, the organisms in this region provide valuable ecosystem goods and services (Gutow et al., 2014; Walag & del Rosario, 2018). In freshwater and estuarine ecosystems, this region of incredible biological activity, as shown in the oxygen curve for lakes and ponds (Smith & Smith, 2003). On the other hand, benthic ecology is the study of the relationships of benthos and their unique and diverse environment. This ecology is essential to the entire aquatic system's functioning, providing myriad purposes (Walag & Canencia, 2016). Even with their innate importance, pollution from various sources has shown to have affected their biology, abundance, diversity, and ecology (Llacuna et al., 2016; Trannum et al., 2019; Trevizani et al., 2016; van Hoey et al., 2010; Walag & del Rosario, 2020; Walag et al., 2018, 2019).

The benthos is primarily dependent on whether the water is moving or static, shallow or deep, and the quality of the substratum. Hard and rocky substrates provide a home to sessile organisms like barnacles and

Ecology and Biodiversity of Benthos
DOI: https://doi.org/10.1016/B978-0-12-821161-8.00002-7

mussels due to their dependence on hard and rough surfaces. The crevices and depressions provide nektonic organisms a source of food and refuge against predators. Muddy, clay, and sandy substrates (soft-bottom substrates), on the other hand, provide both food and habitat to burrowing animals (Lalli & Parsons, 1997). The substratum quality and fineness reflect, in part, the nature and abundance of the parent material. Its ability to stay in place or remain approximately where it settled is highly dependent on the level of scouring and the frequency of wave action, and the shear forces it is subjected to. Of these various environmental characteristics, water depth is commonly highlighted as the most critical ecological correlate of the structure of benthic communities (Pilditch et al., 2015).

The benthic substrate varies as a function of the depth of the ocean and the benthic zone's relationship to neighboring land areas and continental shelves (Smith & Smith, 2003). The benthic zone near the coastal area derives its sediments from weathering and erosion of land areas along with organic matter from marine life. The sediments of the deep-water benthic zone, on the other hand, vary based on the types of marine organisms associated with it.

In the deeper portions of this region, no photosynthesis occurs; thus the bottom community is strictly heterotrophic. Despite this darkness and depth, these communities support rich biodiversity. On the other hand, in shallow benthic regions, the most reported species are polychaete worms and pericarid crustaceans. One hypothesis about this rich biodiversity in deep benthic zones is the absence or lack of widespread disturbance or environmental extremes, like the temperature is almost constant, and storms and typhoons do not stir this region.

As in the pelagic region, the benthic region is also divided by vertical gradients of temperature, light, and salinity, which determines distinctly the different groups of organisms it can support. The seafloor's ecological subdivisions showing depth, topography, and vertical gradients may have clear boundaries from each other; others are more arbitrary, presenting a unique habitat for a distinct group of species. Fig. 1.1 shows the rich biodiversity in benthic ecosystems.

1.1.1 Infauna

The benthic infauna comprises a wide array of taxa and varying abundance and sizes, although large species are rarely recorded (Powell & Mann, 2016). Benthic infauna is the assemblage of organisms that live within the sediments

Figure 1.1 The coral reef ecosystem is dominated mainly by hard corals in Siquijor Island, Philippines.

of the seafloor. These organisms may live wholly or partly within the substrate. These animals range from gastropods, amphipods, polychaetes, and other invertebrates. Infaunal species often dominate communities of a soft substrate and are most abundant in subtidal regions. Although uncommon and few, infaunal species are also found to inhabit hard substrate communities as well. A good example of this is rock-boring clams. The quality and composition of benthic sediment have even been considered the critical determinants of the marine benthic infaunal community (Somerfield et al., 2019). Several current studies on infaunal species are summarized in Table 1.1.

Infaunal ecosystems play an essential role in the cycling of organic matter within seagrass meadows. Any changes in the organic matter input from the benthic environment make this ecosystem susceptible (Johnson et al., 2020). The reason for this is that most infaunal species are detritivores. Fig. 1.2 shows an infaunal species partly burrowed in a seagrass bed. When seagrass and algal biomass are removed, the primary production within this ecosystem will decrease due to this reduction. Several small grazers and megagrazers consume large amounts of seagrass, which in turn causes sizeable physical disturbance in the environment and may reduce the infaunal abundance. More so, a change in the infaunal detritivore community within a meadow could also affect the cycling of nutrients and the amount of available organic matter (Hemminga & Duarte, 2000; MacKenzie et al., 2015). The loss of seagrass also affected the benthic infaunal community structure and the secondary production of this system (Seitz & Lewis, 2018).

Table 1.1 Studies on marine benthic infauna.

Location	Goal	References
Northeast Chukchi Sea, Alaska	Determined abundance, distribution, biomass, diversity of benthic infauna, and their relationship with marine mammals and environmental variables	Schonberg et al. (2014)
Krabbekreek mudflat, Netherlands	Explored the effects of marine vegetation structural complexity on abundance, species richness, size, and body type of infaunal species	González-Ortiz et al. (2016)
Frænfjor, Møre and Romsdal County, Norway	Studied the effects of submarine mine tailings on benthic fauna	Trannum et al. (2019)
Southern Namibia and Namqualand	Characterized the infaunal benthic communities from the inner shelf of Southwestern Africa	Steffani et al. (2015)
Mai Po Tidal Marsh, Deep Bay, Pearl River Estuary, China	Modeled the temporal dynamics of benthic fauna biomass with environmental factors	Yang et al. (2018)
Northeast of Colonsay, Scotland	Assessed the direct impact of scallop dredging on the sediment bed and inhabiting infauna	O'Neill et al. (2013)
York River, Chesapeake Bay, United States	Examined how predation, recruitment, and food availability affect the soft-bottom benthic infauna	Seitz (2011)
Northern Gulf of Mexico	Examined the effects of oil spill on deep-sea soft-bottom infauna	Reuscher et al. (2017)

The benthic infauna can be used as a biological indicator in evaluating environmental stresses as any impacts due to the alteration of the benthos may directly or indirectly affect them (Ali et al., 2018). The macroinfauna, particularly, is also considered helpful in monitoring the effects of pollution since it can be sampled quantitatively. It is regarded as part of the indicators, together with other biological and physical elements of quality, of environmental status in the European Water Framework Directive (van

Figure 1.2 A garden eel (Family: Congridae) partly burrowed in a seagrass bed dominated by *Halophila ovalis* in Caticugan, Siquijor Island, Philippines.

Hoey et al., 2010). In a particular study, the total abundance of infauna species increased in the most impacted area of mine tailings due to the dominance of opportunistic species (Trannum et al., 2019).

Benthic communities, particularly the infaunal community, play an essential role in organic matter remineralization. In one study, the diversity of benthic infauna, both species richness and functional richness, have essential roles in controlling benthic flux rates and organic matter remineralization (Belley & Snelgrove, 2017). More so, their findings suggest that more infaunal species per number of individuals explained a significant portion of the phosphate and silicate fluxes in enriched incubations. At the same time, this benthic community also plays an essential role in recycling fresh phytodetritus. Thus the accelerated anthropogenic impacts toward infaunal diversity have potentially adverse consequences on the ecosystem functioning of the continental shelf sediments. Activities like bottom trawling alter the functional composition of benthic communities and result in lower matter remineralization (Olsgard et al., 2008). In addition, not only interspecific diversity affects ecosystem functioning. In one study, they found out that rapid changes toward the intraspecific diversity of these organisms alter communities and ultimately the functioning of ecosystems (Nicastro et al., 2020). Furthermore, in one study, benthic infaunal diversity increased ammonium concentration and the total inorganic nitrogen consumption rate (Satoh et al., 2007). In addition, infaunal species stimulate nitrogen cycling by extending the oxic-anoxic interfaces with more nutrients through burrow ventilation, particle reworking, irrigation, and excretion (Chen & Gu, 2017). These studies highlight how infaunal diversity impact different ecological characteristics of the benthos.

The dynamics of benthic infauna were also noted to be affected by land reclamation. In a study in Mai Po Tidal Marsh in China, a significant decrease in seasonal biomass variation was noted after land reclamation (Yang et al., 2018). This highlights the substantial ecological disturbance land reclamations can induce to infaunal species. More so, tracking sediment mobilization is vital since this affects the release of nutrients, suspension of phytoplankton cyst and copepod eggs, and the diversity of benthic infauna (Brown et al., 2013). Besides, dredging and sand placement often result in the mortality of the benthic infaunal community (Rosov et al., 2016).

Not just the quality of water and sediments, but marine plants also significantly affect benthic infauna diversity. In an experimental study, above-ground and below-ground vegetation mimics were found to increase the density and the number of macrobenthic infaunal species (González-Ortiz et al., 2016). Also, infaunal species select habitats by migrating laterally from highly unstable sediment to more protected and stable ones, such as those where plants thrive (González-Ortiz et al., 2014). The grazing of sea turtles on seagrass meadows was also found to affect the abundance and community composition of benthic infauna but not their diversity (Johnson et al., 2020). This poses the potential important ecosystem services provided by these kinds of habitats to benthic infaunal species.

Infaunal communities were also found to be resilient over changes in season. An infaunal community in the intertidal mudflats of the Bay of Fundy was observed in terms of community change and the factors that drive this change (Gerwing et al., 2015). They noted that the shift in infaunal community structure during winter seasons was small and not consistent. A weak correlation was also noted between the taxa density or community structure and winter stressors. On the other hand, higher density and diversity were noted in macrobenthic infaunal communities in premonsoon seasons (Kundu et al., 2010). The best environmental variable that correlates with infaunal species abundance is water depth (Schonberg et al., 2014), although, in another report, bottom water temperature, dissolved oxygen, and salinity were highly correlated (Cusson et al., 2007; Kundu et al., 2010). More so, sediment type and composition and the abundance of predators also affect the density of infaunal communities (Lawless & Seitz, 2014). This suggests that benthic distribution and diversity are related to a combination of abiotic and biotic factors.

Since benthic infauna is less mobile than other benthic organisms, they are often stressed with low dissolved oxygen. This hypoxic environment often alters the benthic community composition, diversity, abundance, and biomass (Rabalais & Baustian, 2020). In this kind of environment, small opportunistic species specializing in surface deposit-feeding remain. These organisms do not burrow deeper than the upper few centimeters since the redox discontinuity layer is near the sediment surface. In this scenario, there is a high possibility of efflux of H_2S and NH_4^+ and PO_4^{-3} nutrients in the hypoxic water column (Rabalais, 2019). This accumulation of efflux and nutrients result in phytoplankton biomass accumulation, which indirectly influences hypoxia conditions. This happens since the phytoplankton accumulation leads to carbon flux and further decline of oxygen concentration (Rabalais & Baustian, 2020). Thus resistant polychaete species are favored in this environment due to their fast recruitment and/or low-oxygen environment tolerance.

1.1.2 Epifauna

The benthic epifauna is animals found to be living on or attached to the seafloor. The bulk, about 80%, of larger animals found in the benthic region belongs to this category. Corals, mussels, barnacles, echinoderms, and sponges are among the few examples of these organisms. This category of organisms is found almost in all substrates, but they inhabit mostly hard substrates. These animals are most diverse and abundant in rocky intertidal areas and coral reefs. Fig. 1.3 shows a healthy marine epifaunal community in the Philippines.

Figure 1.3 A patch of healthy coral reef teeming with marine life in Bantayan, Dumaguete City, Philippines.

The morphology of benthic macroalgae has often been positively associated with the abundance and diversity of epifauna (Gan et al., 2019; Torres et al., 2015). Even in seagrass meadows, the abundance and diversity of the epifauna were not related to the biomass of seagrass but that of the algae present as epiphytes (Tano et al., 2016). Anthropogenic activities have continued to flatten the reef structure from rich macroalgae and live branching corals into low-profile turfing algae (Fraser, Stuart-Smith, et al., 2020). This destruction and degradation of coralline ecosystems threaten the epifaunal species residing in them. This is because these species are often associated with habitats with a high degree of branching and a sufficient level of hardness for attachment. In addition, the macroalgae type, whether native or nonnative, possess differing epifaunal assemblages (Lutz et al., 2019). Native species host higher species richness compared to nonnative. Thus changes in the macroalgae community can also affect the richness of the epifauna. It has also been predicted that the transformation of epifaunal species is expected when large erect macroalgae and coralline ecosystems will collapse. Similarly, epifaunal species also play an essential role in the removal of sediments from the coral colony surface. This highlights the complex and essential relationship in this ecosystem (Stella et al., 2011). Not just live branching corals, juvenile coral nurseries can support a range of epifaunal species and function as tools to conserve these threatened mobile species (Wee et al., 2019). More so, the biomass of seagrasses above-ground was noted to have significant attribution to epifaunal species composition, diversity, and productivity (Leopardas et al., 2014; Whippo et al., 2018). Although seagrass meadows offer significant benefits to epifaunal communities, the need to understand the spatial connection between these meadows should be considered more than just determining their present condition (Whippo et al., 2018). In addition, tropical seaweed beds were found to host high mobile epifaunal abundance, biomass, and diversity (Tano et al., 2016).

Epifaunal species are important in the balance of ecosystems as they are extremely prolific in coastal and shallow reef ecosystems. They serve as a link between benthic and pelagic ecosystems by providing a high proportion of their biomass to large predators (Fraser, Lefcheck, et al., 2020; Griffiths et al., 2017). These organisms also facilitate nutrient cycling, provide nursery areas, and modify biochemical regimes, to name a few (Murillo et al., 2020). In addition, benthic macrofauna plays a vital role in the detritus-based food web of seagrasses by serving as a link between primary producers and higher trophic-level predators (Lin et al., 2018).

These critical roles of these species highlight their significant role in the ecosystem. As suggested in the insurance hypothesis of biodiversity, the maintenance of high diversity and redundancy in functional traits contribute to the stability of biotic assemblages and their associated ecological processes and thus help ensure ecosystem recovery from disturbance (Teichert et al., 2017). This concept is particularly relevant as this can be used to evaluate when fishing and other human activities have significant adverse impacts on benthic ecosystem functioning.

In another study, epifaunal bivalves, like the estuarine oyster *Crassostrea virginica*, form dense assemblages of individuals in reefs attached to hard surfaces, including spent shells from previous bivalve generations. This three-dimensional structure present in reefs and the activity of dense populations possesses a vital role in reef-associated organisms and ecosystem energetics (Dame, 2016). An extensive review of the impacts of epifaunal and infaunal bivalves in marine and freshwater ecosystems is presented by Vaughn and Hoellein (2018).

Epifaunal communities are also continuously threatened by disturbances caused by bottom trawls (Tiano et al., 2020). Several reports have indicated an increase in the vulnerability of large-bodied invertebrate epifauna of this kind of fishing activity. This removal of species can cause the ecosystem's instability through a series of food web changes (Shephard et al., 2014). Also, scallop dredging has been noted to be responsible for the most damaging effects to benthic communities and in emerging epifaunal communities in hard substrates (Boulcott et al., 2014; Kaiser et al., 2006). This is alarming since the damage in highly complex epifaunal communities could impair stock recovery and recruitment of exploited species. For this reason, establishments of protective reserves must be pushed since sufficient evidence shows that it can encourage seafloor habitats to recover (Howarth et al., 2015).

In general, there are very few studies on epifaunal response to disturbances, both the composition and functional attributes (Fields et al., 2019). This is vital since most epifaunal species feed directly on sinking material or those that recently deposited organic matter on the surface. Thus this makes them excellent candidates as bioindicators of chemical contamination (Jørgensen et al., 2011). It has also been noted that this group of species are particularly vulnerable to mine tailings as this chronically disturbs the sediment surface and builds up (Trannum et al., 2019). Polymetallic nodules are important to preserve abyssal epifaunal communities but mining them has been found to negatively affect the hard substrate epifaunal communities (Vanreusel et al., 2016). Furthermore, benthic epifauna is considered to be a more sensitive tool in environmental monitoring and assessment of ecological status, especially when biological trait analyses are included in the monitoring. Epifaunal megabenthic organisms in the Antarctic were also found to be suitable biomonitors for the impacts of climate change (Moon et al., 2015). Table 1.2 summarizes the recent studies on marine benthic epifauna.

1.1.3 New advances in the study and monitoring of the Benthos

The benthic ecosystems, particularly those that are greater than 50 m in depth, have been long undersampled (Piechaud et al., 2019). Traditional

Table 1.2 Studies on marine benthic epifauna.

Location	Goal	References
Southern California Bight	Assessed the ecology of foraminiferal species inhabiting oxygen-deficient habitats	Venturelli et al. (2018)
Frisian Front	Analyzed the effects of in situ fisheries disturbance toward the benthic ecosystem using a combination of techniques	Tiano et al. (2020)
North Sea	Determined the epifaunal assemblages associated with subsea pipelines using remotely operated vehicles	Lacey and Hayes (2020)
North Sea	Identified the link between epibenthic communities and environmental conditions and fishing effort	Neumann et al. (2016)
West Antarctic Peninsula	Assessed the impacts of climate change on nearshore marine benthic assemblages	Moon et al. (2015)
Southern Philippines	Examined the effect of taxonomic identity of seagrasses toward benthic epifauna and infauna	Leopardas et al. (2014)
San Jorge Gulf, Argentina	Characterized the spatial distribution of benthic biodiversity	Kaminsky et al. (2018)
Belgium	Investigated macrobenthic fouling community in concrete foundations of wind turbines	De Mesel et al. (2015)
New South Wales, Australia	Investigated the association of epifaunal assemblages to their native or nonnative macroalgal hosts	Lutz et al. (2019)

methods in studying shallow marine benthic environments for biological parameters have not changed drastically since surveys began. While these methods are proven effective and useful, a number of them tend to be limited in scale and are often nonrepeatable. In our effort to employ effective ecosystem management programs, understanding the world of benthos and the physical variables around them is of prime importance. Several technological advances and innovations now have paved the way in getting a better picture of the benthic ecosystems more than ever (Mumby, 2017). Most of these technologies include packages to track population, trajectories, environmental patterns, management interventions, and current ecosystem health status (Bayley & Mogg, 2019). More so, the advancement of computing power allowed scientists to automate and expand on a scale impossible decades ago. Now, the traditional range of monitoring has been scaled up to appropriately and rapidly address the large-scale disturbance the world is currently experiencing (Tittensor et al., 2014). And the advances in monitoring methods provide invaluable contributions and are crucial to this effort of incorporating reliable ecological indicators to policy development and broad-scale threat mitigation (Edgar et al., 2016; La Salle et al., 2016). These methodologies are outlined in Table 1.3.

Table 1.3 Modern methods in monitoring the benthos.

Method	Feature	Advantage	Limitation	References
Remote sensing	Range of techniques utilizing both aircraft and satellite platforms to map sea surface and seafloor. It also includes the use of Lidar, hyperspectral and multispectral radiometry, photo imagery, and radar, among others.	Rapid, economical, and effective in extensive assessments, especially hard to reach areas.	Requires to be coupled with in situ measur ements which is labor-intensive	Strong and Elliott (2017), Colefax et al. (2018), Geller et al. (2017)

(*Continued*)

Table 1.3 (Continued)

Method	Feature	Advantage	Limitation	References
Acoustic-based methods	Use of sound systems to characterize the bathy metry and topography of the seafloor. Capable of measuring the size of objects up to the meter-level resolution along with depth and hardness by transmitting sound pulses.	Useful in describing a wide range of biological parameters and known to be both rapid and accurate mapping technique by integrating object-based image analysis	It is limited to a relatively coarse resolution of objects 1 m or larger. Limited depth range to shallow depths or very deep depths and require a highly trained operator.	Brown and Blondel (2009), Bayley and Mogg (2019), Brown et al. (2011)
Chemistry-based methods	Estimation of health and function of the benthic environment using the net rate of calcification rate such as coral reefs.	Estimates the response of the benthic ecosystem to major environmental changes and ocean warming and acidification.	High temporal and spatial variability poses a challenge in predicting future trajectories. Further studies are required to provide explicitly the gross coral dissolution value.	Perry and Morgan (2017), Langdon et al. (2010), Cohen et al. (2017), Gómez Batista et al. (2020)
Optics-based methods	Use of light-based imagery underwater directly or remotely for marine monitoring assessments.	Can be used remotely and study benthos for a prolonged period of time and at greater depths. A nondestructive method with high definition imagery outputs.	The field of view and the angle of the line of sight varies between and every sampling survey.	Bayley and Mogg (2019), Watson et al. (2010), Bailey et al. (2007)

1.2 Conclusion

The benthic zone is home to diverse groups of organisms that provides both ecological and economic services. This region is in itself diverse and dependent on various environmental parameters. The infaunal and epifaunal communities possess complex and vital roles that, when disturbed, affect the whole balance of the ecosystem. For this reason, most of these organisms are considered bioindicators of physical and chemical disturbance, pollution, and exploitation. Although considerable studies have been made on their ecology, distribution, and biodiversity, their role as bioindicators, their relationship between different marine and estuarine ecosystems, to name a few, much attention should be given to this highly complex region as various anthropogenic activities continuously threaten them. With the help of modern technology and advances in marine research, our understanding and appreciation of this diverse taxon can still be further improved.

References

Ali, T. S., Al-Dawood, S., & Al-Dawood, F. (2018). Univariate analysis of benthic infaunal biodiversity in the kingdom of Bahrain. *Biological and Applied Environmental Research, 2*(2), 111–129.

Bailey, D., King, N., & Priede, I. (2007). Cameras and carcasses: Historical and current methods for using artificial food falls to study deep-water animals. *Marine Ecology Progress Series, 350*, 179–191. Available from https://doi.org/10.3354/meps07187.

Bayley, D. T. I., & Mogg, A. O. M. (2019). *New advances in benthic monitoring technology and methodology. World seas: An environmental evaluation* (pp. 121–132). Elsevier. Available from https://doi.org/10.1016/B978-0-12-805052-1.00006-1.

Belley, R., & Snelgrove, P. V. R. (2017). The role of infaunal functional and species diversity in short-term response of contrasting benthic communities to an experimental food pulse. *Journal of Experimental Marine Biology and Ecology, 491*, 38–50. Available from https://doi.org/10.1016/j.jembe.2017.03.005.

Boulcott, P., Millar, C. P., & Fryer, R. J. (2014). Impact of scallop dredging on benthic epifauna in a mixed-substrate habitat. *ICES Journal of Marine Science, 71*, 834–844. Available from https://doi.org/10.1093/icesjms/fst197.

Brown, C. J., & Blondel, P. (2009). Developments in the application of multibeam sonar backscatter for seafloor habitat mapping. *Applied Acoustics, 70*, 1242–1247. Available from https://doi.org/10.1016/j.apacoust.2008.08.004.

Brown, C. J., Smith, S. J., Lawton, P., & Anderson, J. T. (2011). Benthic habitat mapping: A review of progress towards improved understanding of the spatial ecology of the seafloor using acoustic techniques. *Estuarine, Coastal and Shelf Science, 92*(3), 502–520. Available from https://doi.org/10.1016/j.ecss.2011.02.007.

Brown, L., Bresnan, E., Summerbell, K., & O'Neill, F. G. (2013). The influence of demersal trawl fishing gears on the resuspension of dinoflagellate cysts. *Marine Pollution Bulletin, 66*(1–2), 17–24. Available from https://doi.org/10.1016/j.marpolbul.2012.11.017.

Chen, J., & Gu, J. D. (2017). Faunal burrows alter the diversity, abundance, and structure of AOA, AOB, anammox and n-damo communities in coastal mangrove sediments. *Microbial Ecology, 74*(1), 140–156. Available from https://doi.org/10.1007/s00248-017-0939-5.

Cohen, S., Krueger, T., & Fine, M. (2017). Measuring coral calcification under ocean acidification: methodological considerations for the 45 Ca-uptake and total alkalinity anomaly technique. *PeerJ, 5*, e3749. Available from https://doi.org/10.7717/peerj.3749.

Colefax, A. P., Butcher, P. A., & Kelaher, B. P. (2018). The potential for unmanned aerial vehicles (UAVs) to conduct marine fauna surveys in place of manned aircraft. *ICES Journal of Marine Science, 75*(1), 1–8. Available from https://doi.org/10.1093/icesjms/fsx100.

Cusson, M., Archambault, P., & Aitken, A. (2007). Biodiversity of benthic assemblages on the Arctic continental shelf: historical data from Canada. *Marine Ecology Progress Series, 331*, 291–304. Available from https://doi.org/10.3354/meps331291.

Dame, R. F. (2016). *Ecology of marine bivalves: An ecosystem approach* (2nd ed.). CRC Press.

De Mesel, I., Kerckhof, F., Norro, A., Rumes, B., & Degraer, S. (2015). Succession and seasonal dynamics of the epifauna community on offshore wind farm foundations and their role as stepping stones for non-indigenous species. *Hydrobiologia, 756*, 37–50. Available from https://doi.org/10.1007/s10750-014-2157-1.

Edgar, G. J., Bates, A. E., Bird, T. J., Jones, A. H., Kininmonth, S., Stuart-Smith, R. D., & Webb, T. J. (2016). New approaches to marine conservation through the scaling up of ecological data. *Annual Review of Marine Science, 8*, 435–461. Available from https://doi.org/10.1146/annurev-marine-122414-033921.

Fields, S., Henkel, S., & Roegner, G. C. (2019). Video sleds effectively survey epibenthic communities at dredged material disposal sites. *Environmental Monitoring and Assessment, 191*, 404. Available from https://doi.org/10.1007/s10661-019-7348-9.

Fraser, K. M., Lefcheck, J. S., Ling, S. D., Mellin, C., Stuart-Smith, R. D., & Edgar, G. J. (2020). Production of mobile invertebrate communities on shallow reefs from temperate to tropical seas. *Proceedings of the Royal Society B: Biological Sciences, 287*(1941). Available from https://doi.org/10.1098/rspb.2020.1798.

Fraser, K. M., Stuart-Smith, R. D., Ling, S. D., Heather, F. J., & Edgar, G. J. (2020). Taxonomic composition of mobile epifaunal invertebrate assemblages on diverse benthic microhabitats from temperate to tropical reefs. *Marine Ecology Progress Series, 640*, 31–43. Available from https://doi.org/10.3354/meps13295.

Gan, S. X., Tay, Y. C., & Huang, D. (2019). Effects of macroalgal morphology on marine epifaunal diversity. *Journal of the Marine Biological Association of the United Kingdom, 99*(8), 1697–1707. Available from https://doi.org/10.1017/S0025315419000900.

Geller, G. N., Halpin, P. N., Helmuth, B., Hestir, E. L., Skidmore, A., Abrams, M. J., Aguirre, N., Blair, M., Botha, E., Colloff, M., Dawson, T., Franklin, J., Horning, N., James, C., Magnusson, W., Santos, M. J., Schill, S. R., & Williams, K. (2017). Remote sensing for biodiversity. In M. Walters, & R. J. Scholes (Eds.), *The GEO handbook on biodiversity observation networks* (pp. 187–210). Springer International Publishing. Available from https://doi.org/10.1007/978-3-319-27288-7_8.

Gerwing, T. G., Drolet, D., Barbeau, M. A., Hamilton, D. J., & Allen Gerwing, A. M. (2015). Resilience of an intertidal infaunal community to winter stressors. *Journal of Sea Research, 97*, 40–49. Available from https://doi.org/10.1016/j.seares.2015.01.001.

Gómez Batista, M., Metian, M., Oberhänsli, F., Pouil, S., Swarzenski, P. W., Tambutté, E., Gattuso, J.-P., Alonso Hernández, C. M., & Gazeau, F. (2020). Intercomparison of four methods to estimate coral calcification under various environmental conditions. *Biogeosciences, 17*(4), 887–899. Available from https://doi.org/10.5194/bg-17-887-2020.

González-Ortiz, V., Alcazar, P., Vergara, J. J., Pérez-Lloréns, J. L., & Brun, F. G. (2014). Effects of two antagonistic ecosystem engineers on infaunal diversity. *Estuarine, Coastal and Shelf Science, 139*, 20–26. Available from https://doi.org/10.1016/j.ecss.2013.12.015.

González-Ortiz, V., Egea, L. G., Jiménez-Ramos, R., Moreno-Marín, F., Pérez-Lloréns, J. L., Bouma, T., & Brun, F. (2016). Submerged vegetation complexity modifies benthic infauna communities: The hidden role of the belowground system. *Marine Ecology, 37*(3), 543–552. Available from https://doi.org/10.1111/maec.12292.

Griffiths, J. R., Kadin, M., Nascimento, F. J. A., Tamelander, T., Törnroos, A., Bonaglia, S., Bonsdorff, E., Brüchert, V., Gårdmark, A., Järnström, M., Kotta, J., Lindegren, M., Nordström, M. C., Norkko, A., Olsson, J., Weigel, B., Žydelis, R., Blenckner, T., Niiranen, S., & Winder, M. (2017). The importance of benthic–pelagic coupling for marine ecosystem functioning in a changing world. *Global Change Biology, 23*(6), 2179–2196. Available from https://doi.org/10.1111/gcb.13642.

Gutow, L., Teschke, K., Schmidt, A., Dannheim, J., Krone, R., & Gusky, M. (2014). *Rapid increase of benthic structural and functional diversity at the alpha ventus offshore test site. Ecological research at the offshore windfarm alpha ventus* (pp. 67–81). Springer Fachmedien Wiesbaden. Available from https://doi.org/10.1007/978-3-658-02462-8_9.

Hemminga, M. A., & Duarte, C. M. (2000). *Seagrass ecology. Seagrass ecology*. Cambridge University Press. Available from https://doi.org/10.1017/CBO9780511525551.

Howarth, L. M., Pickup, S. E., Evans, L. E., Cross, T. J., Hawkins, J. P., Roberts, C. M., & Stewart, B. D. (2015). Sessile and mobile components of a benthic ecosystem display mixed trends within a temperate marine reserve. *Marine Environmental Research, 107*, 8–23. Available from https://doi.org/10.1016/j.marenvres.2015.03.009.

Johnson, R. A., Hanes, K. M., Bolten, A. B., & Bjorndal, K. A. (2020). Simulated green turtle grazing affects benthic infauna abundance and community composition but not diversity in a Thalassiatestudinum seagrass meadow. *Journal of Experimental Marine Biology and Ecology, 522*, 151266. Available from https://doi.org/10.1016/j.jembe.2019.151266.

Jørgensen, L. L., Renaud, P. E., & Cochrane, S. K. J. (2011). Improving benthic monitoring by combining trawl and grab surveys. *Marine Pollution Bulletin, 62*(6), 1183–1190. Available from https://doi.org/10.1016/j.marpolbul.2011.03.035.

Kaiser, M. J., Clarke, K. R., Hinz, H., Austen, M. C. V., Somerfield, P. J., & Karakassis, I. (2006). Global analysis of response and recovery of benthic biota to fishing. *Marine Ecology Progress Series, 311*, 1–14. Available from https://doi.org/10.3354/meps311001.

Kaminsky, J., Varisco, M., Fernández, M., Sahade, R., & Archambault, P. (2018). Spatial analysis of benthic functional biodiversity in san jorge gulf, Argentina. *Oceanography, 31*(4), 104–112. Available from https://doi.org/10.5670/oceanog.2018.414.

Kundu, S., Mondal, N., Lyla, P. S., & Ajmal Khan, S. (2010). Biodiversity and seasonal variation of macro-benthic infaunal community in the inshore waters of Parangipettai Coast. *Environmental Monitoring and Assessment, 163*(1–4), 67–79. Available from https://doi.org/10.1007/s10661-009-0817-9.

Lacey, N. C., & Hayes, P. (2020). Epifauna associated with subsea pipelines in the North Sea. *ICES Journal of Marine Science*, 77(3), 1137–1147. Available from https://doi.org/10.1093/icesjms/fsy196.

La Salle, J., Williams, K. J., & Moritz, C. (2016). Biodiversity analysis in the digital era. *Philosophical Transactions of the Royal Society B: Biological Sciences, 371*(1702), 20150337. Available from https://doi.org/10.1098/rstb.2015.0337.

Lalli, C. M., & Parsons, T. R. (1997). *Benthos. Biological oceanography: An introduction* (pp. 177–195). Elsevier. Available from https://doi.org/10.1016/B978-075063384-0/50063-3.

Langdon, C., Gattuso, J.-P., & Andersson, A. (2010). Measurements of calcification and dissolution of benthic organisms and 13 communities. In U. Riebesell, V. J. Fabry, L. Hansson, & J.-P. Gattuso (Eds.), *Guide to best practices for ocean acidification research and data reporting*. Publications Office of the European Union.

Lawless, A. S., & Seitz, R. D. (2014). Effects of shoreline stabilization and environmental variables on benthic infaunal communities in the Lynnhaven River System of Chesapeake Bay. *Journal of Experimental Marine Biology and Ecology*, *457*, 41–50. Available from https://doi.org/10.1016/j.jembe.2014.03.010.

Leopardas, V., Uy, W., & Nakaoka, M. (2014). Benthic macrofaunal assemblages in multispecific seagrass meadows of the southern Philippines: Variation among vegetation dominated by different seagrass species. *Journal of Experimental Marine Biology and Ecology*, *457*, 71–80. Available from https://doi.org/10.1016/j.jembe.2014.04.006.

Lin, J., Huang, Y., Arbi, U. Y., Lin, H., Azkab, M. H., Wang, J., He, X., Mou, J., Liu, K., & Zhang, S. (2018). An ecological survey of the abundance and diversity of benthic macrofauna in Indonesian multispecific seagrass beds. *Acta Oceanologica Sinica*, *37* (6), 82–89. Available from https://doi.org/10.1007/s13131-018-1181-9.

Llacuna, M. E. J., Walag, A. M. P., & Villaluz, E. (2016). Diversity and dispersion patterns of echinoderms in Babanlagan, Talisayan, Misamis Oriental, Philippines. *Environmental and Experimental Biology*, *14*(4), 213–217. Available from https://doi.org/10.22364/eeb.14.28.

Lutz, M. L., Minchinton, T. E., & Davis, A. R. (2019). Differences in architecture between native and non-indigenous macroalgae influence associations with epifauna. *Journal of Experimental Marine Biology and Ecology*, *514–515*, 76–86. Available from https://doi.org/10.1016/j.jembe.2019.03.006.

MacKenzie, R. A., Dionne, M., Miller, J., Haas, M., & Morgan, P. A. (2015). Community structure and abundance of benthic infaunal invertebrates in maine fringing marsh ecosystems. *Estuaries and Coasts*, *38*(4), 1317–1334. Available from https://doi.org/10.1007/s12237-015-9977-8.

Moon, H.-W., Wan Hussin, W. M. R., Kim, H.-C., & Ahn, I.-Y. (2015). The impacts of climate change on Antarctic nearshore mega-epifaunal benthic assemblages in a glacial fjord on King George Island: Responses and implications. *Ecological Indicators*, *57*, 280–292. Available from https://doi.org/10.1016/j.ecolind.2015.04.031.

Mumby, P. J. (2017). Trends and frontiers for the science and management of the oceans. *Current Biology*, *27*(11), R431–R434. Available from https://doi.org/10.1016/j.cub.2017.04.049.

Murillo, F. J., Weigel, B., Bouchard Marmen, M., & Kenchington, E. (2020). Marine epibenthic functional diversity on Flemish Cap (north-west Atlantic)—Identifying trait responses to the environment and mapping ecosystem functions. *Diversity and Distributions*, *26*(4), 460–478. Available from https://doi.org/10.1111/ddi.13026.

Neumann, H., Diekmann, R., & Kröncke, I. (2016). Functional composition of epifauna in the south-eastern North Sea in relation to habitat characteristics and fishing effort. *Estuarine, Coastal and Shelf Science*, *169*, 182–194. Available from https://doi.org/10.1016/j.ecss.2015.12.011.

Nicastro, K. R., McQuaid, C. D., Dievart, A., & Zardi, G. I. (2020). Intraspecific diversity in an ecological engineer functionally trumps interspecific diversity in shaping community structure. *Science of the Total Environment*, *743*, 140723. Available from https://doi.org/10.1016/j.scitotenv.2020.140723.

Olsgard, F., Schaanning, M. T., Widdicombe, S., Kendall, M. A., & Austen, M. C. (2008). Effects of bottom trawling on ecosystem functioning. *Journal of Experimental Marine Biology and Ecology*, *366*(1–2), 123–133. Available from https://doi.org/10.1016/j.jembe.2008.07.036.

O'Neill, F. G., Robertson, M., Summerbell, K., Breen, M., & Robinson, L. A. (2013). The mobilisation of sediment and benthic infauna by scallop dredges. *Marine Environmental Research*, *90*, 104–112. Available from https://doi.org/10.1016/j.marenvres.2013.06.003.

Perry, C. T., & Morgan, K. M. (2017). Bleaching drives collapse in reef carbonate budgets and reef growth potential on southern Maldives reefs. *Scientific Reports*, 7, 40581. Available from https://doi.org/10.1038/srep40581.

Piechaud, N., Hunt, C., Culverhouse, P., Foster, N., & Howell, K. (2019). Automated identification of benthic epifauna with computer vision. *Marine Ecology Progress Series*, *615*, 15–30. Available from https://doi.org/10.3354/meps12925.

Pilditch, C. A., Leduc, D., Nodder, S. D., Probert, P. K., & Bowden, D. A. (2015). Spatial patterns and environmental drivers of benthic infaunal community structure and ecosystem function on the New Zealand continental margin. *New Zealand Journal of Marine and Freshwater Research*, *49*(2), 224–246. Available from https://doi.org/10.1080/00288330.2014.995678.

Powell, E. N., & Mann, R. (2016). How well do we know the infaunal biomass of the continental shelf? *Continental Shelf Research*, *115*, 27–32. Available from https://doi.org/10.1016/j.csr.2016.01.001.

Rabalais, N. N. (2019). *Ocean deoxygenation: Everyone's problem. Causes, impacts, consequences and solutions*. IUCN, https://doi.org/10.2305/iucn.ch.2019.13.en.

Rabalais, N. N., & Baustian, M. M. (2020). Historical shifts in benthic infaunal diversity in the northern gulf of Mexico since the appearance of seasonally severe hypoxia. *Diversity*, *12*(2), 49. Available from https://doi.org/10.3390/d12020049.

Reuscher, M. G., Baguley, J. G., Conrad-Forrest, N., Cooksey, C., Hyland, J. L., Lewis, C., Montagna, P. A., Ricker, R. W., Rohal, M., & Washburn, T. (2017). Temporal patterns of Deepwater Horizon impacts on the benthic infauna of the northern Gulf of Mexico continental slope. *PLoS One*, *12*(6), e0179923. Available from https://doi.org/10.1371/journal.pone.0179923.

Reynolds, C. S. (2013). *Benthic ecology based in part on the article "Benthic Ecology" by Colin S. Reynolds, which appeared in the Encyclopedia of Environmetrics. Encyclopedia of environmetrics*. John Wiley & Sons, Ltd. Available from https://doi.org/10.1002/9780470057339.vab010.pub2.

Rosov, B., Bush, S., Briggs, T. R., & Elko, N. (2016). The state of understanding the impacts of beach nourishment activities on infaunal communities. *Shore and Beach*, *84* (3), 51–55.

Satoh, H., Nakamura, Y., & Okabe, S. (2007). Influences of infaunal burrows on the community structure and activity of ammonia-oxidizing bacteria in intertidal sediments. *Applied and Environmental Microbiology*, *73*(4), 1341–1348. Available from https://doi.org/10.1128/AEM.02073-06.

Schonberg, S. V., Clarke, J. T., & Dunton, K. H. (2014). Distribution, abundance, biomass and diversity of benthic infauna in the Northeast Chukchi Sea, Alaska: Relation to environmental variables and marine mammals. *Deep sea research part II: Topical studies in oceanography*, *102*, 144–163. Available from https://doi.org/10.1016/j.dsr2.2013.11.004.

Seitz, R. D. (2011). Gradient effects on structuring of soft-bottom benthic infauna: Macomabalthica and predation, recruitment, and food availability. *Journal of Experimental Marine Biology and Ecology*, *409*(1–2), 114–122. Available from https://doi.org/10.1016/j.jembe.2011.08.014.

Seitz, R. D., & Lewis, C. J. E. (2018). Loss of seagrass results in changes to benthic infaunal community structure and decreased secondary production. *Bulletin of Marine Science*, *94*(4), 1273–1292. Available from https://doi.org/10.5343/bms.2017.1011.

Shephard, S., Minto, C., Zölck, M., Jennings, S., Brophy, D., & Reid, D. (2014). Scavenging on trawled seabeds can modify trophic size structure of bottom-dwelling fish. *ICES Journal of Marine Science*, *71*(2), 398–405. Available from https://doi.org/10.1093/icesjms/fst134.

Smith, R. L., & Smith, T. M. (2003). *Elements of ecology*. San Francisco *(Fifth)*. Pearson Education Limited.

Somerfield, P. J., McClelland, I. L., McNeill, C. L., Bolam, S. G., & Widdicombe, S. (2019). Environmental and sediment conditions, infaunal benthic communities and biodiversity in the Celtic Sea. *Continental Shelf Research*, *185*, 23–30. Available from https://doi.org/10.1016/j.csr.2018.09.002.

Steffani, N., Sedick, S., Rogers, J., & Gibbons, M. J. (2015). Infaunal benthic communities from the inner shelf off southwestern Africa are characterised by generalist species. *PLoS One*, *10*(11), e0143637. Available from https://doi.org/10.1371/journal.pone.0143637.

Stella, J. S., Pratchett, M. S., Hutchings, P. A., & Jones, G. P. (2011). Coral-associated invertebrates: Diversity, ecological importance and vulnerability to disturbance. *Oceanography and marine biology: An annual review*, *49*, 43–104.

Strong, J. A., & Elliott, M. (2017). The value of remote sensing techniques in supporting effective extrapolation across multiple marine spatial scales. *Marine Pollution Bulletin*, *116*(1–2), 405–419. Available from https://doi.org/10.1016/j.marpolbul.2017.01.028.

Tano, S., Eggertsen, M., Wikström, S. A., Berkström, C., Buriyo, A. S., & Halling, C. (2016). Tropical seaweed beds are important habitats for mobile invertebrate epifauna. *Estuarine, Coastal and Shelf Science*, *183*(Part A), 1–12. Available from https://doi.org/10.1016/j.ecss.2016.10.010.

Teichert, N., Lepage, M., Sagouis, A., Borja, A., Chust, G., Ferreira, M. T., Pasquaud, S., Schinegger, R., Segurado, P., & Argillier, C. (2017). Functional redundancy and sensitivity of fish assemblages in European rivers, lakes and estuarine ecosystems. *Scientific Reports*, 7(1), 17611. Available from https://doi.org/10.1038/s41598-017-17975-x.

Tiano, J. C., van der Reijden, K. J., O'Flynn, S., Beauchard, O., van der Ree, S., van der Wees, J., Ysebaert, T., & Soetaert, K. (2020). Experimental bottom trawling finds resilience in large-bodied infauna but vulnerability for epifauna and juveniles in the Frisian Front. *Marine Environmental Research*, *159*, 104964. Available from https://doi.org/10.1016/j.marenvres.2020.104964.

Tittensor, D. P., Walpole, M., Hill, S. L. L., Boyce, D. G., Britten, G. L., Burgess, N. D., Butchart, S. H. M., Leadley, P. W., Regan, E. C., Alkemade, R., Baumung, R., Bellard, C., Bouwman, L., Bowles-Newark, N. J., Chenery, A. M., Cheung, W. W. L., Christensen, V., Cooper, H. D., Crowther, A. R., & Ye, Y. (2014). A mid-term analysis of progress toward international biodiversity targets. *Science*, *346*(6206), 241–244. Available from https://doi.org/10.1126/science.1257484.

Torres, A. C., Veiga, P., Rubal, M., & Sousa-Pinto, I. (2015). The role of annual macroalgal morphology in driving its epifaunal assemblages. *Journal of Experimental Marine Biology and Ecology*, *464*, 96–106. Available from https://doi.org/10.1016/j.jembe.2014.12.016.

Trannum, H. C., Borgersen, G., Oug, E., Glette, T., Brooks, L., & Ramirez-Llodra, E. (2019). Epifaunal and infaunal responses to submarine mine tailings in a Norwegian fjord. *Marine Pollution Bulletin*, *149*, 110560. Available from https://doi.org/10.1016/j.marpolbul.2019.110560.

Trevizani, T. H., Figueira, R. C. L., Ribeiro, A. P., Theophilo, C. Y. S., Majer, A. P., Petti, M. A. V., Corbisier, T. N., & Montone, R. C. (2016). Bioaccumulation of heavy metals in marine organisms and sediments from Admiralty Bay, King George Island, Antarctica. *Marine Pollution Bulletin*, *106*(1–2), 366–371. Available from https://doi.org/10.1016/j.marpolbul.2016.02.056.

van Hoey, G., Borja, A., Birchenough, S., Buhl-Mortensen, L., Degraer, S., Fleischer, D., Kerckhof, F., Magni, P., Muxika, I., Reiss, H., Schröder, A., & Zettler, M. L. (2010). The use of benthic indicators in Europe: From the Water Framework Directive to the Marine Strategy Framework Directive. *Marine Pollution Bulletin*, *60*(12), 2187–2196. Available from https://doi.org/10.1016/j.marpolbul.2010.09.015.

Vanreusel, A., Hilario, A., Ribeiro, P. A., Menot, L., & Arbizu, P. M. (2016). Threatened by mining, polymetallic nodules are required to preserve abyssal epifauna. *Scientific Reports*, *6*, 26808. Available from https://doi.org/10.1038/srep26808.

Vaughn, C. C., & Hoellein, T. J. (2018). Bivalve impacts in freshwater and marine ecosystems. *Annual Review of Ecology, Evolution, and Systematics*, *49*, 183–208. Available from https://doi.org/10.1146/annurev-ecolsys-110617-062703.

Venturelli, R. A., Rathburn, A. E., Burkett, A. M., & Ziebis, W. (2018). Epifaunal foraminifera in an infaunal world: Insights into the influence of heterogeneity on the benthic ecology of oxygen-poor, deep-sea habitats. *Frontiers in Marine Science*, *5*, 344. Available from https://doi.org/10.3389/fmars.2018.00344.

Walag, A. M. P. (2017). Bioactivities of extracts from different marine organisms around the world (2000 to present). *Biomedical Journal of Scientific & Technical Research*, *1*(7), 1–3. Available from https://doi.org/10.26717/BJSTR.2017.01.000600.

Walag, A. M. P., & del Rosario, R. M. (2018). Proximate biochemical composition and brine shrimp lethality assay of selected sea stars from Goso-on and Vinapor, Carmen, Agusan del Norte, Philippines. *Malaysian Journal of Biochemistry and Molecular Biology*, *21*(3), 11–18.

Walag, A. M. P., & Canencia, M. O. P. (2016). Physico-chemical parameters and macrobenthic invertebrates of the intertidal zone of Gusa, Cagayan de Oro City, Philippines. *AES Bioflux*, *8*(1), 71–82. Available from http://www.aes.bioflux.com.ro/docs/2016.71-82.pdf.

Walag, A. M. P., & del Rosario, R. M. (2020). Initial evaluation of metal content of Acanthasterplanci and Linckia laevigata collected from Carmen, Agusan del Norte, Philippines. *Malaysian Journal of Biochemistry and Molecular Biology*, *20*(3), 1–7.

Walag, A. M. P., del Rosario, R. M., & Canencia, O. P. (2019). Zoochemical composition of selected sea stars collected from the coastal waters of Carmen, Agusan Del Norte, Philippines. *Asian Journal of Biological and Life Sciences*, *8*(2), 53–62. Available from https://doi.org/10.5530/ajbls.2019.8.10.

Walag, A. M. P., Layaog, A. G., & Garcia, G. U. (2018). Survey of echinoderms in the intertidal zone of Goso-on and Vinapor, Carmen, Agusan del Norte, Philippines. *Environmental and Experimental Biology*, *16*(1), 31–38. Available from https://doi.org/10.22364/eeb.16.04.

Watson, D. L., Harvey, E. S., Fitzpatrick, B. M., Langlois, T. J., & Shedrawi, G. (2010). Assessing reef fish assemblage structure: How do different stereo-video techniques compare? *Marine Biology*, *157*, 1237–1250. Available from https://doi.org/10.1007/s00227-010-1404-x.

Wee, S. Y. C., Sam, S. Q., Sim, W. T., Ng, C. S. L., Taira, D., Afiq-Rosli, L., Kikuzawa, Y. P., Toh, T. C., & Chou, L. M. (2019). The role of in situ coral nurseries in supporting mobile invertebrate epifauna. *Journal for Nature Conservation*, *50*, 125710. Available from https://doi.org/10.1016/j.jnc.2019.125710.

Whippo, R., Knight, N. S., Prentice, C., Cristiani, J., Siegle, M. R., & O'Connor, M. I. (2018). Epifaunal diversity patterns within and among seagrass meadows suggest landscape-scale biodiversity processes. *Ecosphere*, *9*(11), e02490. Available from https://doi.org/10.1002/ecs2.2490.

Yang, Y., Chui, T. F. M., Shen, P. P., Yang, Y., & Gu, J. D. (2018). Modeling the temporal dynamics of intertidal benthic infauna biomass with environmental factors: Impact assessment of land reclamation. *Science of the Total Environment*, *618*, 439–450. Available from https://doi.org/10.1016/j.scitotenv.2017.10.325.

CHAPTER 2

Benthic biotopes: abiotic and biotic factors in the sediment

D.B. Arya[1], Salom Gnana Thanga Vincent[1] and Prince S. Godson[2]
[1]Department of Environmental Sciences, University of Kerala, Thiruvananthapuram, Kerala, India
[2]Centre for Marine Science and Technology, Manonmaniam Sundaranar University, Tamil Nadu, India

2.1 Introduction

Benthic communities are the organisms that reside on the sediments of aquatic habitats. They are very diverse and also, they can occur in a wide range of habitats. According to their size, they are subdivided into macrofauna, meiofauna, and microfauna. Aquatic habitats possess a diverse range of benthic communities and they have an important role in maintaining the food web and primary production (Alongi & Christoffersen, 1992). Hence studies related to benthic communities are very important for proper resource management (Raut et al., 2005). The distribution of benthic communities is directly influenced by the physicochemical and biological processes in their habitat (Swami et al., 2000; Varadharajan et al., 2010). Therefore it is important to understand the factors influencing the occurrence and activity of benthic organisms and interactions existing among the organisms. The biodiversity and distribution of organisms within an ecosystem are due to both abiotic (nonliving) and biotic (living) factors. Examples of abiotic factors include depth and light intensity, bottom water and sediment temperature, and sediment grain size.

2.2 Benthic biotopes

A benthic biotope is defined as an area uniform in environmental conditions with characteristic species composition and uniform distribution of benthic communities (Fig. 2.1). In particular, the physical characteristics of water masses near the seabed, seabed topography, and seabed substrate are often the main determining factors in shaping benthic communities (Buhl-Mortensen et al., 2020).

Ecology and Biodiversity of Benthos
DOI: https://doi.org/10.1016/B978-0-12-821161-8.00009-X

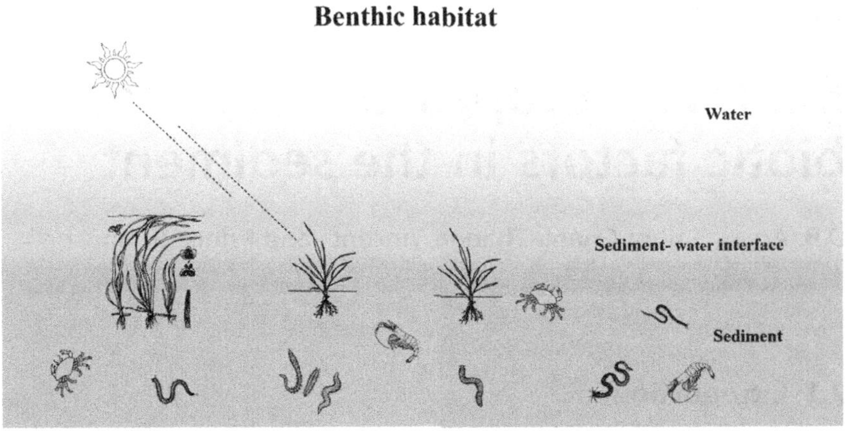

Figure 2.1 Benthic habitat.

2.3 Abiotic factors

2.3.1 Depth and light intensity

In aquatic habitats, the availability of sunlight is the primary source of energy that determines the benthic primary production (Gattuso et al., 2006). The main drivers of biological carbon pumps in aquatic sediments are the sunlight and the organisms which live in the aquatic habitat (Fig. 2.2). It includes complex processes such as photosynthesis, respiration, feeding, and the production of waste products. A small fraction of light that incidents on the surface is reflected and a large portion enters into the water. The distribution of light mainly depends on the absorption and scattering property of water. If the scattering to absorption ratio is high, the light will evenly distribute in all directions (Kirk, 1994). The availability of light changes with respect to space and time that depends on the cloud formation and concentration of dissolved nutrients (Kirk, 1994; Wright, 1997). Four important parameters affect variation in the light distribution, which includes a variation of daily solar irradiance, variation of cloud pattern (Mumby et al., 2001), and the concentration of dissolved solids, which affect the transmittance of water (Van Duin et al., 2001), and variations in tide patterns, which in turn, affect the depth and the amount of light transmitted to the seafloor (Anthony et al., 2004). Light can influence the distribution pattern and survival activity of aquatic inhabitants.

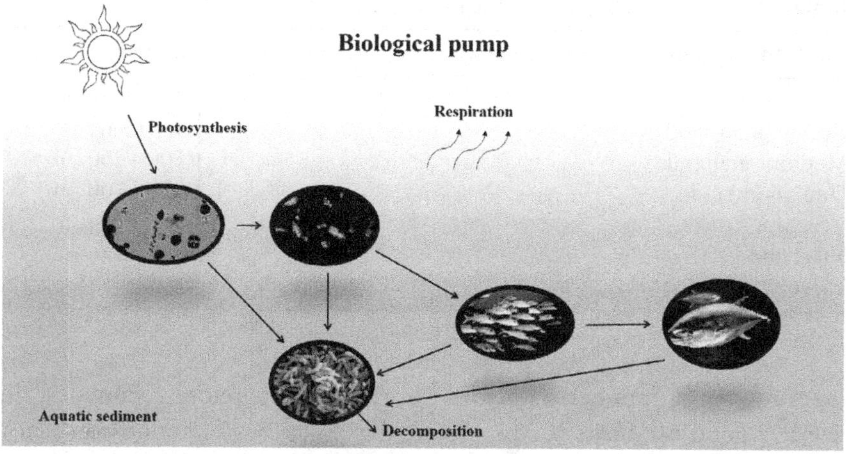

Figure 2.2 Biological pump driven by sunlight.

2.3.2 Bottom water and sediment temperature

Benthic organisms are very diverse and they are present in tropical as well as temperate regions. Seasonal variation in abundance and species composition of benthic communities are outlined in many studies (Heip et al., 1985; Hicks & Coull, 1983). Temperature is one of the important factors which can determine the distribution of benthic organisms (Varadharajan et al., 2010). Temperature can affect species composition directly or indirectly by the temperature-induced controls on anoxic depth level, disturbance, bioturbation, and food type (Coull, 1999).

2.3.2.1 Sediment grain size

Sediment quality is very important in shaping the structure and function (Meyers et al., 1987) of benthos. Grain-size data of the aquatic sediments are of prime importance in differentiating various depositional microenvironments and are also useful for assessing the textural characteristics of the habitat in benthic ecological studies (Hossain et al., 2014) (Table 2.1). Previous studies suggest that sediment grain size can influence the spatial distribution of benthos (Galope-Bacaltos & San, 2002). Various studies by Gaston et al. (1988) and Simboura et al. (2000) also revealed that textural composition is an important parameter that can determine species diversity. Sediment grain size can affect sediment permeability, which is the ease with which water and dissolved oxygen can enter and move through the sediment particles. The availability of large interstitial space can also

Table 2.1 Sediment category and their diameter.

Sediment category	Diameter
Very coarse sand	2–1 mm
Coarse grain sand	1–0.5 mm
Medium grain sand	500–250 μm
Fine grain sand	250–100 μm
Very fine sand	100–63 μm
Silty sand	<63 μm
Clayey sand	<2 μm

support a large number of benthic organisms. Therefore sediment grain size is a good candidate for predicting the diversity of benthic communities (Strayer et al., 1997). Among the various environmental factors, the sediment grain size was identified as the key factor in determining the communities of meiofauna in the sediments (Kim et al., 2020).

Benthic fauna can readily burrow the sediment with a mean grain size of less than 125 μm. Benthic organisms prefer sandy sediments over muddy sediments due to the presence of interstitial space which helps to increase oxygen availability, as they require oxygen to perform their metabolic activities. So, benthos may be present mostly in the surface layer than the bottom layer. The availability of interstitial space depends upon the grain size, sorting, shape, packing, porosity, pore size, and permeability (McLachlan & Turner, 1994). In muddy sediments, benthic invertebrates are restricted to a narrow 2–3 cm oxic zone and in sandy sediments, they may be present up to 10 cm depth.

Sediment grain size is influenced by physical, chemical, and biological factors which include tidal current, wave action, climatic condition, salinity fluctuation, bioturbation, and various anthropogenic activities (Gray & Elliott, 2009; Venkatramanan et al., 2013). Besides these physical factors, biological particles also have an important role in determining sediment properties. Benthic burrowers such as crabs, polychaetes, Nematode, and gastropods in muddy sediments can decrease the sediment compaction by burrowing activity and thus affects the textural composition of the sediment (Davis, 1993). In contrast, tube-dwelling polychaetes such as *Diopatracuprea* stabilize the bottom sediment from erosion and also enhance the accumulation of sediment (Sarkar et al., 2005).

Sediment grain size possesses many ecological applications that strongly influence the distribution of benthic communities (Alongi, 1990). Finer particles such as silt and clay with higher organic matter and low hydrodynamic conditions favor and nourish the infaunal organisms and show high density, diversity, and richness (Day et al., 2012; Gray & Elliott, 2009). Moreover, fine sediment particles like clay have a large relative surface area than coarse sediment like clay, hence can absorb dissolved organic matter and form sedimentary complexes. Therefore the ratio of sand, silt, and clay be useful to determine the faunal distributions (Hossain, 2011). In the case of estuarine sediments, fine-grained sediments are the major sink of riverine heavy metals (Flemming et al., 1996) and they are also bioavailable to benthic organisms present in the sediment and thus enter into the aquatic food chain (Munksgaard & Parry, 2002). In sediments, the availability of oxygen generally decreases toward the bottom and also depends upon the nature of the sediment. In fine-grained sediments, the oxygen concentration is restricted only to top layers because of their low permeability. In the case of coarse-grained sediments, oxygen saturation will occur up to 1 m depth. The oxygen availability in the interstitial habitat is very important to determine the redox potential of the nutrients, the existence of redox condition of the system, and also the activity of interstitial communities (Brown & McLachlan, 2010).

Generally, coarse-grained sediments detain fast and they have poor water, oxygen, or organic matter retaining capacity because they have more interstitial space. These sediments are mobile as compared to fine-grained sediments. Therefore they are inappropriate habitats for those organisms which are unable to tolerate mobile habitats, these types of sediments are only inhabited by those who can tolerate these mobile sediments such as fast burrowing organisms. The higher organic matter content in the mud harbors higher species number and biomass of benthic fauna, thus providing food for higher organisms in the food chain. On the other hand, fine-grained sediments are tightly packed and they can retain large organic matter and water and poor oxygen availability. The presence of high organic matter in turn increases the oxygen demand due to the activity of microbes.

Besides these sandy and muddy sediments, intermediate silty sediments are more favorable for the existence of benthic communities (Gray and Elliott, 2009) Jayaraj et al. (2007) reported that filter feeder benthos prefers medium size sediment and clayey sediments are not suitable substrata for such organisms. A study by Thrush and Townsend (1986) showed that benthic communities in the sublittoral sediment of Ireland exist in two different habitats such as coarse sediment and fine-grained sediment. The study revealed that deposit-feeding spionid polychaete *Pseudopolydora pulchra* dominated in muddy sediment and serpulid polychaete *Pomatoceros* sp. and gastropod *Onobasemicostrata* dominated in coarse sediment. A study by Leduc et al. (2012) in New Zealand reported a greater abundance of nematodes in coarse-grained sediment and also suggested that different feeding group prefers different grain-size preferences. Predatory carnivores prefer sandy bottoms due to the prevalence of their prey organisms in the sand interstitial spaces.

2.4 Biotic factors

Living organisms such as bacteria, microalgae, fish, and other high trophic level organism represents major environmental factors that affect the structure and function of benthic organisms (Giere, 2009). The higher trophic level organisms possess positive, negative effects on the lower trophic levels. The negative interactions include various mechanical disturbances, reduction by sediment ingestion, and alteration of chemical habitat. The mechanical disturbance is mainly caused by the bioturbation activities, for example, disturbance and digging by crabs, sea urchins, and mollusks make disturbances in the sediments to the low trophic level organism. The meiobenthic organisms are less sensitive than macrobenthic organisms to physical and chemical disturbances (Alongi, 1985; Giere, 2009). Deposit feeders ingest the sediments, which directly affects the meiofauna which resides on the sediments, and hence they are forced to migrate downward, thus leading to reduction of the benthic organism in the upper layers (Bell, 1980). The accumulation of fecal matter of higher trophic level organism and algal matter on the upper sediments layers leads to oxygen depletion and cause the reduction of benthic communities in bottom layers (Reise, 2002). The competition of benthic organisms for identical resources can cause niche segregation and lead to amensalism or exclusion.

There are different types of interactions relating to the food chain among benthos and other organisms, for example, predatory interaction like juvenile fish eating small benthos; microphytobenthos are eaten up by nematodes. The polychaete tube produced by annelids protects other fauna from predators (Coull, 1999). Many different studies show that there is an increase in benthic abundance if predators are avoided (Coull & Bell, 1979; Ólafsson & Moore, 1990). Previous studies by Paine (1974) suggest that predation and other physical disturbance have an important role in maintaining the species diversity and also reducing the dominant species to an equilibrium level. Reise (1978) revealed that small benthic predators are the major determinants of the changes in diversity and species abundance pattern. Predation is the major factor that can control the abundance and diversity of benthic communities and also affect the community structure (Gee et al., 1985). The frequency of predator feeding activity depends on factors such as habitat complexity, prey exposure gradient, tidal flow, and prey-predator interactions. Benthic fauna acts as prey for higher trophic level organisms such as juvenile fish. Even if predation occurs continuously, due to the short life span and large reproduction capacity they can compensate easily (Coull, 1990).

2.4.1 Effect of bioturbation and bioirrigation

Bioturbation is the process of the biological makeover of sediments by benthic organisms (Giere, 2009) (Fig. 2.3). Benthic burrowers are the major drivers of bioturbation, especially crustaceans and annelids for macrobenthos; ostracods, nematodes, and copepods for meiofauna. Bioturbation effects can extend up to 20 cm depth in the sediments. Bioturbation possesses beneficial as well as detrimental effects on the ecosystem. Positive effects include the supply of sufficient oxygen and organic matter whereas the negative impacts include disturbance and destabilization (Green & Chandler, 1994). By the process of bioturbation, benthic fauna contributes to the biogeochemical cycling of nutrients in the sediments and sediment-water interface. The sediment mixing depth is also increased by bioturbation or bioirrigation, thus increasing the flux of nutrients from the sediment to the overlying water column.

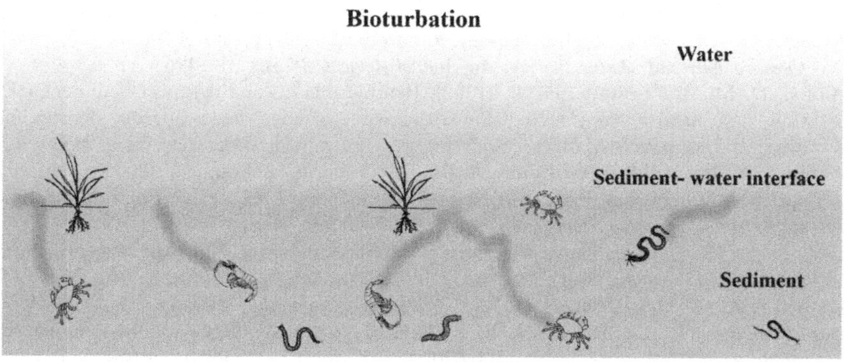

Figure 2.3 Bioturbation activities by benthic organisms.

Various sediment processing activities by insects, mollusks, oligochaetes, crustaceans, and tube dwellers can cause the exchange of particles in the interstitial space, disturbance in chemical equilibria, and changes in the characteristics of organic constituents (Krantzberg, 1985). In marine systems, deposit-feeding benthic organisms can affect the transport of particles present in the sediment (Krantzberg, 1985). Bioirrigation is also dependent on the grain size. The mean distance between the sediment grains changes due to the feeding and burrowing activities of benthic organisms, thus when the grain-to-grain distance increases, the quantity of water between the grains also increases which leads to a reduction of shear strength of the sediment. Also, the vertical sediment structural binding increases due to the mucous secretion of some organisms (McCall et al., 1979).

2.5 Conclusion

Benthic habitats are a very important ecosystem that is the platform for benthic primary production. Various complex interactions exist between the benthic, abiotic, and biotic components. Benthic organisms are ecologically very significant as they can act as the pollution indicators, act as food for higher trophic level organisms and they can enhance organic matter mineralization and nutrient generation. Hence it is very important to understand the various factors influencing, interactions, and activities that exist among them.

References

Alongi, D. M. (1985). Microbes, meiofauna, and bacterial productivity on tubes constructed by the polychaete *Capitella capitata*. Marine ecology progress series. *Oldendorf, 23*(2), 193–196.

Alongi, D. M. (1990). The ecology of tropical soft-bottom benthic ecosystems. *Oceanography and Marine Biology: An Annual Review, 28*(3), 381–496.

Alongi, D. M., & Christoffersen, P. (1992). Benthic infauna and organism-sediment relations in a shallow, tropical coastal area: Influence of outwelled mangrove detritus and physical disturbance. Marine ecology progress series. *Oldendorf, 81*(3), 229–245.

Anthony, K. R., Ridd, P. V., Orpin, A. R., Larcombe, P., & Lough, J. (2004). Temporal variation of light availability in coastal benthic habitats: Effects of clouds, turbidity, and tides. *Limnology and Oceanography, 49*(6), 2201–2211.

Bell, S. S. (1980). Meiofauna-Macrofauna interactions in a high salt marsh habitat. *Ecological Monographs, 50*(4), 487–505.

Brown, A. C., & McLachlan, A. (2010). *The ecology of sandy shores*. Elsevier.

Buhl-Mortensen, P., Dolan, M. F. J., Ross, R. E., Gonzalez-Mirelis, G., Buhl-Mortensen, L., Bjarnadottir, L. R., & Albretsen, J. (2020). Classification and mapping of benthic biotopes in Arctic and sub-Arctic Norwegian Waters. *Frontiers in Marine Science*, 7, 271.

Coull, B. C. (1990). Are members of the meiofauna food for higher trophic levels? *Transactions of the American Microscopical Society*, 233–246.

Coull, B. C. (1999). Role of meiofauna in estuarine soft-bottom habitats. *Australian Journal of Ecology*, *24*(4), 327–343.

Coull, B. C., & Bell, S. S. (1979). *Perspectives of marine meiofaunal ecology. Ecological processes in coastal and marine systems* (pp. 189–216). Boston, MA: Springer.

Davis, W. R. (1993). The role of bioturbation in sediment resuspension and its interaction with physical shearing. *Journal of Experimental Marine Biology and Ecology*, *171*(2), 187–200.

Day, J. W., Jr., Yanez-Arancibia, A., Kemp, W. M., & Crump, B. C. (2012). Introduction to estuarine ecology. *Estuarine Ecology*, *2*, 1–19.

Flemming, H. C., Schmitt, J., & Marshall, K. C. (1996). *Sorption properties of biofilms. Sediments and toxic substances* (pp. 115–157). Berlin, Heidelberg: Springer.

Galope-Bacaltos, D. G., & San, M. D. M. (2002). Composition and spatial distribution of infauna in a river estuary affected by fishpond effluents. *Marine Pollution Bulletin*, *44*(8), 816–819.

Gaston, G. R., Lee, D. L., & Nasci, J. C. (1988). Estuarine macrobenthos in Calcasieu Lake, Louisiana: community and trophic structure. *Estuaries*, *11*(3), 192–200.

Gattuso, J. P., Gentili, B., Duarte, C. M., Kleypas, J. A., Middelburg, J. J., & Antoine, D. (2006). Light availability in the coastal ocean: impact on the distribution of benthic photosynthetic organisms and their contribution to primary production. *Biogeosciences*, *3*(4), 489–513.

Gee, J. M., Warwick, R. M., Davey, J. T., & George, C. L. (1985). Field experiments on the role of epibenthic predators in determining prey densities in an estuarine mudflat. *Estuarine, Coastal and Shelf Science*, *21*(3), 429–448.

Giere, O. (2009). *The biotope: factors and study methods. Meiobenthology: The microscopic motile fauna of aquatic sediments* (pp. 7–62). Springer.

Gray, J. S., & Elliott, M. (2009). *Ecology of marine sediments: From science to management*. Oxford University Press.

Green, A. S., & Chandler, G. T. (1994). Meiofaunal bioturbation effects on the partitioning of sediment-associated cadmium. *Journal of Experimental Marine Biology and Ecology*, *180*(1), 59–70.

Heip, C., Vincx, M., & Vranken, G. (1985). *The ecology of marine nematodes*. Aberdeen University Press.

Hicks, G. R. F., & Coull, B. C. (1983). The ecology of marine meiobenthic harpacticoid copepods. *Oceanography and Marine Biology Annual Review*, *21*, 67–175.

Hossain, M. B. (2011). *Macrobenthic community structure from a tropical estuary* (p. 84) LAP Publishing Company.

Hossain, M. B., Marshall, D. J., & Venkatramanan, S. (2014). Sediment granulometry and organic matter content in the intertidal zone of the Sungai Brunei estuarine system, northwest coast of Borneo. *Carpathian Journal of Earth and Environmental Sciences*, *9*(2), 231–239.

Jayaraj, K. A., Jayalakshmi, K. V., & Saraladevi, K. (2007). Influence of environmental properties on macrobenthos in the northwest Indian shelf. *Environmental Monitoring and Assessment*, *127*(1–3), 459–475.

Kim, H., Song, S. J., Bae, H., Noh, J., Lee, C., Kwon, B., Lee, J., Ryu, J., & Khim, J. S. (2020). Natural and anthropogenic impacts on long-term meiobenthic communities in two contrasting nearshore habitats. *Environment International*, *134*, 105200. Available from https://doi.org/10.1016/j.envint.2019.105200.

Kirk, J. T. (1994). *Light and photosynthesis in aquatic ecosystems*. Cambridge university press.

Krantzberg, G. (1985). The influence of bioturbation on physical, chemical and biological parameters in aquatic environments: A review. *Environmental Pollution Series A, Ecological and Biological, 39*(2), 99–122.

Leduc, D., Rowden, A. A., Probert, P. K., Pilditch, C. A., Nodder, S. D., Vanreusel, A., Duineveld, G. C. A., & Witbaard, R. (2012). Further evidence for the effect of particle-size diversity on deep-sea benthic biodiversity. *Deep Sea Research Part I: Oceanographic Research Papers, 63*, 164–169.

McCall, P. L., Tevesz, M. J., & Schwelgien, S. F. (1979). Sediment mixing by *Lampsilis radiata* siliquoidea (Mollusca) from western Lake Erie. *Journal of Great Lakes Research, 5* (2), 105–111.

McLachlan, A., & Turner, I. (1994). The interstitial environment of sandy beaches. *Marine Ecology, 15*(3–4), 177–212.

Meyers, M. B., Fossing, H., & Powell, E. N. (1987). Microdistribution of interstitial meiofauna, oxygen and sulfide gradients, and the tubes of macro-infauna. *Marine Ecology Progress Series*, 223–241.

Mumby, P. J., Chisholm, J. R., Edwards, A. J., Andrefouet, S., & Jaubert, J. (2001). Cloudy weather may have saved Society Island reef corals during the 1998 ENSO event. *Marine Ecology Progress Series, 222*, 209–216.

Munksgaard, N. C., & Parry, D. L. (2002). Metals, arsenic and lead isotopes in near-pristine estuarine and marine coastal sediments from northern Australia. *Marine and Freshwater Research, 53*(3), 719–729.

Ólafsson, E., & Moore, C. G. (1990). Control of meiobenthic abundance by macroepifauna in a subtidal muddy habitat. *Marine Ecology Progress Series, 65*, 241–249.

Paine, R. T. (1974). Intertidal community structure. *Oecologia, 15*(2), 93–120.

Raut, D., Ganesh, T., Murty, N. V. S. S., & Raman, A. V. (2005). Macrobenthos of Kakinada Bay in the Godavari delta, East coast of India: Comparing decadal changes. *Estuarine, Coastal and Shelf Science, 62*(4), 609–620.

Reise, K. (1978). Experiments on epibenthic predation in the Wadden Sea. *Helgoländer Wissenschaftliche Meeresuntersuchungen, 31*(1–2), 55–101.

Reise, K. (2002). Sediment mediated species interactions in coastal waters. *Journal of Sea Research, 48*(2), 127–141.

Sarkar, S. K., Bhattacharya, A., Giri, S., Bhattacharya, B., Sarkar, D., Nayak, D. C., & Chattopadhaya, A. K. (2005). Spatiotemporal variation in benthic polychaetes (Annelida) and relationships with environmental variables in a tropical estuary. *Wetlands Ecology and Management, 13*(1), 55–67.

Simboura, N., Nicolaidou, A., & Thessalou-Legaki, M. (2000). Polychaete communities of Greece: An ecological overview. *Marine Ecology, 21*(2), 129–144.

Strayer, D. L., May, S. E., Nielsen, P., Wollheim, W., & Hausam, S. (1997). Oxygen, organic matter, and sediment granulometry as controls on hyporheic animal communities. *Archiv für Hydrobiologie, 140*(1), 131–144.

Swami, B. S., Suryawanshi, U. G., & Karande, A. A. (2000). Water quality status of Mumbai (Bombay) harbour-an update. *Indian Journal of Geo-Marine Sciences, 29*(2), 111–115.

Thrush, S. F., & Townsend, C. R. (1986). The sublittoral macrobenthic community composition of Lough Hyne, Ireland. *Estuarine, Coastal and Shelf Science, 23*(4), 551–573.

Van Duin, E. H., Blom, G., Los, F. J., Maffione, R., Zimmerman, R., Cerco, C. F., Dortch, M., & Best, E. P. (2001). Modeling underwater light climate in relation to sedimentation, resuspension, water quality and autotrophic growth. *Hydrobiologia, 444* (1), 25–42.

Varadharajan, D., Soundarapandian, P., Gunalan, B., & Babu, R. (2010). Seasonal abundance of macro benthic composition and diversity along the southeast coast of India. *European Journal of Applied Sciences, 2*(1), 1–5.

Venkatramanan, S., Ramkumar, T., & Anithamary, I. (2013). Distribution of grain size, clay mineralogy and organic matter of surface sediments from Tirumalairajanar Estuary, Tamilnadu, East coast of India. *Arabian Journal of Geosciences*, *6*(5), 1371–1380.

Wright, W. J. (1997). Tropical–extratropical cloud bands and Australian rainfall: I. Climatology. *International Journal of Climatology: A Journal of the Royal Meteorological Society*, *17*(8), 807–829.

CHAPTER 3

Collection, preservation, and documentation of estuarine and marine benthic invertebrates

R. Ravinesh[1,2] and A. Biju Kumar[2]
[1]Coastal and Marine Ecology Division, Gujarat Institute of Desert Ecology, Bhuj (Kachchh), Gujarat, India
[2]Department of Aquatic Biology & Fisheries, University of Kerala, Kariyavattom, Thiruvananthapuram, Kerala, India

3.1 Introduction

The marine ecosystem, especially the marine benthic habitats, are exceptionally rich and diverse in terms of the complexity of substratum and other physicochemical factors and climatic regimes, including the deepest trenches and extreme environments such as hydrothermal vents as well as chemosynthetic habitats (Griffiths et al., 2017; McIntyre, 2010). The marine invertebrates also extend a wide array of ecosystem services such as water purification/filtering, nutrient recycling, and maintenance of marine food webs (Collier et al., 2016; Gosner, 1971; Palumbi et al., 2009). Marine invertebrates such as crustaceans and molluscs often form economically valuable food forms (Khan & Liu, 2019; Lindberg, 2001) and supply several commercial products such as pearls (Zhu et al., 2019). Coral reef ecosystems made primarily with the contribution of invertebrates offers protection from natural hazards (Ferrario et al., 2014; Stella et al., 2011), provide climate change resilience and protection from extreme weather (Cochrane et al., 2016; Duffy et al., 2013), and offers a series of novel bioactive compounds of medicinal importance to mankind (Gouda et al., 2016; Lekshmi et al., 2020; Manomi et al., 2020). Benthic animals also serve as biological indicators of environmental conditions in the estuarine and seafloor habitats (Dreujou et al., 2020; Pratt et al., 2014). Despite all these values, most parts of the marine habitats are not fully explored for documenting invertebrate biodiversity (Collier et al., 2016; Henseler et al., 2019; Luypaert et al., 2020). In the international Union for Conservation of Nature (IUCN) Red List of Threatened Species, which is one the world's well-accepted classification of global risk assessment of

Ecology and Biodiversity of Benthos
DOI: https://doi.org/10.1016/B978-0-12-821161-8.00007-6

species, marine invertebrates are least represented; it is only 2967 against marine vertebrates, which are 11,617 (Chen, 2021).

Biodiversity surveys are one of the most potent tools in assessing and asserting the frontiers of our information into these unknown regions (Purvis et al., 2019; Thomsen and Willerslev, 2015). Biodiversity documentation, including identification and assessment of occurrence, relative abundance, and population dynamics, are mandatory for enriching our knowledge on construction and organization of communities, the effective functioning of the ecosystems and ecosystem services, habitat typologies, and the level of endemism of any geographic region (Fajardo et al., 2014), or to prevent and manage alien species (Di Camillo et al., 2018). Overall, taxonomic identification is essential for biodiversity monitoring and understanding vectors of ecological changes (Boero, 1994, 2001, 2010; Yeates et al., 2011). The systematic inventories and collections would serve as a tangible scientific record for documenting the patterns of diversity across the marine realm and would provide the baseline data for planning and prioritizing conservation, besides discovering and describing new species (Mikkelsen & Cracraft, 2001), not to speak of the values of such documentation in impressing the policymakers toward national, international commitments in conserving the life under water.

Collecting specimens can only be supported when there is a real educational or scientific need, and casual collecting with inadequate documentation can never be justified (Winker et al., 2010). At the same time, the number of professional morphological taxonomists working in museums, research institutions, and universities is gradually declining (Coleman, 2015; Gemeinholzer et al., 2010, 2020; Godfray, 2002; Hopkins & Freckleton, 2002; Hutchings, 2020; Leather, 2009; Ng, 2017). The unplanned and indiscriminate fieldwork may often prove deleterious to the habitat itself and often seriously disturb or damage the ecosystem and impact biodiversity (Nakano, 2020). Properly preserved materials are essential for museum collections, supporting research and academics in universities and research and development institutes, and molecular studies (Ponder et al., 2019, 2020; Suarez & Tsutsui, 2004). The collection methods of invertebrates vary based on taxa, habitats they occupy, and the purpose of collection (Olson, 1981). In marine and estuarine regions, the most frequently encountered benthic invertebrate groups are polychaetes, crustaceans, molluscs, echinoderms, and members of the few groups are poorly lesser known animal phyla (Lalli & Parsons, 1997). Larger marine exploratory surveys are usually generated by numerous marine animals and supportive data samples. New collections, particularly the huge variety of organisms, impose an important load on resources at research institutions. Samples must be processed and documented to reach virtuous conditions and can be incorporated professionally interested in museum collections and databases. Most of the research institutions or museums will generally not accept samples preserved in poor condition or poorly documented. In the current research, DNA analysis is performed intensively by many scientists. Besides whole specimens, many museums collect and store tissue samples for DNA analysis. After the invertebrates samples are collected, all the specimens should be properly preserved for the reference collection. Estuarine and marine invertebrates occur from the sea surface to the seafloor and into the substrate. More than 90% of all animals are invertebrates, and they are classified into at least 33 major groups or phyla. Most of the groups of benthic invertebrates have listed in Table 3.1.

3.2 Collection permits

Any collection from protected areas such as marine reserves requires permission from the respective authorities (Kelleher, 1999). Collectors should be aware of the regulations that prohibit or restrict the collection of live specimens in many parts of the world. For example, many molluscs such as *Tridacna* spp. and *Hippopus hippopus* are legally protected around the world. Most of the collectors or researchers are now aware of the critical importance of obtaining proper permits for collecting the animals. Some animals are listed in the Convention on International Trade in Endangered Species (CITES, available online at https://www.cites.org), International Union for Conservation of Nature (IUCN), and Wildlife Protection Act. Threatened and endangered species cannot be collected or moved across international borders without special permits (Saarman et al., 2018; Costello et al., 2017; Martin, 2016).

While collecting specimens or working with specimens from foreign countries, the researchers should check the prevailing regulations in the respective countries regarding the import and export of specimens and rules related to the Convention on Biological Diversity (CBD). The Nagoya Protocol on Access to Genetic Resources and the Fair and Equitable Sharing of Benefits Arising from their Utilization (ABS) to the CBD offers the legal framework for the access to biological resources and sharing of benefits.

Table 3.1 List of major estuarine and marine benthic invertebrate phyla, subgroups and common names/representatives and ecology.

Phylum	Subgroups	Common names/representatives	Ecology
Retaria (Fig. 3.1A–F)	Foraminifera	Forams or hole bearers	Fine-grained sediments, hard rock substrates, attached to seaweeds, sediment surface, shallow water, and deep sea
Bryozoans (Fig. 3.2A and B)		Ectoprocta or moss animals	Mostly encrusting forms, rock substrates, and fouling
Porifera (Fig. 3.3A–B)		Sponges	Rocks and soft sediments
Cnidaria	Hydrozoa (Fig. 3.4A and B)	Hydroid polyps	Rocky crevices, channels, tide pools, or under the macroalgal canopy, shallow, and deep-sea soft and hard substrates
	Anthozoa (Fig. 3.5A–G)	Sea anemones, black coral, stony corals or hard corals, zoanthids, soft corals and gorgonians, sea pens, sea feathers, sea pansies, and tube-dwelling anemones	Shallow and deep-sea soft and hard substrates
Platyhelminthes (Fig. 3.6A–D)	Turbellaria	Flatworms	Hardy substrates and rock crevices.
Nematoda		Roundworms	Hard and soft sediments
Nemertea (Fig. 3.7A–E)		Ribbon worm or bootlace worm	Under stones, rock crevices
Annelida (Fig. 3.8A–G)	Polychaeta	Bristle worms or polychaetes	Burrows or build tubes in the sediment and commensals, including shallow and deep waters
Sipuncula (Fig. 3.9A and B)		Sipunculid worms or peanut worms	Burrowing sandy and muddy substrates under stones and rock crevices
Echiura (Fig. 3.10A and B)		Echiurids (spoon worms)	Burrowing sandy and muddy substrates under stones and rock crevices

Mollusca	Gastropoda (Fig. 3.11A–L)	Snails and sea slugs	Coral reefs, rocky shores, seagrass, mangroves, muddy or sandy bottom including shallow and deep waters
	Bivalvia (Fig. 3.12A–P)	Clams, oysters, scallops, geoducks, mussels	Coral reefs, rocky shores, seagrass, mangroves, muddy or sandy bottom including shallow and deep waters
	Polyplacophora (Fig. 3.13A and B)	Chitons	Coral reefs, rocky shores, and hardy substrates
	Scaphopoda (Fig. 3.13C and D)	Tusk-shells	Soft sandy and muddy substrates
	Cephalopoda (Fig. 3.13E and F)	Octopus	Coral reefs, rocky shores, and rubbles bottom
Brachiopoda (Fig. 3.14A and D)		Lamp shells	Rocky projections, crevices, caves, and steep slopes of continental shelves
Arthropoda (Class Crustacea)	Ostracoda	Ostracods	Silty or sandy and muddy bottoms and associated with algae
	Copepoda	Cyclopoids; harpacticoids	Hard and soft substrates, caves, and parasitic
	Tanaidacea	Tanaids	Bottom dwellers
	Isopoda (Fig. 3.15A–C)	Isopods	Silty or sandy and muddy bottoms and associated with algae
	Amphipoda (Fig. 3.15C–G)	Amphipods	Silty or sandy and muddy bottoms and associated with algae
	Cirripedia (Fig. 3.15H–J)	Barnacles	Hard substrates
	Decapoda (Fig. 3.16A–G and Fig. 3.17A–G)	Shrimps, spiny lobsters, slipper lobsters, mud shrimp, ghost shrimp, or burrowing shrimps, stomatopod, crabs, and hermit crabs	Coral reefs, hardy substrates, rocky cervices, silty or rocky shores, tidal pools, muddy and sandy, including shallow and deep waters

(*Continued*)

Table 3.1 (Continued)

Phylum	Subgroups	Common names/representatives	Ecology
Echinodermata	Asteroidea (Fig. 3.18A–C)	Starfish	Coral reefs, rocky shores, tidal pools, mud, and sand to kelp forests, seagrass meadows
	Ophiuroidea (Fig. 3.18D and F)	Brittle stars	Associated with sponges, corals, seaweeds, rocks, under the sand or mud
	Holothuroidea (Fig. 3.18G–J)	Sea cucumbers	Coral reefs, rocky shores, tidal pools, mud, and sand to kelp forests, seagrass meadows
	Echinoidea (Fig. 3.18K–N)	Sand dollars and sea urchins	Coral reefs, rocky shores, seagrass meadows, tidal pools, and sandy, including shallow and deep waters
	Crinoidea (Fig. 3.18O)	Feather stars; sea lilies, or comatulids	Hardy substrates and artificial structures

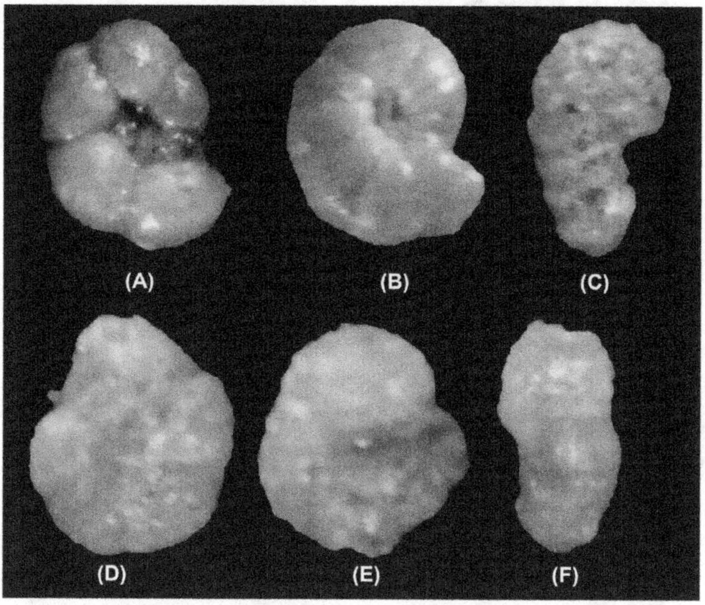

Figure 3.1 [(A)–(F)] Representatives of common foraminifera from the sea.

3.3 Risky animals

Researchers should be aware of the risky animals that they might encounter during the fieldwork. The invertebrate venoms cause mild irritation to immediate death in affected organisms (Williamson et al., 1996; Caras, 1976). Some toxic sponges are found worldwide, and they cause painful skin irritations, sometimes persisting for many hours. Most cnidarians, including hydroids, fire corals, hard corals, soft corals, and sea anemones, may also produce toxins. It causes nettle-like rashes lasting for several days (Rifkin et al., 1993; Sagi et al., 1987). The box jellyfish is a dangerous, highly venomous animal that can cause death within minutes (Lau et al., 2019). However, certain predatory gastropods are equipped with poison glands and means of delivering the poison, which is used to subdue prey. The most infamous of these is the cone shells (family Conidae; Fig. 3.19A); researchers strongly recommend leaving live cone shells alone. All the cones have a poisonous gland derived from the salivary gland in the mouth (Halstead et al., 1990; Hinegardner, 1958). Most nudibranchs produce toxic chemicals or stinging cells from their prey into their skin (Behrens, 2005). Another group of the most dangerous marine molluscs common around the world is the blue-ringed octopus. In this

Figure 3.2 [(A) and (B)] Representatives of common bryozoans.

case, the poison is administered through a bite from beak-like jaws. Not all octopuses have a dangerous bite on humans, but the blue rings are especially dangerous and have been responsible for several fatalities. If stung from these animals, seek professional medical attention immediately, and the best way to avoid the risk is by not handling live specimens (Williamson, 1987). Few echinoderms are also dangerous to humans. The most common slight damages are scratches or wounds acquired from the spines or skin of echinoderms crown of thorns starfish (*Acanthaster planci*; Fig. 3.19B) and sea urchins (Smith, 1977).

Stingrays usually cause injuries by accidentally stepping on while partially buried on a mudflat or in the sand (Fig. 3.20A). The sting is caused

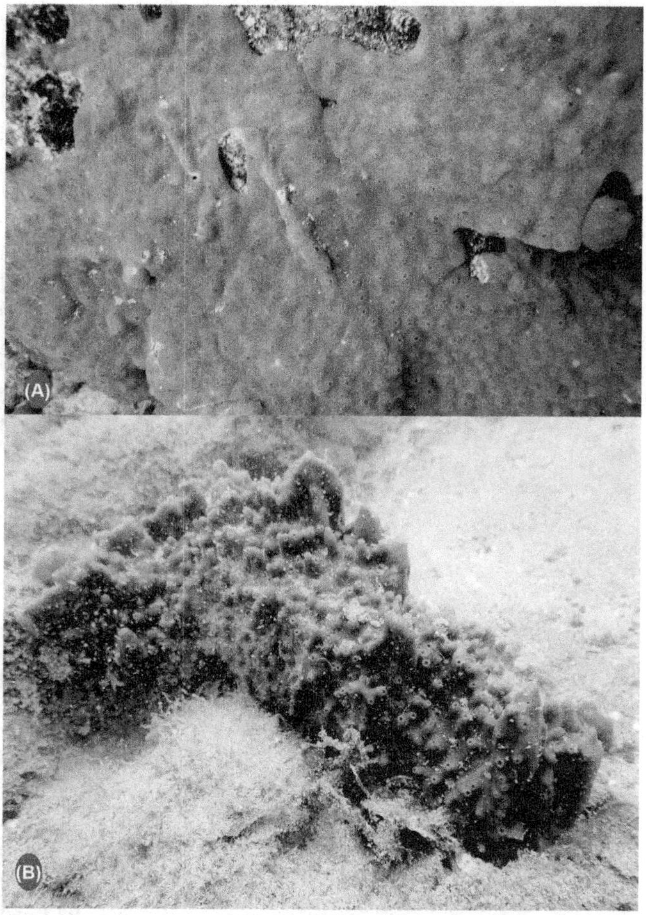

Figure 3.3 [(A) and (B)] Representatives of Porifera.

by a serrated spine situated on a whip-like tail (Myatt et al., 2018). The stonefish is an ugly-looking, sluggish fish that is highly dangerous and usually deadly; injuries are caused by the long caudal spines that eject venom (Fig. 3.20B). It is generally encountered in muddy localities or among reef coral, where its camouflage makes it blend with the natural surroundings of coral, stone, or mud. Scorpionfish are colorful, spinose fish frequently found in shallow water, lurking among or under the coral (Fig. 3.20C). Most injuries are caused by punctures from their venomous spines while lifting coral rocks with bare hands. Stings are painful but not fatal (Rensch & Lavoie, 2021).

Figure 3.4 [(A) and (B)] Representative of hydrozoan.

3.4 Collection methods and apparatus

3.4.1 Hand-picking

Hand-picking is one of the very good sample collection methods used for our bare hands, mainly used rocky shores during low tide, seagrass, sandy shores, mangroves, and tide pools. All the animals were collected directly, and smaller can be removed by using smaller tools such as forceps, knives and brushes, hammer/chisel, can be used for the rocks, coral boulders, dead corals associated animals, and yabby pumps are used for the sandy or muddy bottom. Also hand collection involves picking up the animals

Figure 3.5 [(A) and (B)] Representatives of sea anemones; (C) and (D) hard corals, (E) soft coral; (F) sea fan or gorgonian; and (G) sea pen.

directly through the hand or with a hand-held sieve, scoop or rake, or sampling substrate such as sand, mud, or weed for later examination, usually under magnification (Fig. 3.21A–F).

Figure 3.6 [(A)–(D)] Representatives of flatworms.

3.4.2 Snorkeling

Snorkeling is also one of the well-known practices of collecting samples on the mangroves, lagoons, coral reefs, seagrass, and rocky intertidal to shallow subtidal. Swimming on while equipped with a dive mask, a shaped breathing apparatus called a snorkel, and usually used swim fins (Fig. 3.22).

3.4.3 Self-Contained Underwater Breathing Apparatus diving

Self-Contained Underwater Breathing Apparatus (SCUBA) is one of the excellent sampling methods for most of the marine invertebrates. SCUBA is not only used for the sample collection that can be used to see the animals through ecological observations such as movement, feeding, ecology, crawling, mating, laying eggs, etc. The advent of SCUBA diving has opened up a completely new world for marine biologists, explorers, and collectors, as divers with moderate experience can reach depths up to 50 m for short periods without special gas mixtures (Munro, 2013; Fig. 3.23A and B).

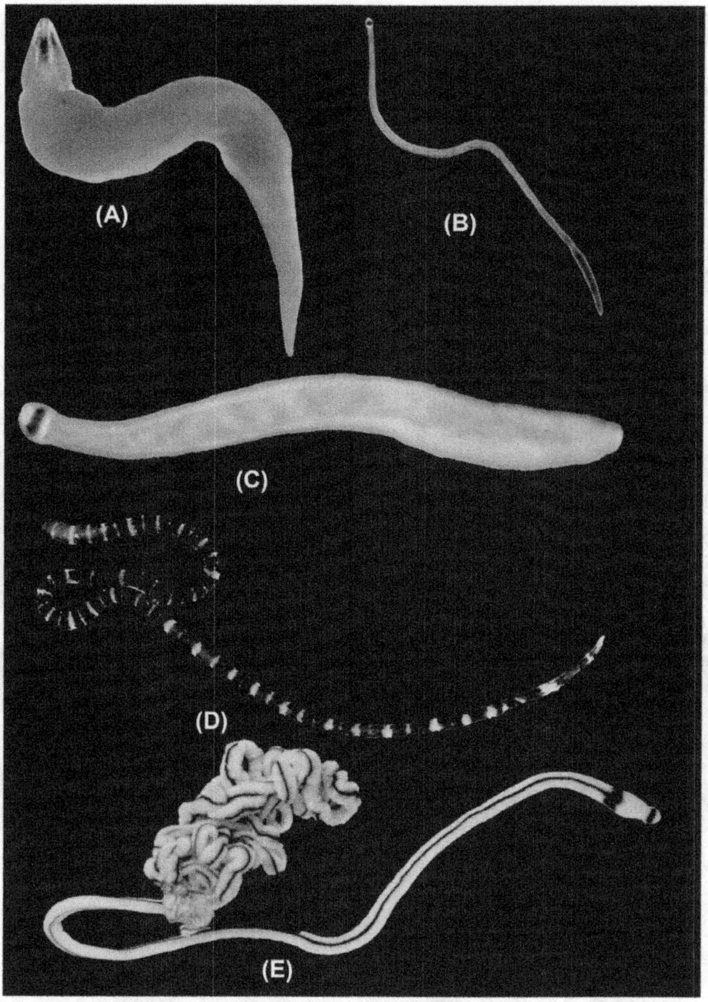

Figure 3.7 [(A)–(E)] Representations of the ribbon worms.

3.4.4 Bottom grabs

Sampling the macro and micro meiobenthic sampling Petersen grab, Hamon, and Van Veen grab may be used to retrieve soil samples from the seabed surface. This is used as seagrass beds, sandy, muddy bottom more than 1000 m depth. Several types of grabs were available, the most common of which are the Shipek, van Veen, and Petersen sediment samplers (Fig. 3.24A and B).

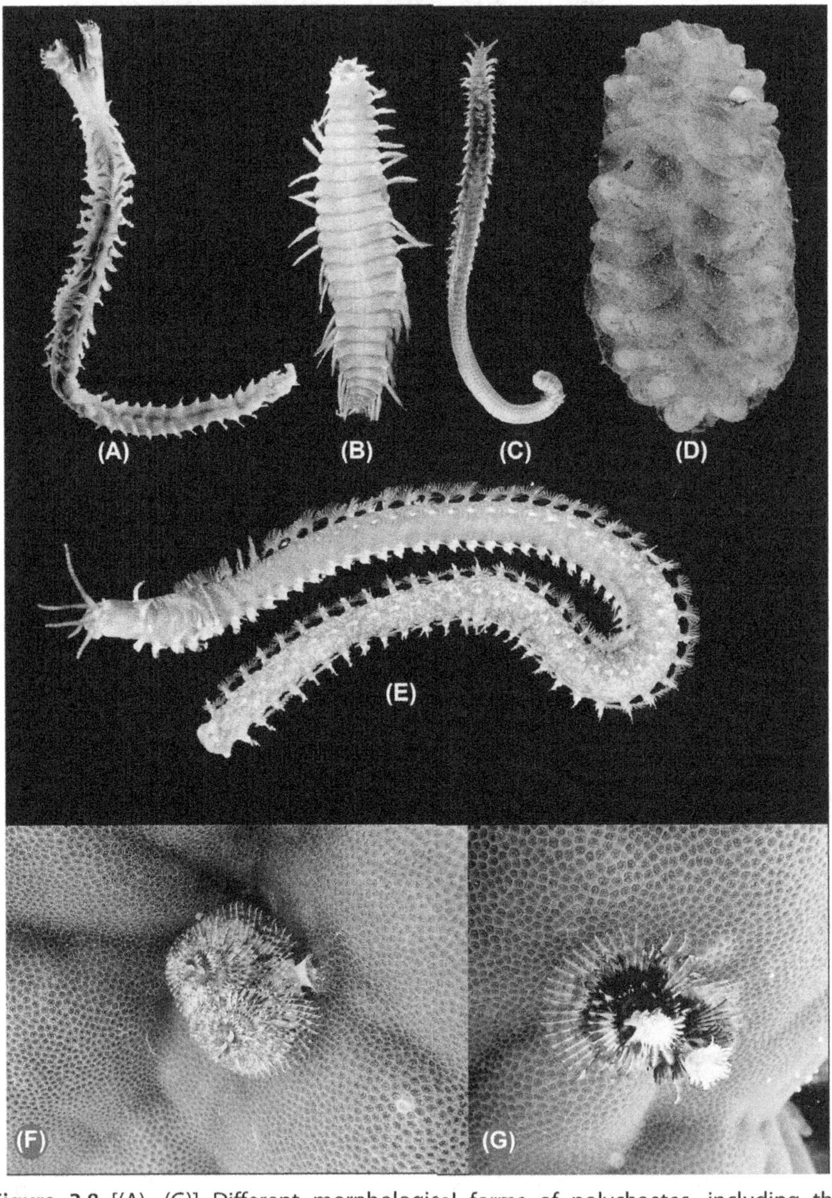

Figure 3.8 [(A)–(G)] Different morphological forms of polychaetes, including the coral reef-associated species.

3.4.5 Core sampling

Core sampling is one of the cheap and simple methods to collect samples from soft sediment regions and mudflats. A core sample is the roughly

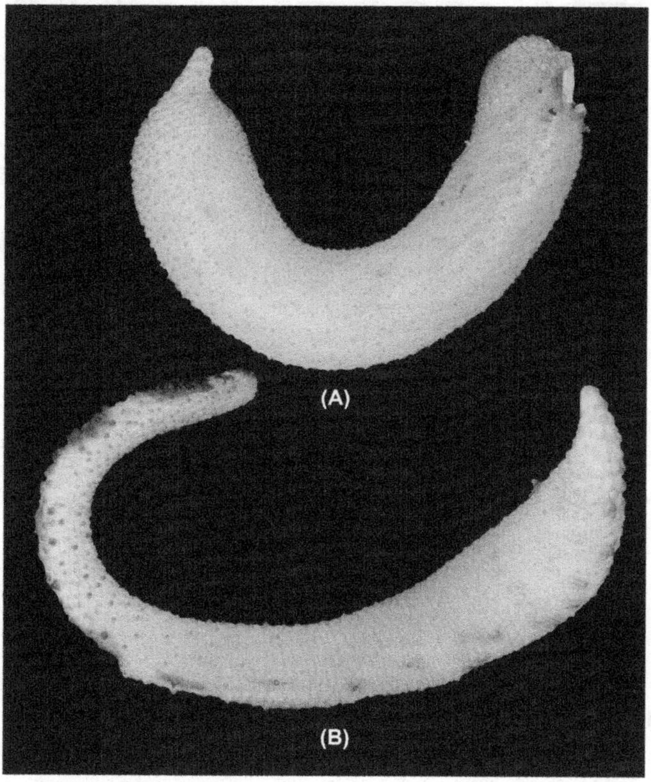

Figure 3.9 [(A) and (B)] Two different morphological forms of the phylum Sipuncula.

cylindrical piece of subsurface material removed by a special drill and brought to the surface for examination. A box corer is similar to a grab though much larger. It is used to collect molluscs and polychaetes that remain buried deep in the bed material. If corers are not available in local situations, a strong PVC pipe driven into the bottom may serve the purpose.

3.4.6 Biological dredge with net

Heavy-duty dredge with the net is ideal for the qualitative sampling of muddy or sandy bottoms. This type is mostly used as the deep-sea exploratory vessels; this is one of the suitable methods to collect deepwater rocky and muddy substrates (Ockelmann, 1964; Fig. 3.25A–D).

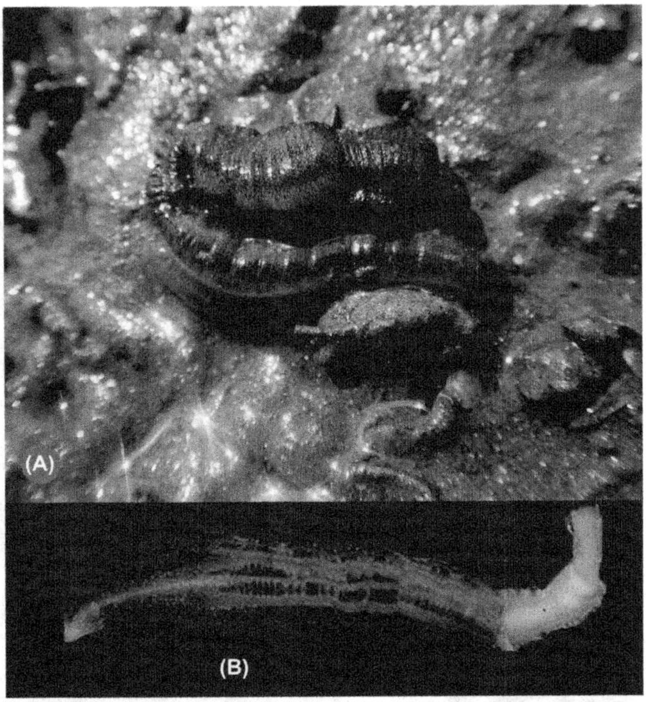

Figure 3.10 [(A) and (B)] The morphological forms of spoon worm.

3.4.7 Trawl

Trawling is one of the traditional methods for collecting large samples of organisms from shallow water to the deep sea. Mainly trawl is used on a sandy or muddy bottom and is used to cover large areas. Even though many different types of trawling operations are used globally, and most of the designs originated from coastal fishing and have either been physically or operationally modified for use in the deep sea, where there are difficulties that do not occur in shallower fishing areas, such as unknown topography and large wire-out to depth ratios. Mainly bottom trawls are used to catch macroinvertebrate communities that live on or near the seafloor. Bottom trawlers use timber or steel trawl boards to keep the mouth of the net open and on the bottom (Ockelmann, 1964; Mare, 1942; Fig. 3.26A–D).

3.4.8 Substratum removal

Active sampling can be done by quickly digging to the sampling depth or turning big stones or rocks using levers at a depth of 20–30 cm or more

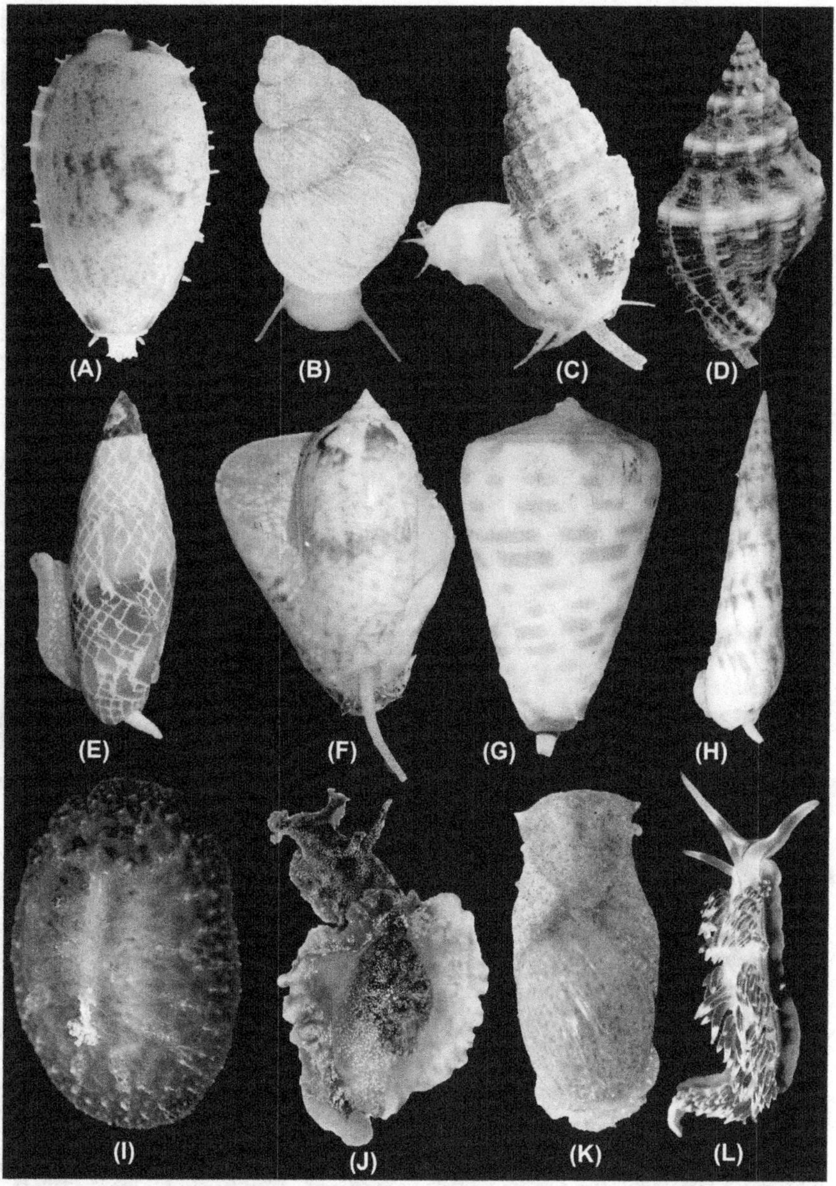

Figure 3.11 [(A)–(L)] Different morphological forms of gastropods and sea slugs.

(Fig. 3.27A). Both activities are arduous, and the latter is risky of injuries and must be undertaken by several people. Underwater rocks, reefs, gravels, finer sediments, and seaweeds are removed manually, and the animals are searched under a microscope/hand lens (Sturm et al., 2006). The bottom of the rocks

Figure 3.12 [(A)–(P)] Different morphological forms of bivalves.

is searched to collect rocky shore-associated animals such as cnidaria, crustacea, mollusca, echinodermata, annelida, porifera, platyhelminthes, and sipuncula (Ravinesh & Biju Kumar, 2013). Seaweeds are one of the important habitats; molluscs, especially sap-sucking slugs (sacoglossans), are found only

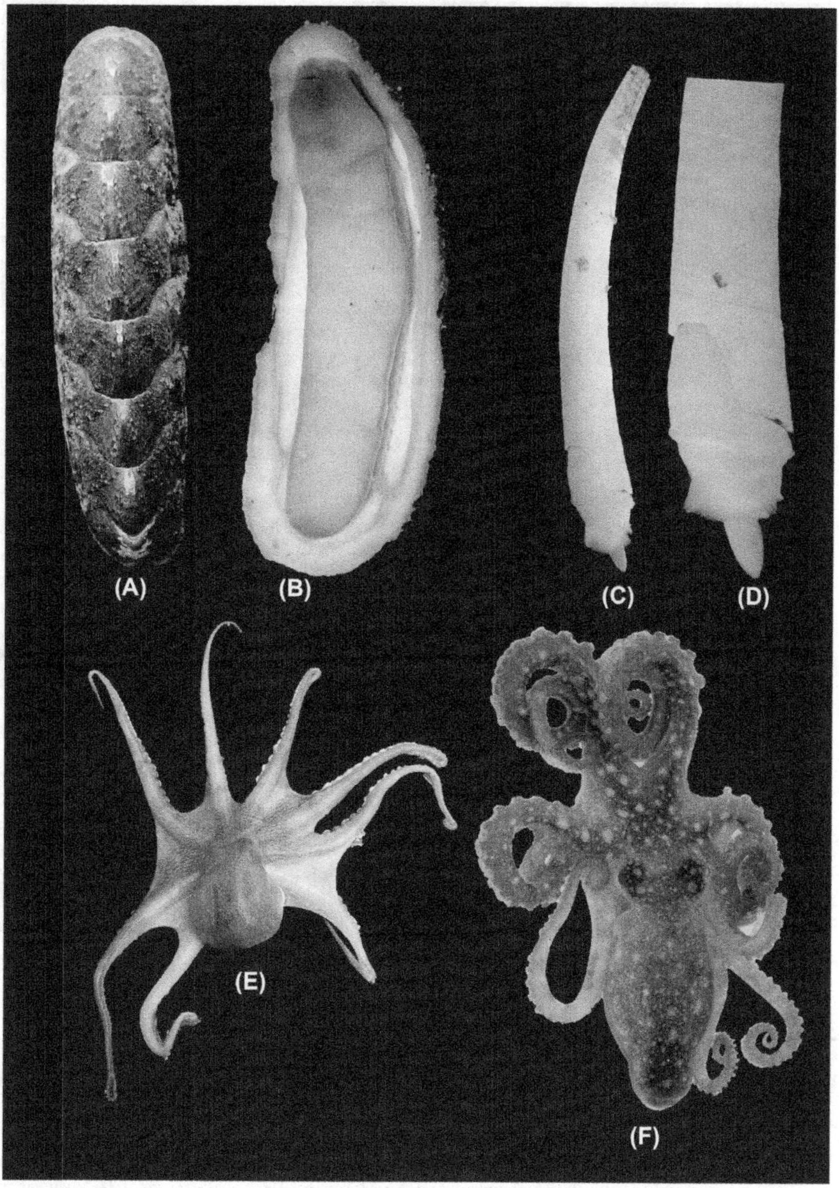

Figure 3.13 [(A) and (B)] Representation of Polyplacophora; [(C) and (D)] Scaphopoda; and [(E) and (F)] Cephalopoda.

on the particular algae and sponges, bryozoans, and hydroids are ascidians one of the important feeding habitats for the nudibranchs and other molluscs (David, 2005; Gosliner et al., 2018; Fig. 3.27B–D).

Figure 3.14 [(A)–(D)] Two different morphological forms of the phylum Brachiopoda.

3.5 Sample processing

Larger expeditions generate a remarkable amount of material to identify. All the species are not always easy to identify species from underwater themselves, and collected samples may not be mass-preserved in the field. Collected samples are best preserved by life itself; as soon as life ceases to exist, decomposition sets in. So that reason for proper processing of the materials databasing, storage of collections, and taxonomic studies (including visits to museum collections) is essential for future research. They may also be picked at the field site, later sorted at the laboratory or field station (Gemeinholzer et al., 2010, 2020) (Table 3.2).

Figure 3.15 [(A)–(C)] Different morphological forms of Isopods; [(D)–(G)] Amphipods, and [(H)–(J)] Barnacles.

There are three major steps to systematically documenting collected animals in specimens or live animal processing: photography, subsampling, and voucher preparation.

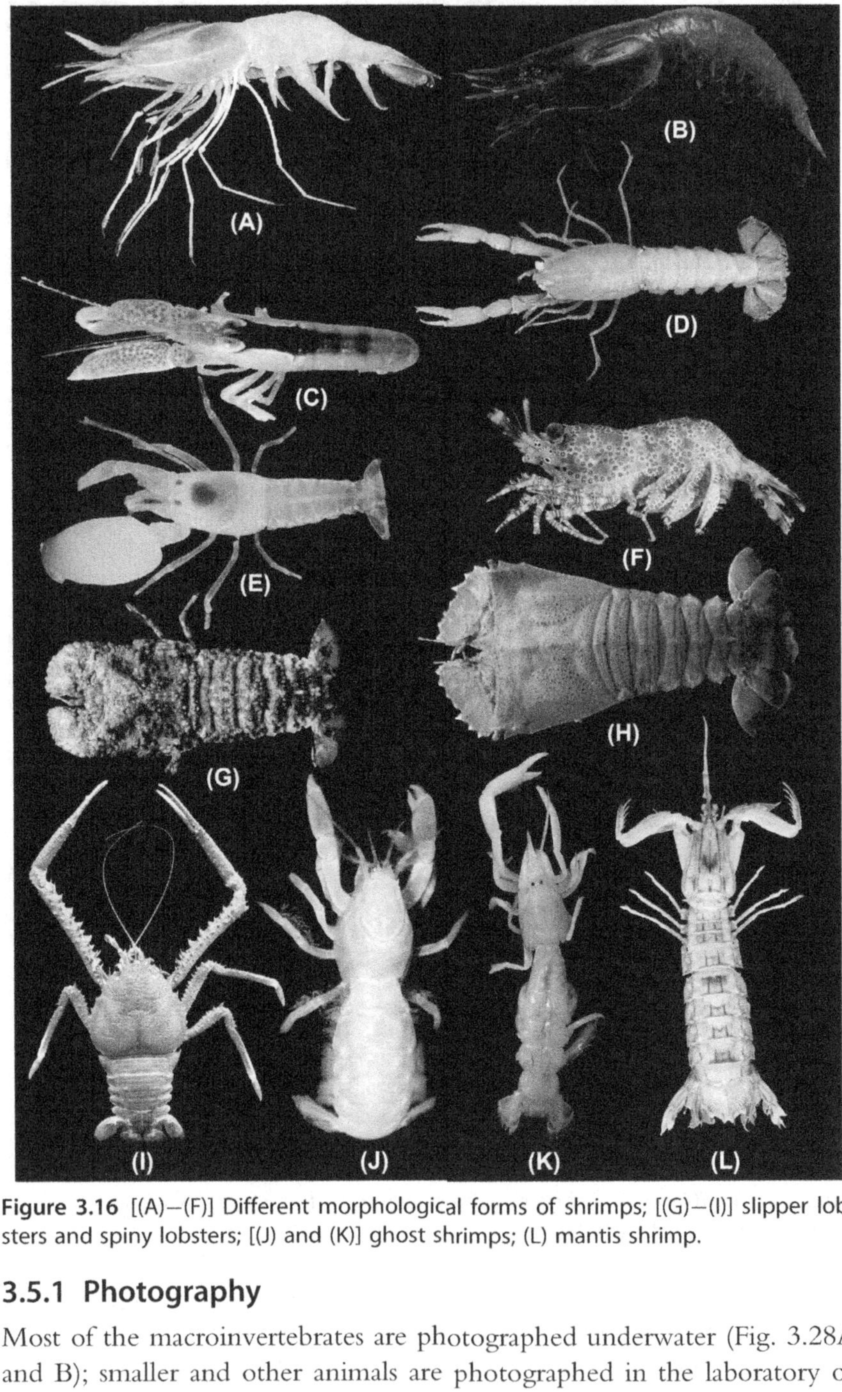

Figure 3.16 [(A)–(F)] Different morphological forms of shrimps; [(G)–(I)] slipper lobsters and spiny lobsters; [(J) and (K)] ghost shrimps; (L) mantis shrimp.

3.5.1 Photography

Most of the macroinvertebrates are photographed underwater (Fig. 3.28A and B); smaller and other animals are photographed in the laboratory or

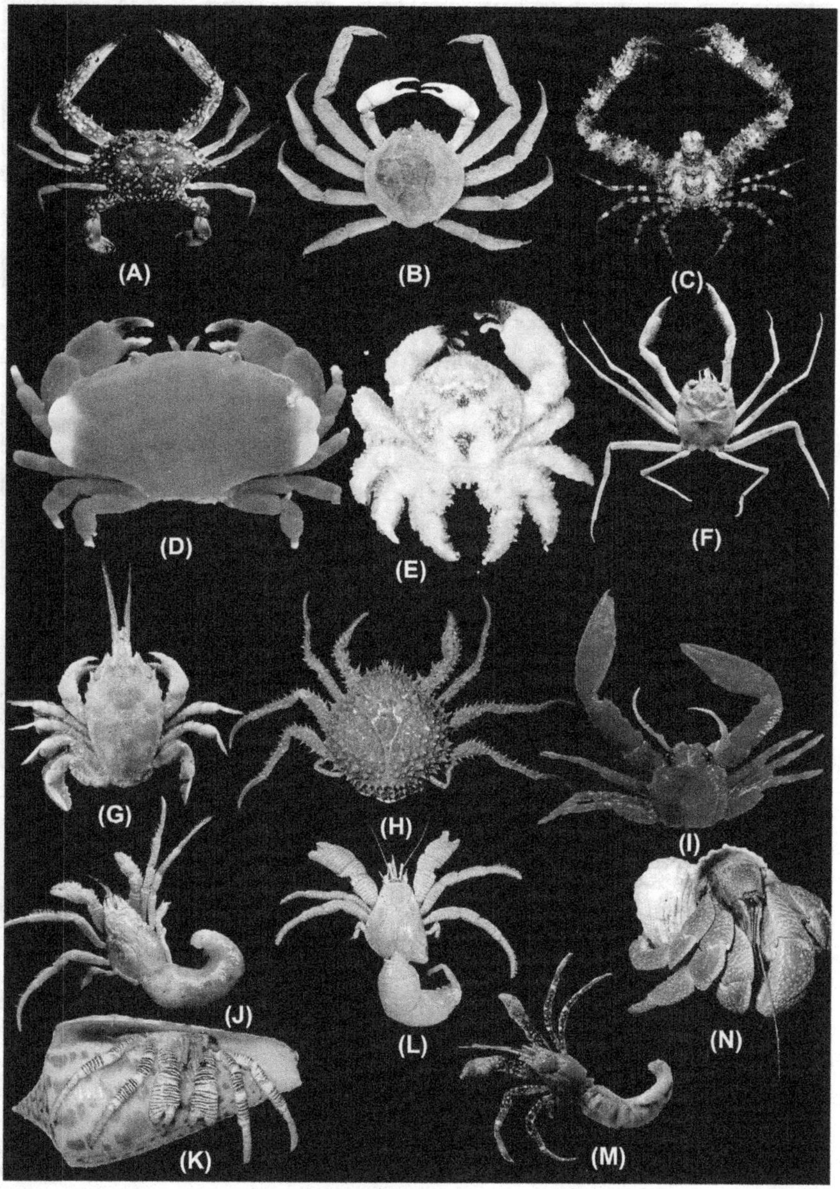

Figure 3.17 [(A)–(G)] Different morphological forms of brachyuran crabs; [(H)–(N)] anomuran crabs.

the field station (Fig. 3.28C and D). For laboratory or field station photography, use a small aquarium tank, sufficient water to cover the whole animal. After prefixation, the photography procedure should be started

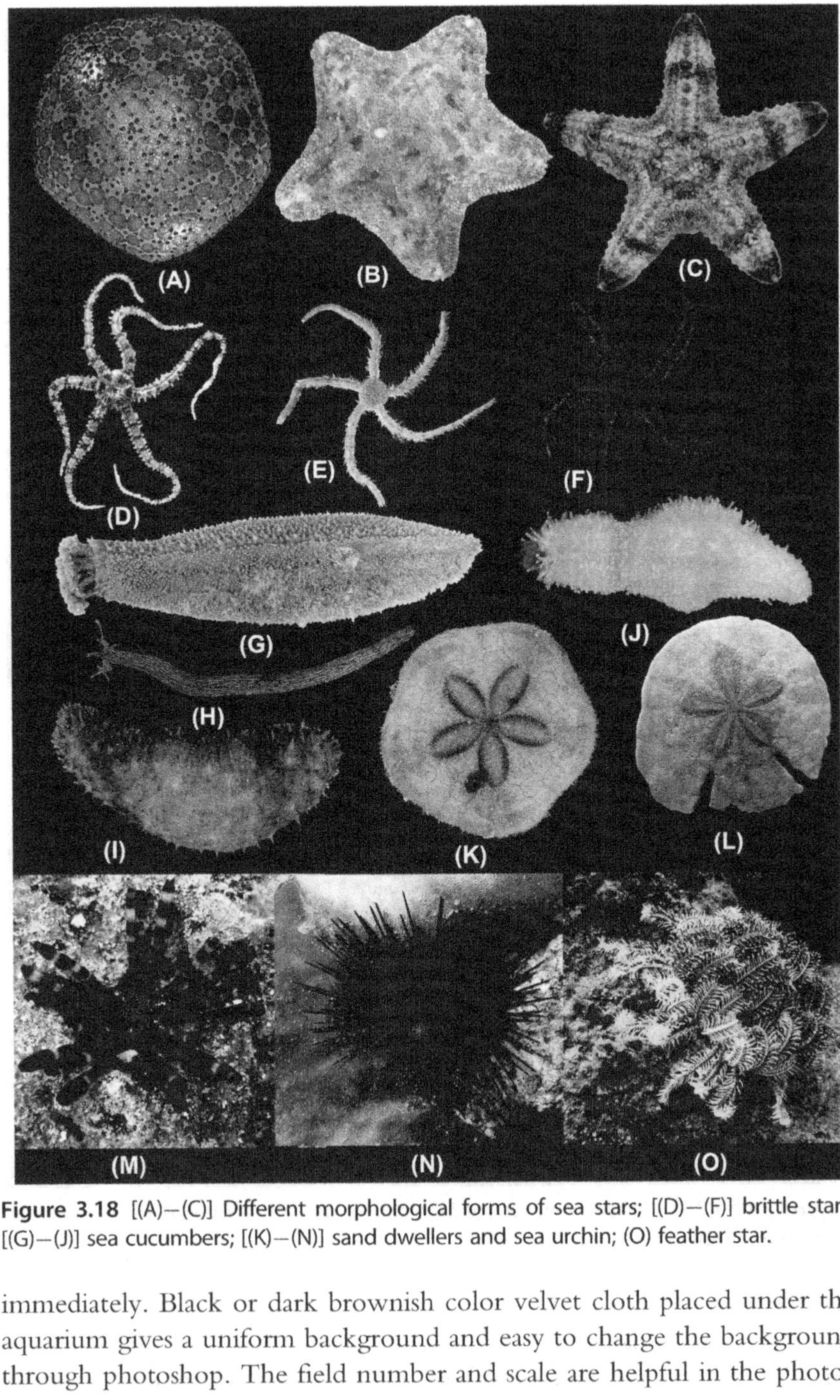

Figure 3.18 [(A)–(C)] Different morphological forms of sea stars; [(D)–(F)] brittle stars; [(G)–(J)] sea cucumbers; [(K)–(N)] sand dwellers and sea urchin; (O) feather star.

immediately. Black or dark brownish color velvet cloth placed under the aquarium gives a uniform background and easy to change the background through photoshop. The field number and scale are helpful in the photograph with the subject in the first photo (Fig. 3.28A–D).

Figure 3.19 The representation of the (A) family Conidae and (B) crown of thorns starfish (*Acanthaster planci*).

3.5.2 Subsampling

Molecular biological research needs a separate subsample of the specimen for genetic (DNA) work in museums, research institutes, and universities. Subsamples should be taken from body parts that are not important for morphological identification, have a low probability of contamination (Evans & Paulay, 2012). Many research institutes collect and store tissue samples for DNA analysis, besides the whole specimens. The genetic subsamples should be placed in roughly 1–2 mm cubes. Tissue subsampling for molecular work requires a 95% or higher ethanol concentration as the preservative (Schander & Hagnell, 2003; Sturm et al., 2006). A very little piece of tissue is needed to get DNA (e.g., end of foot in a gastropod, tube feet from a sea star) and place it directly in 95%–100% ethanol in a small tube or 1.2–2.0 mm plastic DNA vials (Fig. 3.29A and B). In the

Figure 3.20 Dangerous fishes: (A) stingray, (B) stonefish, and (C) lionfish.

molecular studies, it is essential to ensure that the alcohol penetrates into the body to preserve the soft tissue inside.

3.5.3 Voucher preparation

Voucher preparation usually involves relaxing, killing, fixing, and preserving the specimen based on taxon-specific protocols. These specimens should be adequately labeled as voucher specimens or reference material, including the name of the verifying authority (Table 3.3). Permanently preserved materials are stored in the research institute or museum, or laboratory for future reference. Primarily alcohol preserved materials should

Figure 3.21 The technique of hand-picking and various groups of invertebrate animals. (B) Juvenile octopus, (C) polychaetes, (D) colorful nudibranch, (E) boring bivalve, and (F) shrimp.

be replaced with new alcohol after the couple for the days or 1 or 2 weeks. Formalin-fixed materials are also transferred to fresh alcohol. Based on the size of the specimens, they are transferred from the small vials to

Figure 3.22 Sample collection method by snorkeling.

larger jars or containers they fit comfortably into without and occupied to the top with the preservative. Gastropods or bivalve shells are kept in dry condition (Fig. 3.30). In situ photos are very useful for all and they should be linked to the voucher specimen (Eymann et al., 2010; Templado et al., 2010; Lincoln & Sheals, 1979).

3.5.4 Preservation

The main procedure of preservation comprises of three major steps, narcotization, killing and fixation, and preservation.

3.5.4.1 Relaxation

Relaxation is one of the important procedures in many groups; since identification is too tough or impossible if animals are preserved in a tight form; in other groups (e.g., sea slugs, crustaceans, and some echinoderms), autotomy might occur if animals are suddenly dropped straight into fixative. Relaxants naturally hit out both muscles and nerves and make fixation more humane as well. So, our primary objective is to collect specimens that are properly anesthetized and unable to respond after being placed in a fixative. Relaxants are mostly group-specific; useful chemicals are listed below (there are many others).

3.5.4.2 Magnesium chloride ($MgCl_2$)

The live animals are kept in a clean tray or Tupperware or 6-well trays or petri dish with seawater and a small quantity of magnesium chloride is sprinkled over the water and covered with a lid. Normally the animals are

Figure 3.23 [(A) and (B)] Underwater photography and the marine benthic animal collection method using SCUBA diving.

narcotized for 4–6 h, but can take more for larger animals. For animals that contract, it can be added gradually to seawater, typically for a final 50–50 concentration.

3.5.4.3 *Menthol*

The live specimens are kept in a clean tray or Tupperware or 6-well trays or petri dish with seawater sprinkling crushed crystals on top or adding drops of concentrated menthol solution prepared in ethanol. Menthol works especially well for cnidarians and ascidians. Menthol and $MgCl_2$

Figure 3.24 [(A) and (B)] Van Veen Grab and muddy sediment collected from grab.

primarily working holothuroids and is the quickest and the most effective technique for relaxing sea cucumbers.

3.5.4.4 Clove oil

A few drops in a bowl will knock out many crustaceans (especially decapods and stomatopods).

3.5.4.5 Propylene phenoxitol

This is one of the excellent relaxants for molluscs, especially bivalves and some other invertebrates. Few drops are added to a bowl, go slowly into solution and rapidly knock out clams.

Figure 3.25 [(A)–(D)] Collection method of and collected organisms from the deep-sea dredging.

3.5.4.6 Chloretone

Chloretone is an alcohol-based preservative and also called chlorobutanol or chlorbutol. This can be used as an excellent sedative and relaxants for molluscs, especially bivalves and some other invertebrates.

3.5.4.7 Cooling/freezing

The suitable and general method for crustaceans defrosted crabs makes excellent photos.

Figure 3.26 [(A)–(D)] The method of bottom trawling and collected animals.

3.5.4.7.1 Killing and fixation

Narcotized animals are transferred to different killing and fixing agents such as 4% neutral formaldehyde solution, 90% alcohol.

3.5.4.7.2 Preservation

Commonly used preservatives include 70% ethanol or 5% formalin seawater after fixation in 10% formalin; 90% alcohol is recommended for fixation. Typically formalin or formaldehyde solution should be handled with care as it is known to be carcinogenic. It must also be neutralized, or it will attack any shell material. Sodium bicarbonate to excess has proven to be a reliable, inexpensive neutralizing agent. The special fixation will be required for the material to be used for histology, critical point drying, or transmission electron microscope (TEM) examination. However, formalin is a reasonably good all-around fixative and even adequate for TEM studies. Microwave fixation is successful with some marine invertebrates, including molluscs (Berg & Adams, 1984). Material required for electrophoretic or DNA studies should be sorted alive and frozen at $< -88°C$ as soon as possible. Fixation in 95% ethanol will also be sufficient for DNA studies (Evans & Paulay, 2012).

Figure 3.27 [(A) and (B)] The method of substratum removal from the intertidal area for the collection of different groups of invertebrates, (C) colorful nudibranch (*Halgerda tessellata*), and (D) common rocky shore gastropod (*Planaxis sulcatus*).

Table 3.2 List of major estuarine and marine benthic invertebrate phyla and the processing methods of the animals which the best suited methods are indicated.

Phylum and subgroups		Collection method	Narcotization	Killing and fixing	Preservation
Retaria (Foraminifera) (Fig. 3.1A–F)		Dredging, Core sampling, SCUBA diving, and snorkeling	Not necessary	Living animals should be placed first in 50% alcohol or 3%–5% neutral formalin solution	70%–90% alcohol or samples can be washed in fresh water and dried immediately after collection
Bryozoa (Fig. 3.2A and B)		Dredging, SCUBA diving, snorkeling, and trawling	Collected specimens in a shallow vessel containing the seawater keep 30 min to 12 h Bouin's fluid	Narcotized animals should be killed first in 50% alcohol or 3%–5% neutral formalin solution	90% alcohol or stored dry after killing in formalin or colonies may be passed through graded alcohol and xylene and mounted in DPX
Porifera (Fig. 3.3A and B)		Hand-picking, SCUBA diving, snorkeling, and trawling	Not necessary	Fix 80% or 95% alcohol	Preserve 70%–90% of alcohol or drying
Cnidaria	Hydrozoa (Fig. 3.4A and B)	Hand-picking, dredging, SCUBA diving, snorkeling, and trawling	Place the specimens in a shallow vessel containing the seawater and sprinkle crystals of menthol, magnesium chloride, magnesium sulfate, or chloral hydrate	Narcotized animals are fixed in corrosive acetate[2] for 12 h	Preserve 5% formalin or 70%–90% of alcohol
	Anthozoa (Fig. 3.5A–G)	Dredging, Hand-picking, SCUBA diving, snorkeling, and trawling	Place the specimens in a shallow vessel containing the seawater and sprinkle crystals of menthol, magnesium chloride, magnesium sulfate, or chloral hydrate	Narcotized animals are fixed in corrosive acetate[2] for 12 h	Preserve 5% formalin or 70%–90% of alcohol

Platyhelminthes (turbellaria) (Fig. 3.6A–D)		Hand-picking, SCUBA diving, and snorkeling	Live animal to crawl and expand on a piece of moistened paper, then placing the paper and animal gently onto frozen formalin	Narcotized animals are fixed in Gilson's fluid	Wash thoroughly and 70%–90% of alcohol
Nematoda		Hand-picking, SCUBA diving, and snorkeling	Place the specimens in a shallow vessel containing the seawater and sprinkle crystals of menthol, magnesium chloride, magnesium sulfate, or chloral hydrate	Fix in 4%–10% formalin or 50% alcohol	85% alcohol or 5% formalin
Nemertea (Fig. 3.7A–C)		Hand-picking and SCUBA diving	Place the specimens in a shallow vessel or trays containing the seawater and magnesium chloride	Fix in 4%–10% formalin or 50% alcohol	85% alcohol or 5% formalin
Annelida (Polychaeta) (Fig. 3.8A–G)		Hand-picking, SCUBA diving, snorkeling, dredging, grab, core sampling, and trawling	–	Narcotized animals should be fixed 70 Bouin's fluid	85% alcohol or 5% formalin
Sipuncula (Fig. 3.9A and B)		Hand-picking, SCUBA diving, snorkeling, dredging, grab, core sampling, and trawling	Place the specimens in a shallow vessel or trays containing the seawater, propylene phenoxetol, or magnesium chloride	Narcotized animals should be fixed 70%–90% alcohol or 4% formalin	85% alcohol or 5% formalin
Echiura (Fig. 3.10A and B)		Hand-picking, SCUBA diving, and snorkeling	Place the specimens in a shallow vessel or trays containing the seawater, crystals of menthol, 7% solution of magnesium chloride, or 1% solution of propylene phenoxetol	Narcotized animals should be fixed 5% neutral formalin	85% alcohol or 5% formalin
Mollusca	Gastropoda (Fig. 3.11A–L)	Hand-picking, SCUBA diving, snorkeling, dredging, grab, core sampling, and trawling	Place the specimens in a shallow vessel or trays allow extending the animal add crystals of	Not necessary	90% alcohol

(Continued)

Table 3.2 (Continued)

Phylum and subgroups		Collection method	Narcotization	Killing and fixing	Preservation
			menthol, 7% solution of magnesium chloride		
	Bivalvia (Fig. 3.12A–P)	Hand-picking, SCUBA diving, snorkeling, dredging, grab, core sampling, and trawling	Place the specimens in a shallow vessel or trays allow extending the animal add crystals of menthol, 7% solution of magnesium chloride	Not necessary	90% alcohol
	Polyplacophora (Fig. 3.13A and B)	Hand-picking, SCUBA diving, snorkeling, dredging, grab, core sampling, and trawling	Place the specimens in a shallow vessel or trays allow extending the animal add crystals of menthol, 7% solution of magnesium chloride or keep the freezer for 30 min to 1 h	Not necessary	90% alcohol
	Scaphopoda (Fig. 3.13C and D)	Hand-picking, SCUBA diving, snorkeling, dredging, grab, core sampling, and trawling	Place the specimens in a shallow vessel or trays allow extending the animal add crystals of menthol, 7% solution of magnesium chloride or keep the freezer for 30 min to 1 h	Not necessary	90% alcohol
	Cephalopoda (Fig. 3.13E and F)	Hand-picking, SCUBA diving, snorkeling, and trawling	Place the specimens in a shallow vessel or trays allow extending the animal add crystals of menthol, 7% solution of magnesium chloride or keep the freezer for 30 min to 1 h	Not necessary	90% alcohol
Brachiopoda (Fig. 3.14A–D)		Hand-picking, SCUBA diving, and snorkeling	Place the specimens in a shallow vessel or trays. Allow extending the animal add crystals of menthol, 7% solution of magnesium	Not necessary	90% alcohol

Arthropoda (Ostracoda, Copepoda, Tanaidacea, Isopoda, Amphipoda, Cirripedia, and Decapoda) (Figs. 3.15A–J, 3.16A–L, 3.17A–N)		Hand-picking, SCUBA diving, snorkeling, dredging, grab, core sampling, and trawling	Place the specimens in a shallow vessel or trays to extend the animal add clove oil or tricaine, or keep the freezer for 30 min to 1 h	Narcotized animals should be fixed or killed 70%–80% of ethanol	90% alcohol
Echinodermata	Asteroidea (Fig. 3.18A–C)	Hand-picking, SCUBA diving, snorkeling, dredging, and trawling	Place the specimens in a shallow vessel or trays and relax with magnesium chloride or chloretone	Narcotized animals should be fixed 80%–95% alcohol	90% alcohol or drying
	Ophiuroidea (Fig. 3.18D–F)	Hand-picking, SCUBA diving, snorkeling, dredging, and trawling	Place the specimens in a shallow vessel or trays and relax with magnesium chloride or chloretone	Narcotized animals should be fixed or killed 80%–95% alcohol	90% alcohol
	Holothuroidea (Fig. 3.18G–J)	Hand-picking, SCUBA diving, snorkeling, dredging, and trawling	Place the specimens in a shallow vessel or trays allow them to relax in magnesium chloride with a pinch of menthol on top, wait until the animal does not contract when touched		90% alcohol.
	Echinoidea (Fig. 3.18L–N)	Hand-picking, SCUBA diving, snorkeling, and trawling	Place the specimens in a shallow vessel or trays. Allow extending the animal add crystals of menthol, 7% solution of magnesium chloride, or keep the freezer for 30 min to 1 h	Narcotized animals should be fixed or killed 80%–95% alcohol	90% alcohol or drying
	Crinoidea (Fig. 3.18O)	Hand-picking, SCUBA diving, and snorkeling	Not necessary	Rapidly push down face first into 95% EtOH, spreading arms out as pushing animal into fixative by hand; the animal will die in 30–90 s in a spread-out position	90% alcohol

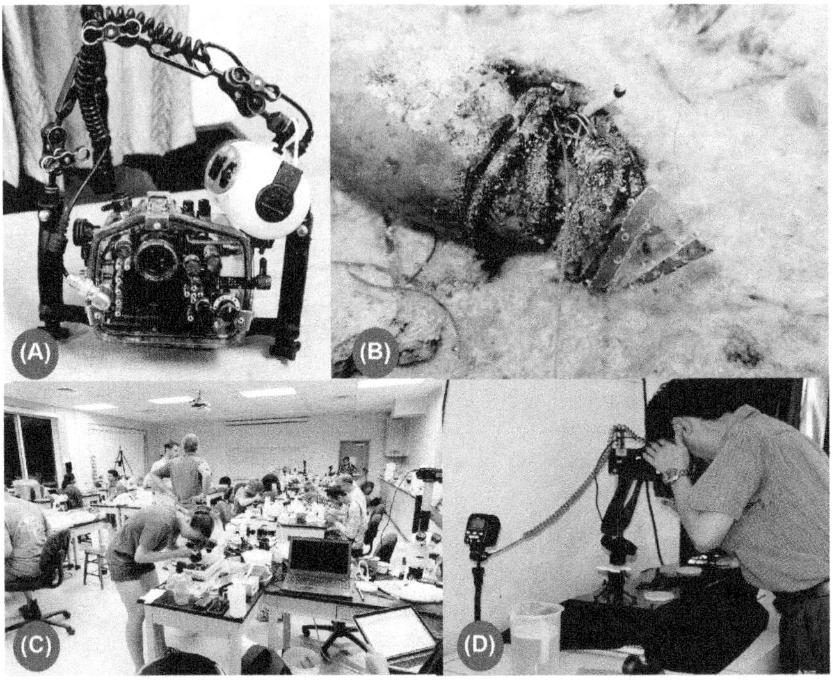

Figure 3.28 [(A) and (B)] Underwater camera and photographed hermit crab; (C) photography for mass sampling, sorting animals, and curation; and (D) photography of an individual specimen.

3.5.5 Storage

The specimens should be stored in tubes, watertight glass jars, or drums, ideally with lids constructed of noncorroding material or jars or drums, depending upon the size of the specimen. Dry specimens should be stored in an acid-free and dry environment. Alcohol preserved specimens should be checked from time to time and replaced as and when necessary. Sometimes more than one specimen is stored in the same bottle, and all the specimens should have their tags and stored (Fig 3.31A-B).

3.5.6 Labeling

All the animals are appropriately labeled with relevant information demonstrating the phylum and other classifications, name of species, habitat, locality, depth, associations, microhabitat, date of collecting, name of the collector, and the number of examples, etc. (Table 3.3).

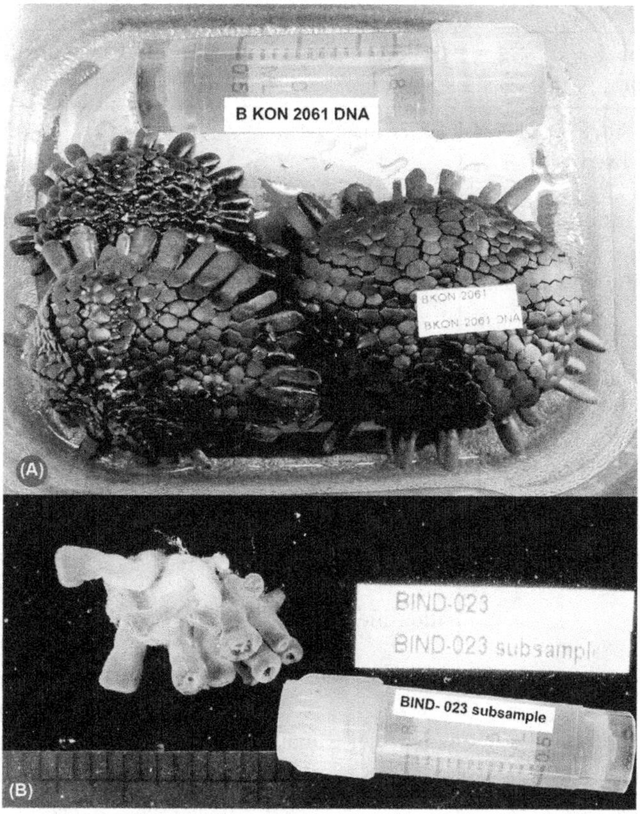

Figure 3.29 [(A) and (B)] Photography of the collected animals and subsample labels.

Table 3.3 Datasheet.

Name of the investigator:	**Collected by:**	**Institute:**	**Sl. no.**
Collection data	**Specimen details**	**Specimen identifiers**	
Date collected (e.g., 01/Jan/2021)	Voucher Type: available/Not available/fixed in Formalin/ethanol/dry	Sample ID Code*: (Eg) IND_ MOLL_001	
Habitat: Coral reef/ mangrove/estuary/ supra tidal/intertidal/ subtidal/oceanic/	Tissue type:	Field number:	

(*Continued*)

Table 3.3 (Continued)

Name of the investigator:	**Collected by:**	**Institute:**	**Sl. no.**
Substrate: Rock/sandy bottom/mud/ seaweed/ …………	Extra info:	Museum ID:	
Other methods of collection: Market/ trader/trawl by catch/………	Sex: Male/female/ hermaphrodite	Collection code:	
Location details: FAO region: Country: State:	Life Stage: Adult/immature/ larva/egg	Gear used: Hand-picking/ beach wash/ scuba diving/ ……	
District:	*Specimen taxonomy*	*Oceanographic details*	
Village:	Identified by:	Wave action: Heavy/ moderate/ low	
Exact place: Latitude (D.d):	Identifier e-mail: Identifier institution:	Currents: Water temperature:	
Longitude (D.d): Co-ordinate source: e.g., Map/GPS	*Identification:* Phylum:	Oxygen: Salinity:	
Co-ordinate accuracy: …3…m, ………km, ……feet, ……yards, …………degrees	Class:	Nutrients:	
Elevation:…0……m, ………km, ………yards, ……feet (negative values for the collection below sea level)	Order:	Meteorology: Rainy/ sunny/ cloudy/ ……	

(*Continued*)

Table 3.3 (Continued)

Name of the investigator:	**Collected by:**	**Institute: Sl. no.**
Elevation accuracy: ......m, ...km, ...feet, ...yards	Family:	Sample ID code: First three letters of Phylum/first two letters of the country/or state/ first three letters of exact locality/ number
Depth: ......m, ...km, ...feet, ...yards	Subfamily:	
Depth accuracy: ......m, ...km, ...feet, ...yards	Genus:	Eg:
No. of organisms collected:	Subgenus:	Phylum: mollusca
No. of organisms counted in the locality:	Species:	Country: India Locality: Kovalam MOL/IND/KOV/1
PHOTOGRAPH: Yes/No If yes picture no.	Local name: Language:	Remarks:
Measurements (mm) depends on phyla	Conservation status: IUCN:........................... Indian:...............	
SL: ....; SW....; SH:... Body length:............ Body width:.........	Ecological notes: Color, cryptic species, etc. Ecological notes	

3.5.6.1 *Image processing*

The image-processing method is essential for taxonomy, identification, and good publications (Figs. 3.32A–H and 3.33A–F). The following steps explain the actual procedure of Adobe Photoshop cs6 software, but basically, any version may be used for this method (Motomura & Ishikawa, 2013).

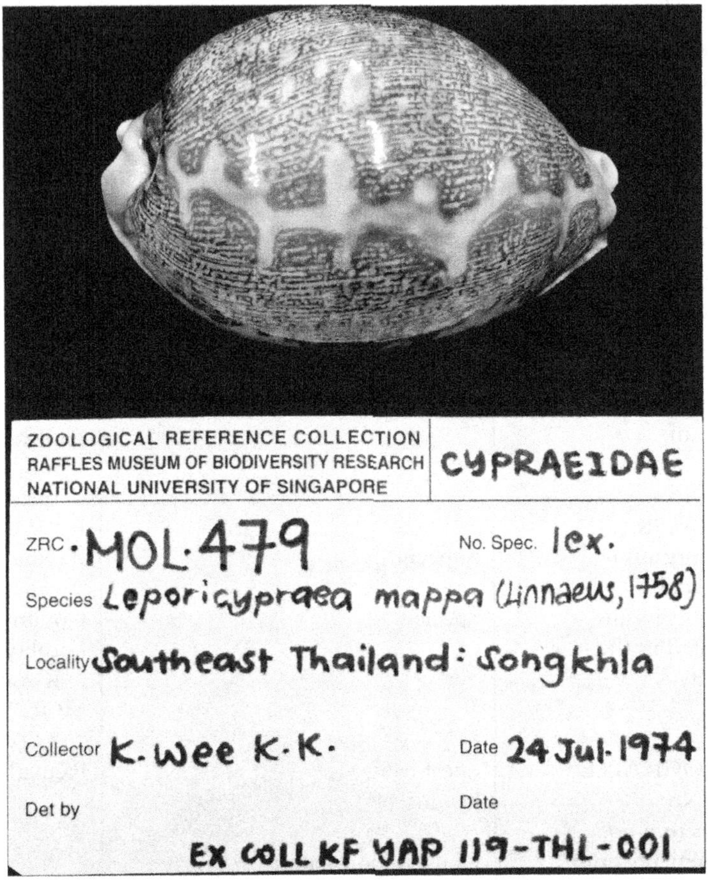

Figure 3.30 Voucher specimen from the zoological reference collection.

3.6 Conclusion

Marine invertebrates live on various habitats, such as sand and rubble bottoms, tide pools, brackish pools, mangroves, shallow reef flats, and deep seas. Biodiversity inventories can become never-ending tasks, and documentation is the process of identifying/recording the animals. Documentation of biodiversity is the groundwork of conservation action. Therefore, it is essential to keep the motivation of participants and stakeholders by carefully planning and executing projects involving biodiversity documentation. Collecting can only be supported when there is a real educational or scientific need, and casual collecting with inadequate documentation can never be justified. Digital imaging or photography is one

Figure 3.31 (A) Adequately preserved and labeled materials; (B) proper storage animals.

of the powerful tools for both amateurs and professionals. Estimating the diversity of cryptic species and limitations of the current morpho taxonomic methods are challenging because molecular surveys that reveal

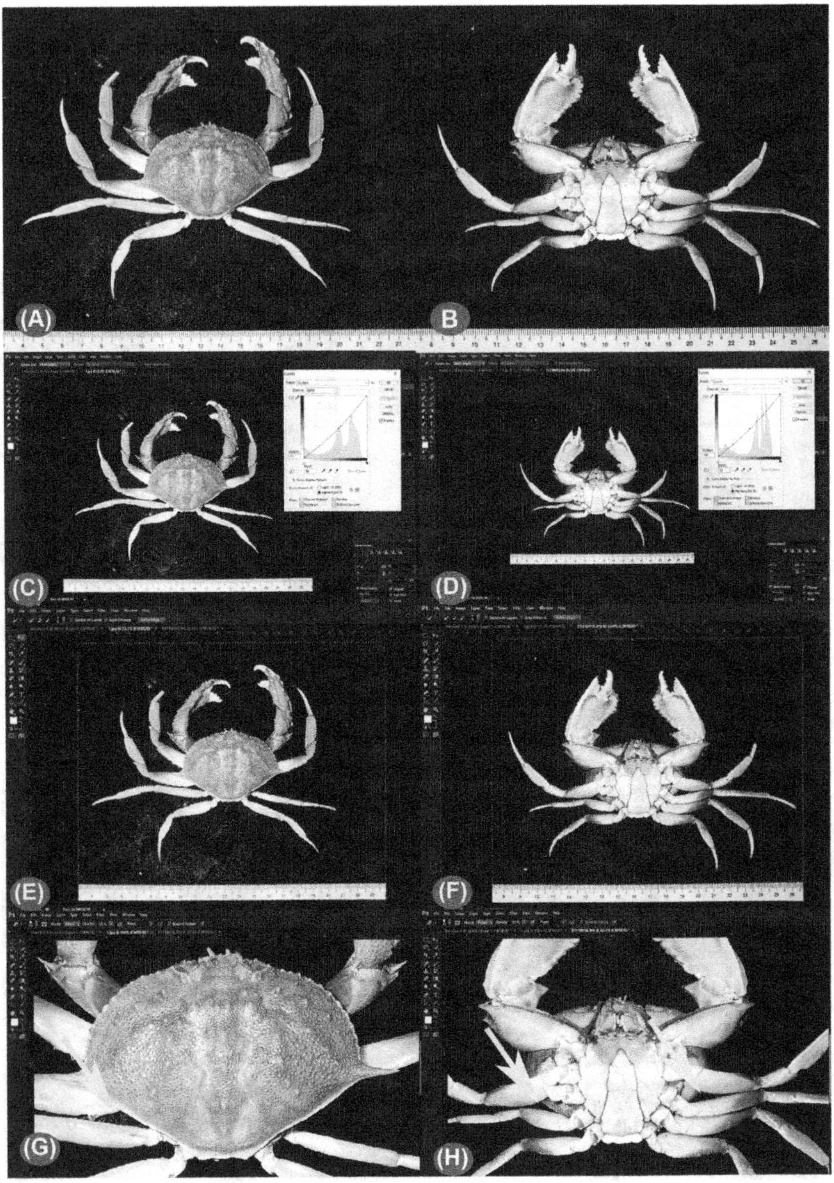

Figure 3.32 [(A) and (B)] Unprocessed original image. [(C) and (D)] Fine-tune the image. Select "Image" → "Adjustments" → "curves." A dialog box will appear, Adjust the curves and click "OK". [(E) and (F)] Roughly select the quick selection tool,→ delete the selected area and change the background black or your preferred color [(G) and (H)] Select "eraser tool" and increase or decrease the diameter of the eraser →erase all the unselected area.

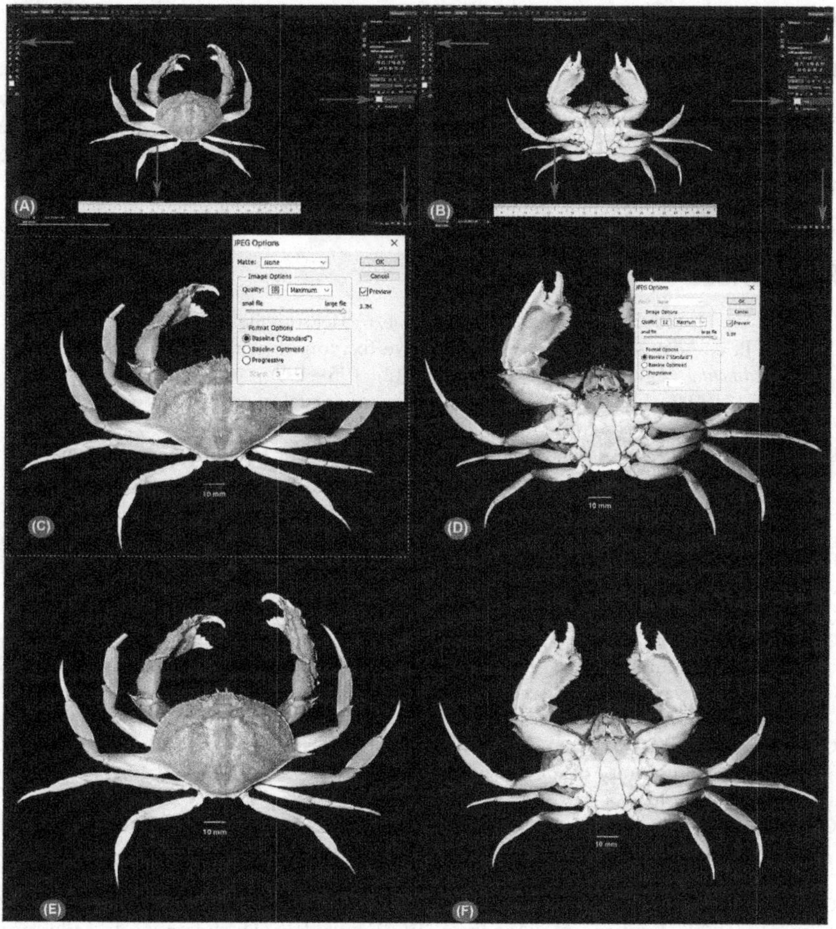

Figure 3.33 (A) and (B) Click the "new layer"→Click "control + R"→measure 10 mm→Click the "brush tool"→press shift key→and draw the line for 10 mm. Finally scale bar for the measurements. (C) and (D) Finally select "File"→"Save as." Set the image quality to 12, and confirm that the compression is set as "Highest (Low Compression Rate)." Select the "Baseline (Standard)" option. [(E) and (F)] Processed image with the black background.

them most readily have been applied to only a fraction of marine diversity. This chapter is helpful for beginners in the field and who may, for some reason required to collect or study biological samples from the seabed and familiarize themselves with the different groups of animals. There is no single comprehensive method incorporating all the study techniques, and different approaches have to be applied depending on the scope of the investigation.

Acknowledgments

The first author wishes to acknowledge Gustav Paulay, Amanda Bemis, and John Slapcinsky, Florida Museum of Natural History, and Mary Hagedorn, Smithsonian Conservation Biology, for the guidance received during the MarineGEO Hawaii Biodiversity Assessment. The authors are also grateful to Patterson Edward, SDMRI, Tuticorin for helping with the biodiversity surveys.

References

Behrens, W. D. (2005). *Nudibranch behavior* (p. 176) Jacksonville: New World Publishing.

Berg, C. J., & Adams, N. L. (1984). Microwave fixation of marine invertebrates. *Journal of Experimental Marine Biology and Ecology*, *74*(2), 195–199.

Boero, F. (1994). Fluctuations and variations in coastal marine environments. *Marine Ecology*, *15*(1), 3–25.

Boero, F. (2001). Light after dark: The partnership for enhancing expertise in taxonomy. *Trends in Ecology and Evolution*, *16*(5), 266.

Boero, F. (2010). The study of species in the era of biodiversity: A tale of stupidity. *Diversity*, *2*(1), 115–126.

Caras, R. A. (1976). *Dangerous to man* (p. 422) London: Barrie & Jenkins Ltd.

Chen, E. Y. (2021). Often overlooked: Understanding and meeting the current challenges of marine invertebrate conservation. *Frontiers in Marine Science*, *8*, 690704.

Cochrane, S. K. J., Andersen, J. H., Berg, T., Blanchet, H., Borja, A., Carstensen, J., Elliott, M., Hummel, H., Niquil, N., & Renaud, P. E. (2016). What is marine biodiversity? Towards common concepts and their implications for assessing biodiversity status. *Frontiers in Marine Science*, *3*, 248.

Coleman, C. O. (2015). Taxonomy in times of the taxonomic impediment—Examples from the community of experts on amphipod crustaceans. *Journal of Crustacean Biology*, *35*(6), 729–740.

Collier, K. J., Probert, P. K., & Jeffries, M. (2016). Conservation of aquatic invertebrates: Concerns, challenges and conundrums. *Aquatic Conservation: Marine and Freshwater Ecosystems*, *26*(5), 817–837.

Costello, M. J., Basher, Z., McLeod, L., Asaad, I., Claus, S., Vandepitte, L., Yasuhara, M., Gislason, H., Edwards, M., Appeltans, W., Enevoldsen, H., Edgar, G. J., Miloslavich, P., Silvia De Monte, S., Pinto, I., Obura, D., & Bates, A. E. (2017). Methods for the study of marine biodiversity. In M. Walters, & R. Scholes (Eds.), *The GEO handbook on biodiversity observation networks*. Springer.

David, W. B. (2005). *Nudibranch Behaviour.* Jacksonville, Florida: New World Publications Inc.

Di Camillo, C. G., Gravili, C., De Vito, D., Pica, D., Piraino, S., Puce, S., & Cerrano, C. (2018). The importance of applying standardised integrative taxonomy when describing marine benthic organisms and collecting ecological data. *Invertebrate Systematics*, *32*(4), 794.

Dreujou, E., McKindsey, C. W., Grant, C., Tréau de Coeli, L., St-Louis, R., & Archambault, P. (2020). Biodiversity and habitat assessment of coastal benthic communities in a subArctic industrial harbor area. *Water*, *12*, 2424.

Duffy, J. E., Amaral-Zettler, L. A., Fautin, D. G., Paulay, G., Rynearson, T. A., Sosik, H. M., & Stachowicz, J. J. (2013). Envisioning a marine biodiversity observation network. *Bioscience*, *63*(5), 350–361.

Evans, N., & Paulay, G. (2012). DNA barcoding methods for invertebrates. In W. Kress, & D. Erickson (Eds.), *DNA barcodes. Methods in molecular biology (methods and protocols)* (858, pp. 47–77). Humana Press.

Eymann, J., Degreef, J., Häuser, Ch, Monje, J. C., Samyn, Y., & Vanden Spiegel, D. (2010). *Manual on field recording techniques and protocols for all taxa biodiversity inventories and monitoring*. Brussels: ABC Taxa. Belgian National Focal Point for the GTI, vol 8.

Fajardo, J., Lessmann, J., Bonaccorso, E., Devenish, C., & Muñoz, J. (2014). Combined use of systematic conservation planning, species distribution modelling, and connectivity analysis reveals severe conservation gaps in a megadiverse country (Peru). *PLoS One*, *9*(12), e114367.

Ferrario, F., Beck, M. W., Storlazzi, C. D., Micheli, F., Shepard, C. C., & Airoldi, L. (2014). The effectiveness of coral reefs for coastal hazard risk reduction and adaptation. *Nature Communications*, *5*, 3794.

Gemeinholzer, B., Rey, I., Weising, K., Grundmann, M., Muellner, A. N., Zetzsche, H., Droege, G., Seberg, O., Petersen, G., Rawson, D., & Weigt, L. (2010). Organising specimen and tissue preservation in the field for subsequent molecular analyses. In J. Eymann, J. Degreef, Ch Häuser, J. C. Monje, Y. Samyn, & D. VandenSpiegel (Eds.), *Manual on field recording techniques and protocols for All Taxa Biodiversity Inventories and Monitoring*. Brussels: ABC Taxa. Belgian National Focal Point for the GTI, vol 8.

Gemeinholzer, B., Vences, M., Beszteri, B., Bruy, T., Felden, J., Kostadinov, I., Miralles, A., Nattkemper, T. W., Printzen, C., Renz, J., Rybalka, N., Schuster, T., Weibulat, T., Wilke, T., & Renner, S. S. (2020). Data storage and data re-use in taxonomy—The need for improved storage and accessibility of heterogeneous data. *Organisms Diversity and Evolution*, *20*, 1–8.

Godfray, H. (2002). Challenges for taxonomy. *Nature*, *417*, 17–19.

Gosliner, T., Valdez, A., & Behrens, W. D. (2018). *Nudibranch and sea slug identification-Indo-Pacific* (2nd ed., p. 452). New World Publications.

Gosner, K. L. (1971). *Guide to identification of marine and estuarine invertebrates* (p. 693) New York, NY: Wiley Interscience.

Gouda, S., Das, G., Sen, S. K., Shin, H.-S., & Patra, J. K. (2016). Endophytes: A treasure house of bioactive compounds of medicinal importance. *Frontiers in Microbiology*, *7*, 1538.

Griffiths, J. R., Kadin, M., Nascimento, F. J. A., Tamelander, T., Törnroos, A., Bonaglia, S., & Winder, M. (2017). The importance of benthic-pelagic coupling for marine ecosystem functioning in a changing world. *Global Change Biology*, *23*(6), 2179–2196.

Halstead, B. W., Auerbach, P. S., & Campbell, D. (1990). *A colour atlas of dangerous marine animals* (p. 192) London: Wolfe Medical Publications Ltd.

Henseler, C., Nordström, M. C., Törnroos, A., Snickars, M., Pecuchet, L., Lindegren, M., & Bonsdorff, E. (2019). Coastal habitats and their importance for the diversity of benthic communities: A species- and trait-based approach. *Estuarine, Coastal and Shelf Science*, *226*, 106272.

Hinegardner, R. T. (1958). The venom apparatus of the cone shell. *Hawaii Medical Journal*, *17*, 533–563.

Hopkins, G., & Freckleton, R. P. (2002). Declines in the numbers of amateur and professional taxonomists: Implications for conservation. *Animal Conservation*, *5*(03), 245–249.

Hutchings, P. (2020). Major issues facing taxonomy—A personal perspective. *Megataxa*, *1*(1), 46–48.

IPBES. (2019). Summary for policymakers of the global assessment report on biodiversity and ecosystem services of the Intergovernmental Science-Policy Platform on Biodiversity and Ecosystem Services. In S. Díaz, J. Settele, E. S. Brondízio, H. T. Ngo, M. Guèze, J. Agard, A. Arneth, P. Balvanera, K. A. Brauman, S. H. M. Butchart, K. M. A. Chan, L. A. Garibaldi, K. Ichii, J. Liu, S. M. Subramanian, G. F. Midgley, P. Miloslavich, Z. Molnár, D. Obura, & ... C. N. Zayas (Eds.), (p. 56). Bonn, Germany: IPBES secretariat.

Kelleher, G. (1999). *Guidelines for marine protected areas*. IUCN, Gland, Switzerland and Cambridge.

Khan, B. M., & Liu, Y. (2019). Marine mollusks: Food with benefits. *Comprehensive Reviews in Food Science and Food Safety, 18*(2), 548–564.

Lalli, C. M., & Parsons, T. R. (1997). *Biological oceanography an introduction* (2nd ed.). Burlington, MA: Elsevier Butterworth-Heinemann.

Lau, M. T., Manion, J., Littleboy, J. B., Oyston, L., Khuong, T. M., Wang, Q. P., Nguyen, D. T., Hesselson, D., Seymour, J. E., & Neely, G. G. (2019). Molecular dissection of box jellyfish venom cytotoxicity highlights an effective venom antidote. *Nature Communications, 10*(1), 1655.

Leather, S. (2009). Taxonomic chauvinism threatens the future of entomology. *Biologist (Columbus, Ohio), 56*(1), 10–13.

Lekshmi, N., Umar, M. D., Dhaneesha, M., Joseph, R., Ravinesh, R., & Sajeevan, T. P. (2020). Endophytic fungi isolated from the marine sponges as a source of potential bioactive compounds. *Journal of Aquatic Biology & Fisheries, 8*, 58–66.

Lincoln, R. J., & Sheals, J. G. (1979). *Invertebrate animals: Collection and preservation* (p. 150) London: British Museum (Natural History), Cambridge University Press.

Lindberg, D. R. (2001). Molluscs. In S. Levin (Ed.), *Encyclopedia of biodiversity* (pp. 235–347). Academic Press.

Luypaert, T., Hagan, J. G., McCarthy, M. L., & Poti, M. (2020). Status of marine biodiversity in the anthropocene. In S. Jungblut, V. Liebich, & M. Bode-Dalby (Eds.), *YOUMARES 9*—The oceans: Our research, our future. Cham: Springer.

Manomi, S., Jayesh, P., Ravinesh, R., Prabhakaran, M. P., Lathika, C. T., Neema, J., & Rosamma, P. (2020). Hidden mycota of marine macroalgae: diversity, molecular phylogeny and antimicrobial profile. *FEMS Microbiology Letters, 367*(11), 1–9.

Mare, M. F. (1942). A study of a marine benthic community with special reference to the microorganisms. *Journal of the Marine Biological Association of the United Kingdom, 25*(03), 517.

Martin, J. W. (2016). Collecting and processing crustaceans: An introduction. *Journal of Crustacean Biology, 36*(3), 393–395.

McIntyre, A. D. (Ed.), (2010). *Life in the world's oceans: Diversity, distribution, and abundance.* Oxford: Wiley-Blackwell.

Mikkelsen, P. M., & Cracraft, J. (2001). Marine biodiversity and the need for systematic inventories. *Bulletin of Marine Science, 69*(2), 525–534.

Motomura, H., & Ishikawa, S. (2013). *Fish collection building and procedures manual. English edition* (p. 70) Kyoto: The Kagoshima University Museum, Kagoshima and the Research Institute for Humanity and Nature.

Munro, C. (2013). Diving, 125–173. In A. Eleftheriou (Ed.), *Methods for the study of marine benthos* (Fourth Edition). John Wiley & Sons, Ltd.

Myatt, T., Nguyen, B. J., Clark, R. F., Coffey, C. H., & O'Connell, C. W. (2018). A prospective study of stingray injury and envenomation outcomes. *The Journal of Emergency Medicine, 55*(2), 213–217.

Nakano, H. (2020). Marine animals. In K. Inaba, & J. Hall-Spencer (Eds.), *Japanese marine life*. Singapore: Springer.

Ng, P. K. L. (2017). Collecting and processing freshwater shrimps and crabs. *Journal of Crustacean Biology, 37*(1), 115–122.

Ockelmann, K. W. (1964). An improved detritus-sledge for collecting meiobenthos. *Ophelia, 1*(2), 217–222.

Olson, D. K. (1981). *Techniques for the study of primate population ecology. By the Subcommittee on Conservation of Natural Populations, Committee on Nonhuman Primates, NRC* (p. 233) Washington: National Academy Press.

Palumbi, S. R., Sandifer, P. A., Allan, J. D., Beck, M. W., Fautin, D. G., Fogarty, M. J. Halpern, B. S., Incze, L. S., Leong, J., Norse, E., Stachowicz, J. J., & Wall, D. H

(2009). Managing for ocean biodiversity to sustain marine ecosystem services. *Frontiers in Ecology and the Environment*, 7(4), 204–211.

Ponder, W. F., Lindberg, D. R., & Ponder, J. M. (2019). *Biology and evolution of the mollusca* (p. 924) Taylor & Francis Group, CRC Press, Vol. 1.

Ponder, W. F., Lindberg, D. R., & Ponder, J. M. (2020). *Biology and evolution of the mollusca* (p. 870) Taylor & Francis Group, CRC Press, Vol. 2.

Pratt, D. R., Lohrer, A. M., Pilditch, C. A., & Thrush, S. F. (2014). Changes in ecosystem function across sedimentary gradients in estuaries. *Ecosystems*, *17*, 182–194.

Purvis, A., Molnar, Z., Obura, D., Ichii, K., Willis, K., Chettri, N., Dulloo, E., Hendry, A., Gabrielyan, B., Gutt, J., Jacob, U., Keskin, E., Niamir, A., Öztürk, B., Salimov, R., & Jaureguiberry, P. (2019). Status and Trends – Nature. In E. S. Brondízio, J. Settele, S. Díaz, & H. T. Ngo (Eds.), *Global assessment report of the Intergovernmental Science-Policy Platform on Biodiversity and Ecosystem Services*. Bonn, Germany: IPBES secretariat.

Ravinesh, R., & Biju Kumar, A. (2013). Comparison of intertidal biodiversity associated with natural rocky shore and sea wall: A case study from the Kerala coast, India. *Indian Journal of Geo-Marine Sciences*, *42*(2), 223–235.

Rensch, G. & Lavoie, H. M. (2021). *Lionfish, scorpionfish, and stonefish toxicity*. [Updated 2021 Jan 11]. In: StatPearls [Internet]. Treasure Island (FL): StatPearls Publishing. Available from: https://www.ncbi.nlm.nih.gov/books/NBK482204/.

Rifkin, J. F., Fenner, P. J., & Williamson, J. A. H. (1993). First aid treatment of the sting from the hydroid *Lytocarpus philippinus*: The structure of and in vitro discharge experiments with its nematocysts. *Journal of Wilderness Medicine*, *4*, 252–260.

Saarman, E. T., Owens, B., Murray, S. N., Weisberg, S. B., Ambrose, R. F., Field, J. C., Nielsen, K. J., & Carr, M. H. (2018). An ecological framework for informing permitting decisions on scientific activities in protected areas. *PLoS One*, *13*(6), e0199126.

Sagi, A., Rosenberg, L., Ben-Meir, P., & Hauben, D. J. (1987). Fire coral *Millepora dichotoma* as a cause of burns: A case report. *Burns: Journal of the International Society for Burn Injuries*, *13*, 325–326.

Schander, C., & Hagnell, J. (2003). Death by drowning degrades DNA. *Journal of Molluscan Studies*, *69*, 387–388.

Smith, M. M. (1977). *Sea and shore dangers: Their recognition, avoidance and treatment* (pp. 16–17). JLB Smith Institute of Ichthyology, Rhodes University.

Stella, J. S., Pratchett, M. S., Hutchings, P. A., & Jones, G. P. (2011). Coral-associated invertebrates: Diversity, ecological importance and vulnerability to disturbance. In R. N. Gibson, R. J. A. Atkinson, & J. D. M. Gordon (Eds.), *Oceanography and marine biology, an annual review* (49, pp. 43–104). CRC Press, Taylor & Francis Group.

Sturm, C. F., Pearce, T. A., & Valdés, A. (Eds.), (2006). *The mollusks, a guide to their study, collection and preservation*. Pittsburgh, Pennsylvania, USA: American Malacological Society.

Suarez, A. V., & Tsutsui, N. D. (2004). The value of museum collections for research and society. *Bioscience*, *54*(1), 66.

Templado, J., Paulay, G., Gittenberger, A. A., & Meyer, C. (2010). Sampling the marine realm. In J. Eymann, J. Degreef, C. Häuser, J. C. Monje, Y. Samyn, & D. VandenSpiegel (Eds.), *Manual on field recording techniques and protocols for all taxa biodiversity inventories* (pp. 273–307). Brussels: ABC Taxa. Belgian National Focal Point for the GTI, Vol 8.

Thomsen, P. F., & Willerslev, E. (2015). Environmental DNA—An emerging tool in conservation for monitoring past and present biodiversity. *Biological Conservation*, *183*, 4–18.

Williamson, J. A., Fenner, P. J., Burnett, J. W., & Rifkin, J. F. (Eds.), (1996). *Venomous and poisonous marine animals: A medical and biological handbook*. Sydney: University of New South Wales Press.

Williamson, J. A. H. (1987). The blue-ringed octopus bite and envenomation syndrome. *Clinics in Dermatology*, *5*, 127–133.

Winker, K., Reed, J. M., Escalante, P., Askins, R. A., Cicero, C., Hough, G. E., & Bates, J. (2010). The importance, effects, and ethics of bird collecting. *The Auk*, *127*(3), 690–695.

Yeates, D. K., Seago, A., Nelson, L., Cameron, S. L., Joseph, L., & Trueman, J. W. H. (2011). Integrative taxonomy, or iterative taxonomy? *Systematic Entomology*, *36*(2), 209–217.

Zhu, C., Southgate, P. C., & Li, T. (2019). Production of pearls. In A. Smaal, J. Ferreira, J. Grant, J. Petersen, & Ø. Strand (Eds.), Goods and services of marine bivalves (pp. 73–93). Springer Nature.

CHAPTER 4

Source and composition of organic matter and its role in sediment benthic communities

J.V. Sinu
P.G. Department of Zoology, Christian College, Chengannur, Kerala, India

4.1 Introduction

Macro, meio, and microbenthic communities, important components linking pelagic and benthic ecosystems, are primary consumers and are found to feed on organic matter (OM). Benthic fauna play a central role in OM retention, remineralization, and benthic-pelagic exchange, both directly through ingestion, biomass production, respiration, and excretion (Middelburg, 2018) and indirectly through bioturbation (Denga et al., 2020). The distribution of benthic communities in the bottom sediment environments varies in response to many variables (El-Serehy et al., 2015). Most benthic organisms being sessile are directly subjected to environmental fluctuations. Both pelagic and benthic zones in aquatic ecosystems are subjected to a variety of stressors that may interact together to produce combined impacts on biodiversity and ecosystem functioning. Fluctuations in the condition, nature, and characteristics of aquatic ecosystems affect benthic community composition, function, diversity, and production. Benthic fauna act as a connecting link between the biotopes of substratum and water column in the aquatic ecosystem and their distribution highly depends on the physical nature of the substratum, sediment OM, nutritive content, degree of stability, and oxygen content. Benthic organisms affect the biogeochemical fluxes between the sediment and water column through their metabolism and behavior (Ehrnsten et al., 2019). Among the benthic communities, tolerant or opportunistic species tend to dominate stressed environments and less tolerant species tend to become increasingly rare or will disappear. Therefore both these species can serve as stress indicators (Belan, 2003). The drastic change in the composition of sediment and the organic carbon content of the benthic

Ecology and Biodiversity of Benthos
DOI: https://doi.org/10.1016/B978-0-12-821161-8.00001-5

environment may bring considerable change in the composition, abundance, and biomass of the species. The successional changes in the structure of the benthic community can be predicted and with the increased perturbation there will be a decline in the diversity of the species, then abundance as well as the biomass (Pearson & Rosenberg, 1978). The spatiotemporal variation of benthic invertebrates will have impacts on other trophic levels which in turn will determine the fate of energy and nutrient flow in the benthic-pelagic zones of the aquatic ecosystems. This chapter aims to provide an account of the sediment composition and OM content and their role in the distribution patterns among benthic groups in the marine and estuarine ecosystems.

4.2 Sediment composition of aquatic ecosystems

Aquatic sediments are derived from and composed of physical, chemical, and biological components and are originated from bed load transport, beach and bank erosion, and land runoff. Sediments range from micron-sized clay particles through silt, sand, gravel, rock, and boulders. Sediment is a naturally found material that is broken down by processes of weathering and erosion and is then transported. For tracing sediment sources, the sediment properties are also used. Sediment sources have fundamental importance in understanding the sediment dynamics of aquatic ecosystems (Juez et al., 2018). Sand and silt can be carried in suspension in river water and on reaching the sea bed get deposited by sedimentation. Like soil in the terrestrial ecosystem, sediments hold some importance in the aquatic ecosystem because they are the source of enormous resources. Substratum provides the anchor to life and food in the form of OM to the bottom fauna. Sediment characteristics play a very important role in deciding the nature of the marine benthic environment. The functional traits of infaunal benthic communities are related to their particle reworking and burrow ventilation capacity, which in turn affect transport conditions and biogeochemistry of the sediment and consequently solute exchange between sediment and overlying water (Kristensen et al., 2014). Sediments are important catalysts of the dynamics of water quality and environmental food cycles. The prevalent hydrodynamic conditions sort the sediments by size naturally. Generally, slow-moving water will contain fine-grained sediments and fast-moving water will contain coarse-grained sediments. In the continental shelf region, the sediment is composed of three different types different in origin—a near-shore sand zone, extending from the shore to water depth of 5–10 m, is succeeded by the mud zone (silt and clay), which extends to a depth of 50–60 m (inner shelf). Beyond 50–60 m (outer shelf) is covered by coarse calcareous sand. Sedimentation processes on the aquatic systems are a function of climate, bathymetry, and hydrography (Hashimi et al., 1981).

In aquatic ecosystems, sediments play a key role and they act as storage reservoirs for all kinds of pollutants which are introduced into the aquatic ecosystem either by anthropogenic or natural processes (Batvari & Mariappan, 2016). Sedimentation is the direct result of erosion from other aquatic areas or land-based areas and can be detrimental or beneficial to aquatic environments. The beneficial effects are the building up of new substrates for the development of marshes and the detrimental effects are related to the burial of organisms that dwell in the bottom. The environmental impacts of sedimentation include the following—loss of important or sensitive aquatic habitats, decrease in fishery resources, loss of recreation attributes, loss of coral reef communities, human health concerns, changes in fish migration, increases in erosion, loss of wetlands, nutrient balance changes, circulation changes, increases in turbidity, loss of submerged vegetation, and coastline alteration. The inorganic and organic sedimentary materials produce dissolved chemicals in the overlying and sediment pore water as well as runoff and groundwater that range from fresh to marine in salinity. The characteristics of sediments are affected by the complex biogeochemical interactions between the inorganic, living, and nonliving organic components. The sediment biotic community includes micro, meso, and macrofauna and flora that are interdependent and their host sediment's biogeochemical characteristics.

In the material cycles of waters, sediments form a natural filter system and buffer. The physical characteristics of the sediments are greatly determined by the depositions of the environment. Sediment texture is one of the key features of aquatic sediments. The study on the textural characteristics of sediments gives valuable information about the grain size which determines the transport and deposition in aquatic systems. A wide range of colors is found in aquatic sediments. The sediment color may vary due to several reasons. The sediment associated with coarse-grained quartz or limestone deposits is usually lighter colored which may be white or pale tint. The darker sediments will have a predominance of mineral substances, such as iron oxide or manganese oxide. Their colors may be red, blue, or green and are usually found in mud. The sediments that are in the form of black are often found in undisturbed areas. Aquatic sediments are studied by collecting bottom samples. The grain size of sediments varies from the finest clay to large boulders. The ocean floor in some areas contains deposits of either organic or inorganic origin predominately of siliceous or calcareous remains. The physical properties of sediments are important variables to understand geological events in the marine and estuarine environments.

The physical and chemical properties of sediment in an aquatic system depend on the physical topography, climate, temperature, vegetation, and microbial activities. It is the physical hydrodynamics that controls the substratum type which, in turn, affect the biological features of these habitats. Particle size, hydrological regime, OM quality, and nutrient availability influence communities in the benthic system. Surface sediments act as an efficient natural trap for diverse substances including contaminants in the aquatic ecosystems. They also act as a natural regulator in the processes that happens in the benthic system (Shah et al., 2019). The nutrients of the sediments including macro and meioinvertebrates would be directly controlling the growth of both plants and soil microorganisms. The nutrients include phosphorus and nitrogen compounds. Sediments play a key role in the nitrogen cycle of aquatic environments and can act either as a source or a sink of biologically available (fixed) nitrogen. In recycling fixed nitrogen in the water, sediments play more or less a fundamental role. The complete understanding of the sediment characteristics by routine sampling and analysis are useful tools for the management of essential nutrients of the soil.

Currently, there are widespread studies associated with the physical properties of aquatic sediments which are used to understand geological events. They are good indicators for the composition, microstructure, and environmental conditions during and after the depositional process (Kim et al., 2001). Sediment bed is also a major site for OM decomposition which is largely carried out by bacteria. Some natural components of sediments could be regarded as pollutants at concentrations high enough to cause adverse biological effects. These contaminants do not ordinarily biomagnify in aquatic food webs, but at extremely high concentrations they can travel up the food chain which leads to a severe reduction in the diversity of bottom-dwelling organisms that live in the aquatic system and adverse effects can spread, via the food chain, from phytoplanktons to zooplanktons, fish, birds, and mammals that feed on contaminated aquatic life.

Sediments usually have an impact on ecosystem quality because of their quality, or their quantity. Compared to the water column, contaminants fixed to the sediment matrix typically create a prolonged hazard for the ecosystem (Kara et al., 2017). When the sediment is subjected to alteration, it will be degraded (Akita et al., 2020). But we can wisely manage this degradation by implementing regular monitoring programs. Bottom sediments serve as reservoirs for pollutants, therefore deserve special consideration in the planning and design of aquatic ecosystem studies. Sediments are indicators of the quality of overlying water and their study is a useful tool in the assessment of environmental pollution status (Kumar & Khan, 2012). The benthic region plays a major role in elementary cycles and acts as a sink for carbon and various pollutants. Multiple stressors in sediment profile can produce an ecosystem imbalance with a global impact, so it is essential to monitor these ecosystems to preserve their pristine form. In addition to the abiotic factors, the floor of the aquatic system is home to a vast array of benthic organisms performing various ecosystem services. The complete understanding of the ecology of the sediment ecosystem is a key part of designing and managing ecosystems in which the long-term fertility and productive capacity of the sediment are maintained or even improved which in turn supports a healthier benthic environment.

Anthropogenic activities, however, have significantly enhanced sedimentation as well as sediment loss. With the intensification of human influence on the natural environment, the regional sedimentary processes are undergoing extensive changes which are illustrated by changes in the hydrological processes and the resulting ecological and environmental effects (Liu et al., 2020). For many decades, the issue of sediment in aquatic environments has been a topic of concern. There are land-based sedimentation activities (like forestry, agriculture, urbanization, construction, and recreation) and water-based sedimentation activities also (such as navigation, port activities, drag fishing, dams, channelization, water diversions, wetlands loss, and other large-scale hydrological modifications). Bank or beach protection activities, water diversions, and many of the other aquatic activities cause sediment impoverishment or loss. Morphological changes such as physical changes over a large area such as the change in the size and shape of a water body will result in new water flow patterns leading to major changes in natural sediment erosion and sedimentation patterns which in turn cause sediment removal from sensitive areas. From a broad perspective, ecosystem recovery and restoration are the major objectives of sediment dynamics (Ouillon, 2018).

An aquatic system contains a certain variety of sediment based upon its geology, topography, precipitation rates, and other natural and anthropogenic factors. The depositional and erosional processes are responsible for the wide variety of sedimentary patterns and profiles. Unlike the water column, the sediment layer is in an immovable and steady condition which disallows rapid re-suspension and biochemical dissolution of pollutants back into the water column (Yu et al., 2016). Sediment has both direct and indirect influences upon both biotic and abiotic ecosystem components. For aquatic plant and animal life, sediments are absolutely necessary. Sediments are indispensable and vital resources if managed properly. If the sediment management is improper, it will lead to the destruction of aquatic habitat. Sediment in rivers gets deposited in a lake/estuary or the sea as the river reaches the estuarine and marine zones. Additionally, health risks posed by sediment from pollutants and pathogens can attach themselves to sediment particles entering the aquatic systems and can very rapidly change the water quality. These chemicals are harmful to both plants and animals.

4.3 Organic matter in sediments

Organic Materials in sediments are derived from multiple sources like biogenic particulate matter from primary production in surface waters, decomposed tissues of aquatic plants and animals, and land-source OM from soils, plant debris, and man made compounds transported by runoff and atmospheric deposition into the aquatic domain and from various point and nonpoint waste water discharges. Consequently sedimentary OM constitutes a heterogeneous and complex mixture of organic compounds (Lu et al., 2020) with a wide range of chemical and physical properties. As the size of sediment mineral particles decreases, the concentration of the OM content increases. The degradation of OM is an important process going on in sediments (Rowe et al., 2020). It links the biological carbon cycle with the geological subsurface cycle (Hulse et al., 2017). Benthic biogeochemical dynamics is linked directly or indirectly to the degradation of OM. This process controls the dissolution of carbonates, the recycling of inorganic carbon and nutrients, the flux of organic carbon to the deep biosphere, and the burial of organic carbon in the sediments. As a result, OM degradation determines the net carbon dioxide removal from, and oxygen input to, the atmosphere. When the oxygen content is low in the sediments the organic carbon burial is at its maximum. However, the mechanism remains enigmatic. Particle size, hydrological regime, OM quality, and nutrient availability influence communities in sediments. As particle size changes, changing redox conditions and the availability of organic carbon substrates determine the benthic community structure. If marine sediments have lower redox potential, it points to the presence of OM in them. Changes in redox potential reflect changes in the community in response to fluctuating oxygen status. Redox potential tends to be more negative in fully saturated sediments, indicating low oxygen status (Reddy & Laune, 2008). The organic fraction of the particle input to the sediment is a potential source of energy for benthic organisms. The main link between "active" pools of carbon in the water bodies, atmosphere, and landmasses and carbon pools that cycle on much longer geological timescales, is the primary long-term sink, that is created by the burial of OM in the sediments (Burdige, 2007). Therefore investigating the processes that control the composition of sedimentary OM that is buried in fresh/brackish/marine sediments is crucial for understanding carbon cycling on a global scale. The environmental factors that affect the accumulation and preservation of OM in surface sediments are the hydrodynamic environment for particle deposition, the water depth, and the redox conditions of the bottom water and sediments (Yang et al., 2018). Sediments of the bottom are the source of nutrients to the water column above them which leads to benthic-pelagic coupling and also influences primary productivity. Recent studies point to the fact that terrigenous OM is heterogeneous in composition which is composed of at least two distinct sources, viz., soils and plant debris.

Sediments accumulated in the lake/seafloor consist of various organic and inorganic materials, mostly utilized as proxies for climatic changes and as historical records of the lake's/sea's connectivity to terrestrial ecosystems and anthropogenic activities. The productivity of aquatic systems is greatly influenced by the quantity and quality of sedimentary OM (Cividanes et al., 2002). The OM preserved in sedimentary deposits is also an important progenitor of fossil fuels, a recorder of Earth's history. As an accumulation of geochemical fossils, the OM content of sediments provides information that is important to interpretations of both natural and human-induced changes in local and regional ecosystems (Dianto et al., 2019). The quantity and character of the OM available in the sediment are also influenced by the sediment quality which in turn supports benthic organisms (Jofre et al., 2001). Sediment OM has an important role in controlling benthic organism metabolism, composition, distribution, benthic nutrient regeneration, and remineralization rates (Pusceddu et al., 2011) and also recovery following disturbances (Hermand et al., 2008). In short, the composition of OM in sediments has been recognized as the major factor affecting the metabolism, distribution, and dynamics of benthic organisms. Land-derived OM is commonly thought of as resistant to change due to its high weight molecular components, while labile fractions of OM, including chlorines (early diagenetic products of chlorophylls), amino acids, proteins, lipids, and carbohydrates are readily digested by benthic organisms and rapidly mineralized (Woulds & Cowie, 2009). The quantity of annual organic carbon buried in lake sediment is almost equal to half of that of ocean sediment (Dean & Gorham, 1998).

4.4 Estuarine sediments: A case study from selected estuaries from Indian subcontinent

Although fresh waters cover only about 2% of the Earth's surface area, they play an important role in the global carbon cycle. Estuarine/lake systems are diverse and the multiple sources and also alterations of OM are geographically and temporally variable (Costa et al., 2021). To understand the autochthonous and allochthonous input as well as to assess the global biogeochemical cycles, the source characterization of OM within the estuarine sediments is of utmost importance (Goni et al., 2003). Organic matter constitutes a minor but important fraction of estuarine sediments and the primary source of OM originates from the complex mixture of lipids, carbohydrates, proteins, and other OM components produced by organisms that have lived in and around the lake (Rullkotter, 2000). Organic matter is then preserved in sediments according to its composition, sedimentary mineralogical composition, and redox conditions in interstitial water (Barber et al., 2017). Within aquatic ecosystems, bottom sediments have important functions such as efficient natural trap for diverse substances including contaminants, and as a natural regulator of the processes that occur in the lake floor (Jorgensen, 1996). They also store huge amounts of OM and also affect the oxygen content of the bottom water. Bottom sediment provides nutrients to the water column above them, leading to benthic-pelagic coupling; thus they influence the primary productivity (Jorgensen, 1996). It is very important to study the source of OM and organic carbon since lakes and water bodies are the major sinks for carbon. The OM content of estuarine sediments provides a variety of indicators, or proxies that can be used to reconstruct paleoenvironments of estuaries and to infer histories of regional climate changes (Lone et al., 2018).

Estuaries are among the most diverse and productive environments that serve as breeding and nursery grounds for various faunal communities

(Bhowmik & Mandal, 2021) and they play an important role in carbon and nutrient biogeochemistry (Bianchi, 2006). Estuarine sediment dynamics are a consequence of the advective and dispersive transport of suspended sediments originating from freshwater and marine sources, and the bed-water exchange processes of erosion and deposition (Mathew & Winterwerp, 2020). The geologic, bathymetric, and hydrologic settings in which the bottom sediments get deposited, determine their texture and composition. Terrigenous detritus, biogenic debris, and pollutants are accumulated in the bottom sediments in the estuarine zone. As a result, estuarine sediments become a unique environment for organisms because they are relatively heterogeneous, comprising detritus, inorganic and organic particles (Bragadeeswaran et al., 2007). Sediments in estuaries are the sites of chemical reactions as they can either temporarily bind nutrients and other contaminants, or more permanently store these contaminants with burial, and that in turn affect the composition of coastal waters, coastal nutrient cycling, and large-scale geochemical cycles. The estuaries trap the sediments and nutrients that are carried from the land by rivers and the ocean tides. Sediment conditions can deteriorate as a result of eutrophication or contamination. Sediment accumulation can be enhanced by macrobenthic species and thus they can protect the habitat from erosion. They can influence benthic-pelagic coupling and perform a crucial role in bioturbation by their physiological activities, which have been the focal point in major ecological studies (Zhang et al., 2019). Healthy sediments support healthy benthic invertebrate communities and they may retain their temporary stores of nutrients longer as they are oxygenated more deeply into the surface layers. Human activities have modified estuarine systems over time, consequently the estuaries are highly dynamic, and the primary biological and physical drivers vary from hours to weeks to annual and multidecadal time scales. To understand the autochthonous and allochthonous input as well as to assess the global biogeochemical cycles, source characterization of OM within the estuarine sediments is important (Goni et al., 2003).

Indian estuarian systems contribute much to the economy of the country due to their enormous fishery resources, transportation uses, esthetic and recreational values, etc. Currently, due to urbanization and industrialization, these valuable resources are deteriorated to a large extent due to reclamation, domestic and sewage inputs, which in turn upset its ecological balance. Hence an attempt was made to evaluate sediment textural composition and OM content and their role in the distribution patterns among benthic groups in the selected estuarine ecosystems along the

Indian coast. Indian estuaries are significantly different from other estuaries in the discharge pattern during the monsoon period (southwest and northeast monsoons) when significant monsoonal precipitation occurs. Based on the review of research conducted by various researchers, three seasonal divisions have been demarcated—premonsoon (February–May), monsoon (June–September), and postmonsoon (October–January). The monsoon precipitation during June–September was more significant, due to its strong spatial variations.

4.4.1 Materials and methods

4.4.1.1 Study area

Estuarine sediment and benthic samples were collected during the year 2018–2019 for the three seasons—premonsoon, monsoon, and postmonsoon from three estuaries—Kochi estuary (a station I), Vembanad Ramsar site (station II), and Ashtamudi Ramsar site (station III) Kerala, India (Fig. 4.1; Table 4.1).

4.4.1.2 Description of the sampling stations

4.4.1.2.1 Station I (Kochi estuary)

Kochi estuary (09° 30′–10°10′ N and 76° 15′–76°25′ E) is a bar-built estuary located on the southwest coast of Kerala, India. The estuary opens to the Arabian Sea at Kochi and receives freshwater from six rivers and several tributaries make this one of the largest wetlands along the west coast of India. Kochi estuary is well-known as a nursery ground for many of the commercially important fishes and shell fishes and is a system of shallow brackish water lagoons along the Kerala coast.

4.4.1.2.2 Station II (Vembanad Ramsar site)

Vembanad wetland ecosystem (09° 30′ N 76° 13′ E to 76° 50′ E)—the longest and the largest Ramsar site on the southwest coast of India, located in the coastal region of Kerala, which is highly populated. It is an agricultural area and also a major hotspot of biological diversity. It is about 65 km long and 0.5–15 km wide. Its depth varies from 1.0 to 12.0 m. Being a Ramsar site as well as an important tourist area, Lake Vembanad has immense socioeconomic and ecological significance.

4.4.1.2.3 Station III (Ashtamudi Ramsar site)

Ashtamudi Wetland (Ashtamudi Ramsar site) (08° 57′ N and 76° 35′ E) located at Kollam, Kerala, India, is the second-largest wetland in Kerala

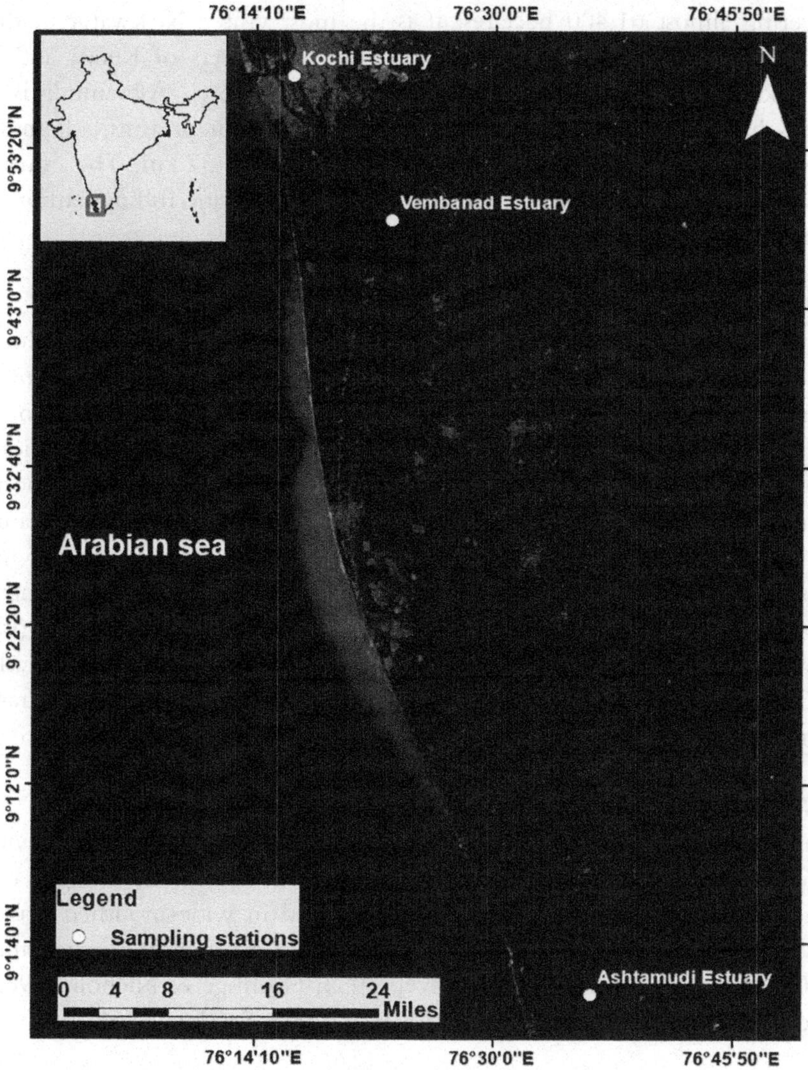

Figure 4.1 Location map of the study area.

Table 4.1 Location of sampling stations.

Station no.	Latitude	Longitude	Sampling stations
1	09° 30′–10°10′ N	76° 15′–76°25′ E	Kochi estuary
2	09° 30′ N	76° 13′ E	Vembanad Ramsar site
3	08° 57′ N	76° 35′ E	Ashtamudi Ramsar site

covering almost 61,400 hectares. It is the most visited backwater in the state and is also called the gateway to the backwaters of Kerala and is well known for its houseboat and backwater resorts. Ashtamudi is a palm-shaped extensive water body with eight prominent arms, adjoining the Kollam town of Kollam district with an area of 57 km. The estuary has its outlet at Neendakara, which is the biggest marine fishing harbor in Kerala.

4.4.1.3 Sampling protocol

Sediment quality along with macrobenthic assemblages was investigated through a regular periodical biomonitoring program. Cruises were performed to evaluate the abiotic and biotic variables to set up a seasonal database. In this monitoring program, measurement and sampling sessions were organized three times a year, once per season—premonsoon, monsoon, and postmonsoon during the year 2018–19. For this purpose, samples for sediment analysis were collected using a Van Veen grab (0.25 m^2 mouth area) and were analyzed using standard methods. The sediment organic carbon content was estimated by the wet oxidation method (El-Wakeel & Riley, 1957) which was then converted into OM (Trask, 1939). Sediment textural analyses for computing sand, silt, and clay fractions were carried out by pipette analysis (Krumbein & Pettijohn, 1938).

4.4.1.4 Macrofaunal core samples

The sediment samples for macrobenthic analyses were anesthetized with 7% $MgCl_2$ sieved on a 500 μm mesh, stained with Rose Bengal and preserved in 70% isopropyl alcohol, and stored in wide-mouthed plastic jar containers for identification using standard keys such as Boggemann, 2005; Day, 1967; FAO, 1998; Fauvel, 1953; Gorringe & Nicholls, 1982; Modayil, 2007; Parulekar et al., 1980; Rao, 2003; Sanders, 1956.

4.4.1.5 Data analysis

Principal component analysis (PCA) for sedimentological variables was computed with Primer version 6. Mean ± standard deviation was computed using SPSS version 20.

4.4.2 Organic matter composition of Indian estuarine systems

Estuaries are important zones of high OM production and processing (Canuel & Hardison, 2016) and are the regions where the mixing of terrigenous and marine OM occurs (Middelburg & Herman, 2007). To understand the processes that influence the sources and fate of OM in estuaries, we should assess the biogeochemical processes that occur within these regions. This assessment is important in quantifying the contributions of carbon from land and rivers to the global carbon budget at the land-ocean margin (Canuel & Hardison, 2016). External inputs and stresses often interact in both synergistic and compensatory ways which further complicates the interpretation of long-term patterns in estuarine biogeochemical processes. The estuarine carbon cycle is closely linked to anthropogenic changes that are occurring on land and in the open ocean and are tied to the biogeochemical cycles (Testa et al., 2018). Estuaries receive OM from multiple sources and the contribution of OM from diverse sources makes the OM cycling in estuaries more complex. Being highly productive zones (Countway et al., 2007) estuaries can act as a sink, or due to extensive processing of OM, they can be the source of atmospheric carbon dioxide. Estuaries process significant quantities of terrestrial and marine OM and play a controlling factor in determining fluxes of dissolved and particulate organic carbon (DOC and POC, respectively) to the coastal ocean (Takasu et al., 2020).

Sediment OM content (Fig. 4.2; Tables 4.2 and 4.3) analysis from three Indian estuaries—Kochi estuary, Ashtamudi, and Vembanad Ramsar sites of international significance revealed that OM content in estuarine systems ranged from 2.65% to 5.87%. Kochi estuary (station I) registered maximum sediment organic carbon of 5.87% during postmonsoon and a minimum of 4.69% during monsoon with a mean value of 5.33 ± 0.596. Compared to other estuaries, Kochi estuary recorded high OM content which is mainly imparted as part of urbanization, industrialization, developmental activities, and other anthropogenic interactions in the region. A study conducted in Kochi estuarine system revealed the presence of both freshly deposited as well as aged sedimentary OM during the premonsoon

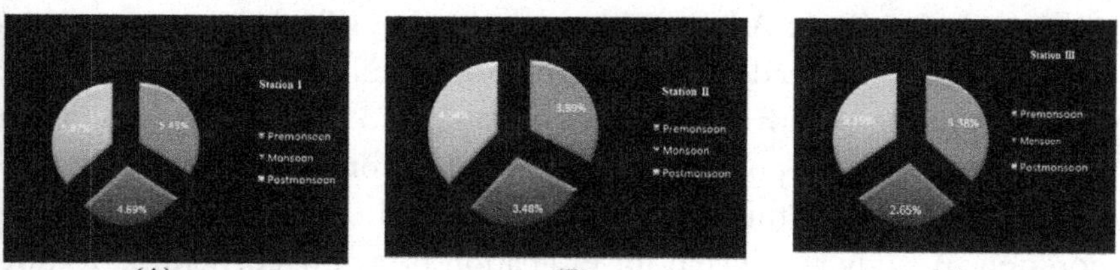

(A) (B) (C)

Figure 4.2 Variation in sediment organic carbon (%) in Indian estuaries. (A) Kochi estuary, (B) Vembanad estuary, (C) Asthamudi estuary.

Table 4.2 Stationwise mean variation in the sediment organic matter and granulometry of Indian estuaries.

Parameters	Station I	Station II	Station III	Total
Sand	56.97 ± 1.48	50.28 ± 4.49	69.33 ± 3.79	58.86 ± 8.90
Silt	30.44 ± 0.36	31.10 ± 4.90	20.3 ± 3.84	27.58 ± 6.32
Clay	12.59 ± 1.23	17.72 ± 0.97	10.36 ± 0.06	13.557 ± 3.36
Organic matter	5.33 ± 0.59	3.97 ± 0.53	3.10 ± 0.37	4.12 ± 0.08

Table 4.3 Seasonwise mean variation in the sediment organic matter and granulometry of Indian estuaries.

Parameters	Premonsoon	Monsoon	Postmonsoon	Total
Sand	56.06 ± 8.88	59.61 ± 11.94	60.92 ± 8.74	58.86 ± 8.90
Silt	29.84 ± 4.87	27.65 ± 9.07	25.25 ± 6.17	27.58 ± 6.32
Clay	14.10 ± 4.05	12.75 ± 3.36	13.82 ± 3.99	13.56 ± 3.36
Organic matter	4.23 ± 1.07	3.61 ± 1.03	4.53 ± 1.34	4.12 ± 0.08

and postmonsoon with enhanced quality of OM content in postmonsoon and consequently the estuarine surface sediment nature varied between mesotrophic, eutrophic, and meso-oligotrophic conditions (Mathew et al., 2019). At Vembanad estuary (station II), the values varied from 3.48% during monsoon to 4.54% during postmonsoon with a mean value of 3.97 ± 0.53. Vembanad wetland ecosystem, aptly called the rice bowl of Kerala, is one of the most populous coastal segments of Kerala, revealed high OM content but it was lesser when compared to Kochi estuary. A detailed investigation in this Ramsar site indicated altered water and sediment quality, leading to accelerated eutrophication with the dominance of opportunistic benthic faunal bioindicators of organic enrichment (Asha et al., 2016). At Ashtamudi estuary (station III), the maximum value of 3.38% was observed during premonsoon and a minimum of 2.65% during monsoon, with a mean value of 3.073 ± 0.37. Seasonal mean values revealed that postmonsoon accounted for high OM content of 4.53 ± 1.34 whereas the low value of 3.61 ± 1.03 was observed during monsoon. A study conducted for evaluation of hydrogeochemical features of the Ashtamudi estuary revealed that there was a wide spatial concentration fluctuation of organic carbon in bottom sediment during monsoon and nonmonsoon seasons and revealed a good correlation with sediment texture and also high deterioration of the physicochemical quality of water during the nonmonsoon season concerning monsoon season (Babu et al., 2009).

4.4.3 Sediment composition in Indian estuarine systems

Sediment grain size is one of the most important of the physical properties of sediments and can reveal important information about the sediment source, transport history, and depositional situation (Oyedotun, 2016). The variation in grain sizes in the distal and proximal sediment sources is simply related to the aggradation/degradation processes which occur in the alluvial and coastal areas (Juez et al., 2018). Grain size is influenced by several physical, chemical, and biological factors including tidal current, wave action, climatic condition, salinity fluctuation, bioturbation, and various anthropogenic activities (Sivaleela & Venkataraman, 2013). Sediments that reach an estuarine area are first suspended and carried back and then eroded before they are finally deposited either temporarily or permanently (Manning et al., 2010). If the speed of the current of a river is faster, it can carry sediment particles that are greater in size. When the velocity of the tidal current decreases, along the course of the estuary, finer sediment particles settle on the bottom (Ralston et al., 2013).

Granulometric analyses of sediments (Figs. 4.3–4.5; Tables 4.2 and 4.3) from the selected estuarine systems of India revealed that sand, silt, and clay fractions showed spatiotemporal fluctuations. The sand was predominant in these estuarine systems. The percentage composition of sand ranged from 72% to 47.25% with a mean of 58.86 ± 8.90. Seasonal mean values revealed that postmonsoon accounted for high sand content of 60.92 ± 8.74 whereas the low value of 56.06 ± 8.88 was observed during premonsoon. Silt percentage in the Indian estuaries ranged from 35.24% to 17.6% with a mean of 27.58 ± 6.32. Seasonal mean values revealed that premonsoon accounted for high silt content of 29.83 ± 4.87 whereas a low value of 25.26 ± 6.18 was observed during postmonsoon. Clay in percentage ranged from 18.36% to 10.3% with a mean of 13.56 ± 3.36. Seasonal mean values revealed that premonsoon accounted for high clay content of 14.11 ± 4.05 whereas a low value of 12.75 ± 3.36 was observed during monsoon.

At Kochi estuary (station I), sand was predominant and showed significant variations during the premonsoon, monsoon, and postmonsoon

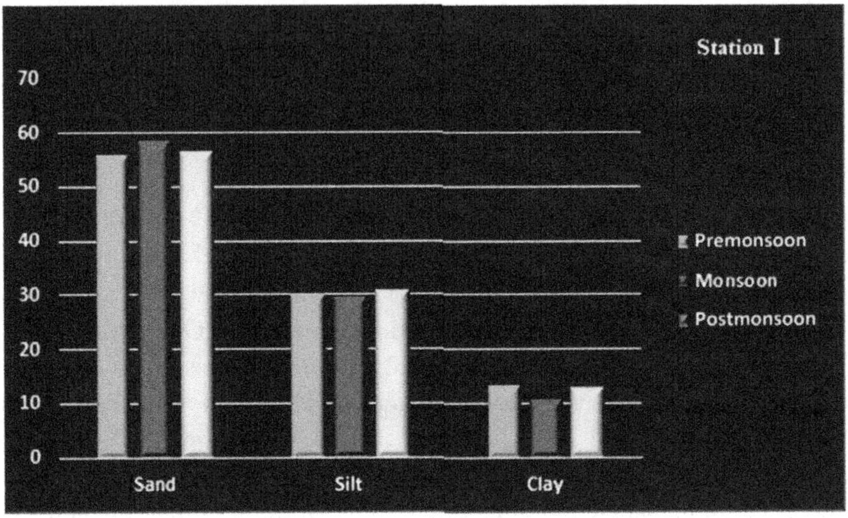

Figure 4.3 Variation in sediment textural composition (%) at Kochi estuary.

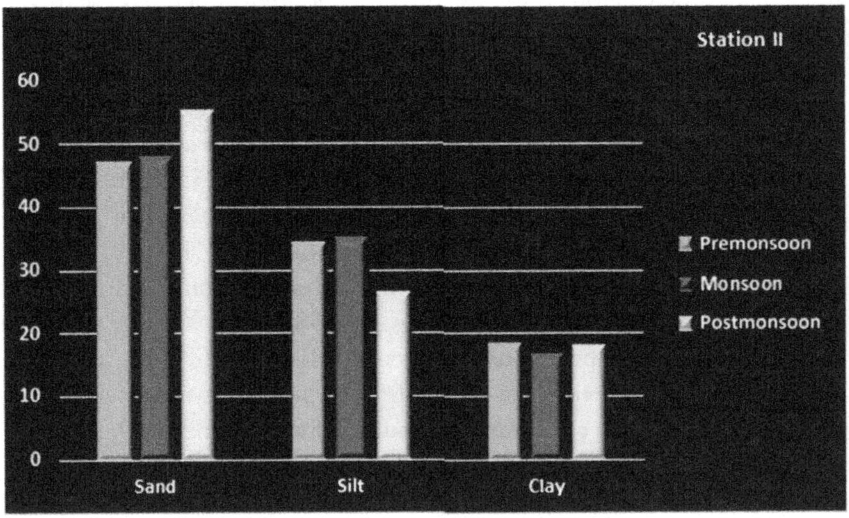

Figure 4.4 Variation in sediment textural composition (%) at Vembanad estuary.

seasons. The sand was high during monsoon (58.66%) and low during premonsoon (55.92%), with a mean value of 56.97 ± 1.478. Even though sand predominated in this region, near-shore marine regions of this estuary were reported to be dominated by clayey sediments and the sediment of that region in the estuary was found to be unfavorable for the thriving

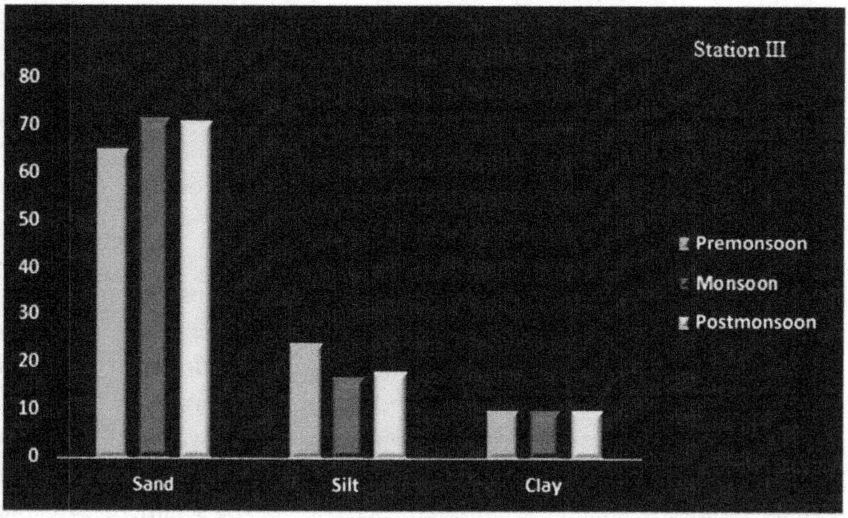

Figure 4.5 Variation in sediment textural composition (%) at Ashtamudi estuary.

of the standing crop (Gandhi et al., 2017). The silt percentage at Kochi estuary, was high during postmonsoon 30.81% whereas the low value of 30.1% was observed during monsoon with a mean value of 30.44 ± 0.36. Clay in percentage was high during premonsoon 13.66% whereas the low value of 11.24% was observed during monsoon with a mean value of 12.59 ± 1.23.

At Vembanad estuary (station II), a high percentage of the sand value of 55.44% was recorded during postmonsoon and a low value of 47.25% during premonsoon with a mean value of 50.28 ± 4.49. Sediment texture analysis conducted in the Vembanad estuary showed the abundance of sand and silt with a minor fraction of clay on nearshore regions (Mohan et al., 2014; Priju & Narayana, 2007). The high percentage of silt value at Vembanad estuary (station II), was recorded during monsoon (35.24%) and a low value of 26.36% during postmonsoon with a mean value of 31.00 ± 4.9. At Vembanad estuary (station II), a high percentage of clay (18.36%) was recorded during premonsoon and a low value (16.6%) during monsoon with a mean of 17.72 ± 0.97. Depth wise textural analysis carried out in the Vembanad estuary also revealed the predominance of sand in nearshore regions (Murty & Veerayya, 1981).

Granulometric composition of surficial sediments from the Ashtamudi lake recorded a high percentage of sand of 72% during monsoon and a minimum of 65% during premonsoon, with a mean value of 69.33 ± 3.79.

Sediment substratum is mainly composed of sand with an admixture of silt and clay. In a study conducted from 2008 to 2010, a higher fraction of sand was observed during the premonsoon period whereas higher silt and clay content was recorded during the monsoon period (Rajan, 2018). But in the current study, monsoon accounted for high sand percentage. At Ashtamudi estuary (station III), the percentage of silt recorded a maximum value of 24.7% during premonsoon and a minimum of 17.6% during monsoon, with a mean value of 20.3 ± 3.84. Percentage of clay recorded a maximum value of 10.4% during monsoon and postmonsoon and a minimum of 10.3% during premonsoon, with a mean value of 10.37 ± 0.05. In a study on the sediments of Ashtamudi lake, sand dominated textural composition was found especially during monsoon season and the data generated from the textual analysis showed that there was no remarkable variation in the granulometric distribution of sediments among different seasons, indicating an almost consistent behavior in the sediment dynamics at different locations (Krishnakumar & Das, 2019).

The percentage composition of each grade (sand, silt, and clay) was calculated and plotted on triangular graphs based on the terminology suggested by Shepard (1954). Regarding the sediment texture of Indian estuaries studied, four textural classes of sediments (Fig. 4.6; Table 4.4) were obtained, viz., sandy loam, loam, sandy clay loam, and loamy sand. Kochi

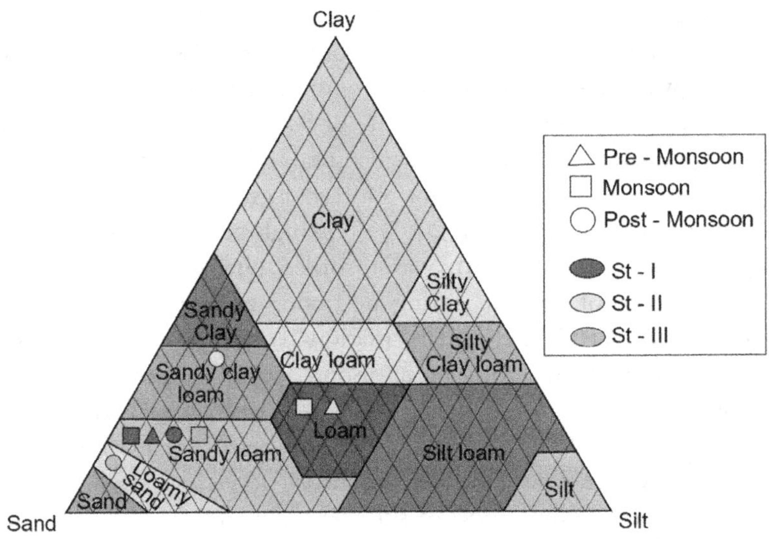

Figure 4.6 Trilinear plot for the seasonal and stationwise distribution of sediment textural classes in Indian estuaries.

Table 4.4 Seasonal and stationwise distribution of sediment textural classes.

	Premonsoon	Monsoon	Postmonsoon
Station I	Sandy loam	Sandy loam	Sandy loam
Station II	Loam	Loam	Sandy clay loam
Station III	Sandy loam	Sandy loam	Loamy sand

estuary (station I) and Ashtamudi Ramsar Site (station III) were characterized by relatively higher percentages of sand content (sandy loam and loamy sand texture). In the trilinear diagram, surficial sediments of nearshore to the marine region of Kochi estuary were reported to be sand, sandy silt, and silt (Gandhi et al., 2017). In the present study, the Vembanad Ramsar site had comparatively higher percentages of loamy texture (loam and also loamy sand). A study conducted by Priju and Narayana (2007) revealed four textural classes of sediments—most of the samples constituted muddy sand, sandy mud, silty sand, and mud in the Vembanad estuary.

4.4.4 Principal component analysis for sedimentological variables

PCA analysis for sedimentological variables (Fig. 4.7; Tables 4.5 and 4.6) with a total of three canonical axes explains a cumulative variance of 99.9%. Of the total variables analyzed PC1 and PC2 explain about 93.7% of the variance. Sand and silt, have a greater contribution to PC1 which accounted for 74.8% of the cumulative variance with an eigen value of 2.99. Clay and OM have a greater contribution with PC2 with 93.7% of the cumulative variance with an eigen value of 0.756 (Table 4.5).

PCA plot illustrated the distribution of sand ordinated to the left of station III which was characterized by high sand content. Silt, clay, and OM were ordinated to the right to stations I and II indicating their greater concentration in those stations. From the PCA plot, it is clear that sand is negatively correlated to silt, clay, and OM as they are positioned oppositely in the PCA plot.

4.4.5 Role of estuarine sediments in macrobenthic group composition and occurrence

Estuarine macrobenthos of Indian waters has been investigated extensively as an attempt to elucidate the role of sediment OM and composition responsible for their community organizations. Macrofauna are important contributors to estuarine ecosystem services including coastal stabilization,

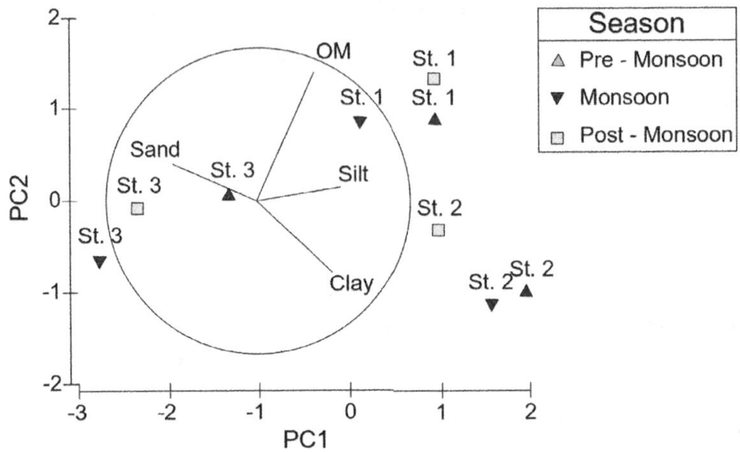

Figure 4.7 Principal component analysis (PCA) ordinations for sedimentological variables.

Table 4.5 PCA values for geochemical variables.

PC	Eigenvalues	%Variation	Cum. %Variation
1	2.99	74.8	74.8
2	0.756	18.9	93.7
3	0.247	6.2	99.9

Table 4.6 Pattern matrix table for geochemical variables.

Variable	PC_1	PC_2	PC_3
Sand	− 0.561	0.243	0.225
silt	0.548	0.087	− 0.615
Clay	0.492	− 0.469	0.661
OM	0.378	0.845	0.366

nutrient cycling, carbon cycling and storage, and also trophic transfer (Janas et al., 2019). Analysis from Indian estuarine systems revealed that macrobenthic invertebrates (Figs. 4.8–4.10; Tables 4.7 and 4.8) consisted of polychaetes, crustaceans, molluscs, and other groups. Macrobenthos are the dominant member of the benthic community which also act as a food source for various demersal predators inhabiting the same ecosystem (Griffiths et al., 2017). Among macrobenthic groups, polychaetes were abundant with high dominance in Kochi estuary, at station I (80 No./m^2)

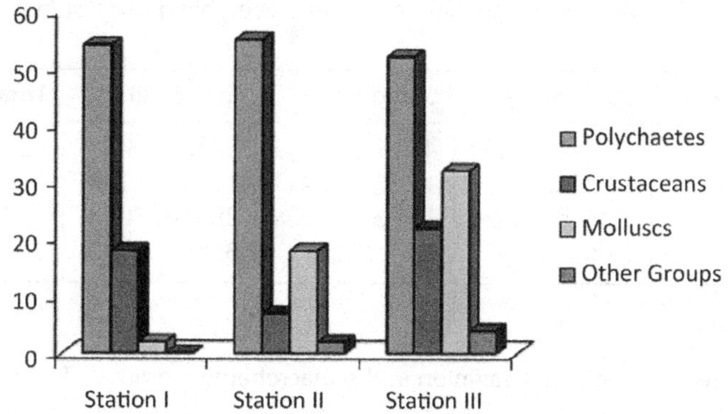

Figure 4.8 Spatiotemporal variation in the composition and distribution of macrobenthic groups (ind./m^2) during premonsoon.

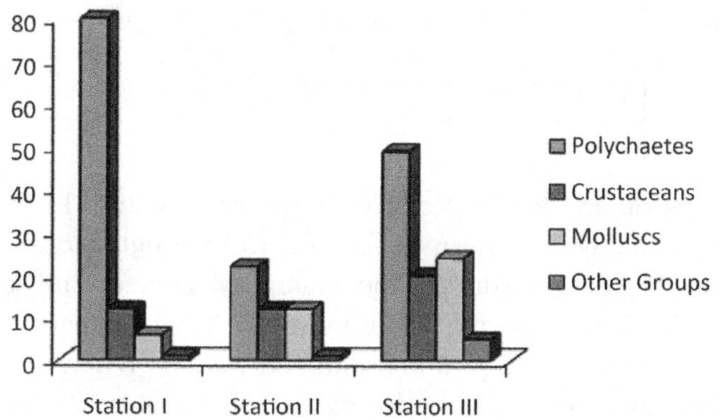

Figure 4.9 Spatiotemporal variation in the composition and distribution of macrobenthic groups (ind./m^2) during monsoon.

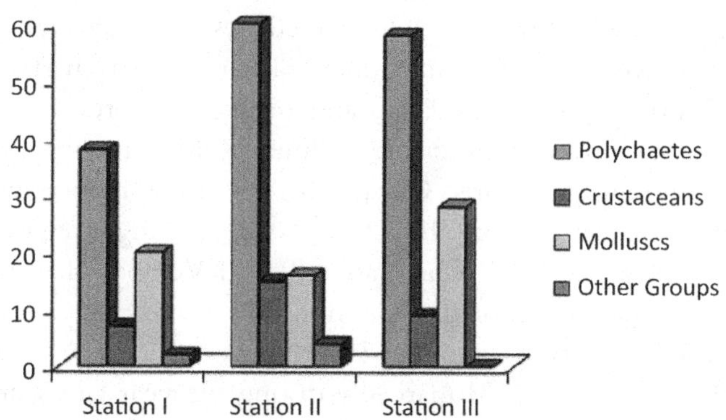

Figure 4.10 Spatiotemporal variation in the composition and distribution of macrobenthic groups (ind./m^2) during postmonsoon.

Table 4.7 Stationwise mean variation in the macrobenthic groups of Indian estuaries.

Macrobenthos	Station I	Station II	Station III	Total
Polychaetes	57.33 ± 21.20	45.67 ± 20.65	53.00 ± 4.58	52.00 ± 15.82
Crustaceans	12.33 ± 5.51	11.33 ± 4.04	17.00 ± 7.00	13.56 ± 5.55
Molluscs	9.33 ± 9.45	15.33 ± 3.06	28.00 ± 4.00	17.56 ± 9.84
Others	1.00 ± 1.00	2.33 ± 1.53	3.00 ± 2.65	2.11 ± 1.83

Table 4.8 Seasonwise mean variation in the macrobenthic groups of Indian estuaries.

Macrobenthos	Premonsoon	Monsoon	Postmonsoon	Total
Polychaetes	53.67 ± 1.53	50.33 ± 29.02	52.00 ± 12.17	52.00 ± 15.82
Crustaceans	15.67 ± 7.77	14.67 ± 4.62	10.33 ± 4.16	13.56 ± 5.55
Molluscs	17.33 ± 15.01	14.00 ± 9.17	21.33 ± 6.11	17.56 ± 9.84
Others	2.00 ± 2.00	2.33 ± 2.31	2.00 ± 2.00	2.11 ± 1.83

during monsoon and low in Vembanad estuary, at station II (22 No./m^2) during monsoon with a mean of 52.00 ± 15.81. High monsoonal polychaete assemblages were due to moderate OM content and sandy loam texture of the Kochi estuary. In contrast, the Kochi estuary had a high percentage of OM during postmonsoon and was also reflected by the low faunal abundance. Low faunal density and abundance of macrobenthic groups during monsoon season were due to loamy sediment texture with fairly high OM content. Macrobenthic species, especially polychaetes, perform invaluable ecosystem services during the process of feeding, tube construction, and bioturbation which mediates the exchange of energy and matter between sediment and water column (Mestdagh et al., 2018).

Macrobenthic group abundance and diversity recorded their highest values at the stations with moderate sediment OM content with a higher proportion of sandy texture. Crustaceans recorded high abundance at Ashtamudi wetlands (station III) (22 No./m^2) during premonsoon and low value (7 No./m^2) at Kochi (station I) and Vembanad wetlands (station II) during postmonsoon and premonsoon, respectively, with a mean of 13.56 ± 5.55. A study conducted by Nair et al. (1984) observed that textural characteristics varied markedly during the year and sandy texture dominated the Ashtamudi estuary with higher benthic faunal population during that period. Molluscs also recorded greater abundance at Ashtamudi estuary (station III) during premonsoon (32 No./m^2) whereas lower value at Kochi estuary (station I) during premonsoon (2 No./m^2) with a mean of 17.56 ± 9.84. Other groups contributed to high abundance at Ashtamudi estuary, station III (5 No./m^2) during monsoon with a mean of 2.11 ± 1.83. In the Indian estuaries, macrobenthic groups exhibited marked changes in the distribution and diversity patterns concerning sediment composition and OM content.

4.5 Marine sediments: a case study from the Arabian Sea, north western extension of tropical Indian Ocean, west coast of Indian subcontinent

The ocean floor covers about 70% of the Earth's surface, is a complex and dynamic component of aquatic ecosystems and is composed of materials deposited through ages, and is the key to ancient and historical biogenic marine environments (Turchyn et al., 2021). Marine sediments are blessed with inexhaustible resources that have not been properly exploited. Marine sediments are of great importance for understanding long-term evolution of the climate system of Earth's environment (Hulse et al., 2017), predicting future environmental and climate changes, exploiting and utilizing new energy sources on the seabed, and studying the extreme microbial diversity of oceans and other biological genetic resources. The contamination of marine environments represents an increasing global concern because of the potential risks to both human health and along the coast heavily affecting the marine ecosystems (Giglioli et al., 2020). Marine sediments are of two major groups. They are pelagic and terrigenous. Pelagic sediments are found far away from shore, predominantly of either organic or inorganic origin while terrigenous sediments are found nearshore and may contain at least some coarse material of terrigenous origin. The constituents of sediments may undergo chemical changes after they are deposited, thus developing a new equilibrium between solid and dissolved components. Hence granulometry of marine sediments is often heterogeneous in nature. Their characteristics also will differ spatially depending on the nature of the adjacent coastline, the dynamics of the water, and the features of the sea bed. Marine sediments can be sensitive indicators for monitoring contaminants in aquatic environments (El-Sikaily et al., 2021).

The distribution of sediments on the seafloor represents the integration of several geological, physical, chemical, and biological processes. The physical and chemical properties of seawater have a large impact on the performance of marine sediments (Wang et al., 2019). Marine sediments may be sand, silt, or clay which gets deposited at the sea bottom. They are composed of dead and living organisms, eroded soil, chemical solids, water, and nutrients. Continental and oceanic crust, volcanoes, chemical processes, plants, animals, and microbes are the main sources of marine sediments. Most sedimentary particles and chemical constituents are transported to the ocean by rivers and deposited in the nearshore environment—the continental margins (Ouillon, 2018). The collective name of the shelf, the slope, and the rise is the continental margin. Sediments of the continental margins include sand and mud from rivers and the changes in sediment load are due to geology, geomorphology, and organic elements. The sediments of biogenous and hydrogenous types are most common on the continental shelf region, slope region, and also in the continental rise region. Sediment transport is not constant and can be affected by other external factors such as water flow, water level, weather events, and human influence.

In marine systems, the benthic environment is a crucial component because it provides ecosystem services, sustains biodiversity, regulates the climate, and therefore it is important to human society. The benthic habitat represents the transitional zone between the biosphere and geosphere; it provides a habitat for organisms from all domains of life; and it modulates biogeochemical cycles of carbon, macro and micronutrients, and trace elements (Lessin et al., 2018). Benthos in marine sediments plays a vital role in the ecosystem processes. They perform sediment reworking by feeding, burrowing, and construction activities. Marine sediments that are near to water surface are perforated by burrow structures ranging from simple unbranched vertical tubes to complex and highly branched networks of funnels and chambers inhabited by organisms. These burrows can be found in different stages of excavation and construction. Organic matter deposited on the seafloor provides food for the benthic communities, either at the sediment surface or upon burial into the sediment layers below (Jorgensen et al., 2019). The benthic communities which are already established in a particular marine environment can modify their surrounding sedimentary regime.

Indian coastal zone is renowned for its amazing beaches, fishery resources, and other biotic and abiotic resources having wide social, economical, and ecological significance. As part of urbanization, many threats from various anthropogenic sources such as local encroachment, pilgrim influx, tourist influx, urbanization, industrialization, aquaculture, dumping of plastics, residential and sewage waste disposal, etc. are adversely affecting these ecosystems. As marine sediments act as a sink for various contaminants, it is essential to assess their composition and OM content which in turn affect the biogeochemical cycles and also the quality of ecosystems. Benthos being sedentary, are greatly influenced by the quality and composition of the sediments. So present study is an attempt to assess the sediment composition and OM content and also its role in macrobenthic groups from the selected coastal regions in the Arabian Sea, the northwestern extension of the tropical Indian ocean.

4.5.1 Materials and methods

4.5.1.1 Study area, sampling sites, periods, and research cruises

The present study was carried out in the Arabian Sea (8–24°N latitudes and 50–77°E longitudes), the northwestern extension of the tropical Indian Ocean, Kerala, India. Seasonal samples for the three seasons—premonsoon, monsoon, and postmonsoon were taken from four stations (stations I, II, III, and IV) from the shoreline of the Arabian Sea, Kerala, India during the year 2018–19. Premonsoon includes February–May, monsoon includes June–September, and postmonsoon includes October–January. Coastal zones were represented by station I (Kochi), station II (Alappuzha), station III (Kollam), and station IV (Thiruvananthapuram). A map showing the study area along with sampling stations is presented in Fig. 4.11.

4.5.1.2 Sampling protocol

For the present study, cruises were performed to analyze the marine sediment samples and macrobenthic faunal assemblages. Sediment samples for geochemical analyses were collected and subsampled and immediately brought to the laboratory for further analyses (Al-Shiwafi et al., 2005). The collected sediment and macrobenthic samples were analyzed as per standard procedures.

4.5.2 Organic matter in marine sediments from Indian subcontinent

Organic matter (OM) in marine sediments is finely dispersed in different concentrations depending largely on the size of the OM source, water depth, and sedimentation rates. Organic and geochemical imprints in the marine sediments are a function of complex interactions between the source area and sedimentation (Choudhary et al., 2018). Burial of organic

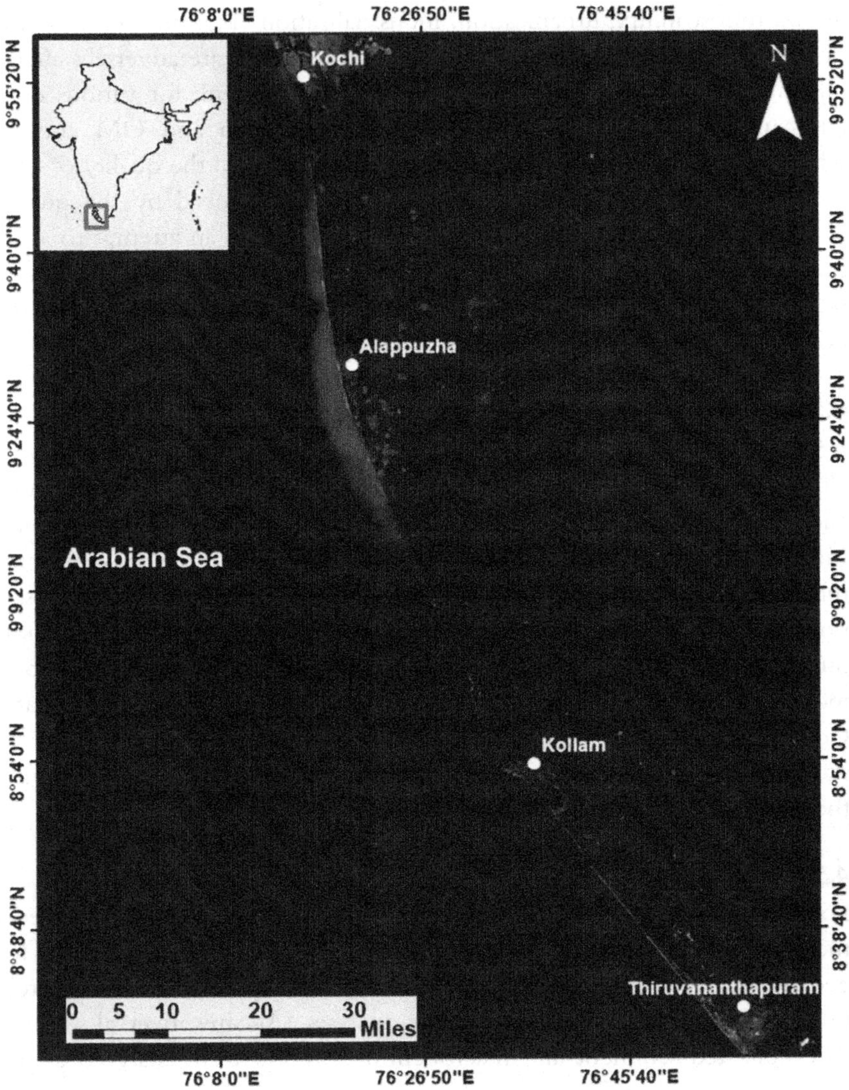

Figure 4.11 Location map of the study area.

carbon (OC) in marine sediment is a long-term sink for atmospheric carbon dioxide (CO_2) as it removes CO_2 from the atmosphere through the biosphere and stores it in the geosphere (Baroni et al., 2020). The OM preserved in sedimentary deposits is also important as a progenitor of fossil fuels, a recorder of Earth history (Zonneveld et al., 2010), and the ultimate source of essentially all atmospheric oxygen. Organic matter preservation in marine sediments directly links the global cycles of carbon, oxygen, and sulfur over geologic time. The biogeochemical cycling of OM strongly influences the global carbon cycle by acting as a sink or the source for carbon dioxide in the atmosphere. It is strongly linked to global climate and thus biogeochemical cycling of the OM becomes a global concern (Rowe et al., 2020). The mineralization of OM has direct as well as indirect relation to many of the reactions and biogeochemical processes that occur in the marine environment. The decomposition of OM promotes the recycling of essential nutrients, balances the amount of oxygen in oceans and sediments, and helps in most early diagenetic processes. However, there is a direct relationship between OM decomposition rates and OM composition or between mineralization rates and OM decomposition pathways (oxic vs anoxic). In recent decades, a significant increase of carbon dioxide in the atmosphere due to fossil fuel burning has been observed, resulting in global warming (IPCC, 2007).

In the retention and permanent removal of nutrients and OM, sediments play a very important role. The degradation of OM is an important process going on in sediments. It links the biological carbon cycle with the geological subsurface cycle. One of the challenges of characterizing within this complex mosaic of habitats is the diverse range of sources from which OM derives (Canuel, 2001). When the oxygen content is low in the marine sediments, the organic carbon burial is at its maximum. However, the mechanism remains enigmatic. The grain size has an immense influence on the distribution of various sedimentary parameters in aquatic environments (Renjith et al., 2011). As particle size changes, changing redox conditions and the availability of organic carbon substrates determine the benthic community structure. If marine sediments have lower redox potential, it points to the presence of OM in them. Particle size, hydrological regime, OM quality, and nutrient availability influence communities in sediments. The complex coastal bathymetry controls the heterogeneity of sediment types and the inhabiting biota, which have large spatial variability within coastal ecosystems. Most of the researchers were focused on the processes in deep, dark, and muddy sediments but not in shallower sandy sediments. These sediments may be with or without sufficient light to support benthic primary producers (McGlathery et al., 2007).

Analysis of sediment OM from the four selected stations of the Arabian Sea, the northwestern extension of tropical Indian Ocean from the Indian subcontinent (Fig. 4.12; Tables 4.9 and 4.10) revealed that

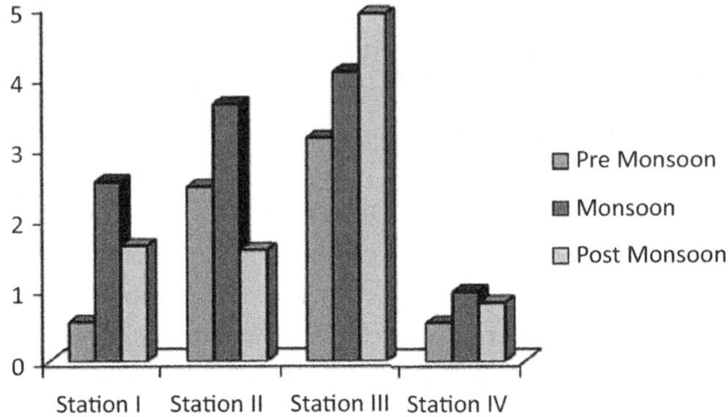

Figure 4.12 Variation in sediment organic carbon (%) in the Arabian Sea.

OM content ranged from 4.90% to 0.54%. A study conducted on the shelf sediments of the southeastern Arabian Sea, west coast of India, indicated that sediment OM in the shelf sediments of the southwest coast of India ranged from 0.95% to 6.29% (mean, 2.67% ± 1.57%). The highest amount of OM was observed where sediment was clayey silt in nature (Ramya et al., 2013). In the present study, at Kochi (station I), maximum sediment OM of 2.53% during monsoon and minimum of 0.55% during premonsoon with a mean value of 1.57 ± 0.99 was observed. At Alappuzha (station II), the values varied from 3.63% during monsoon to 1.57% during postmonsoon with a mean value of 2.55 ± 1.03. Compared to the Kochi coast, the OM content of Alappuzha, was slightly high which was mainly due to tourists and pilgrim influx as the coast is renowned for its esthetic beauty and popular pilgrim centers. At Kollam (station III), the maximum value of 4.90% was observed during postmonsoon and a minimum of 3.15% during premonsoon, with a mean value of 4.04 ± 0.88. Kollam coast accounted for high OM content in the Arabian Sea and it was greatly imparted by various Anthropogenic interventions. The coast is noted for its harbors—Neendakara and Shakthikulangara, of which former is the largest fishing harbor in Kerala. The region is densely populated and marine fishing is the major activity and livelihood of people residing in this region. Research studies on organic carbon stored in coastal sediments showed clearly that DOM is rapidly remineralized and incorporated into the biosphere. On the other hand, POM is mostly trapped in sediments within the silt-clay size fraction. (Harris, 2020). At Thiruvananthapuram (station IV), maximum sediment organic carbon of 0.99% during monsoon and minimum of 0.54% during premonsoon with

Table 4.9 Stationwise mean variation in the sediment organic matter and granulometry of marine sediments from the Indian subcontinent.

Parameters	Station I	Station II	Station III	Station IV	Total
Sand	80.2 ± 7.89	85.3 ± 9.27	71.46 ± 5.22	78.86 ± 2.80	78.95 ± 7.75
Silt	15.21 ± 5.83	13.7 ± 8.84	26.75 ± 5.04	19.22 ± 2.79	18.72 ± 7.37
Clay	4.59 ± 2.93	1.00 ± 0.46	1.78 ± 0.19	1.91 ± 0.35	2.32 ± 1.91
Organic matter	1.57 ± 0.99	2.55 ± 1.03	4.04 ± 0.88	0.79 ± 0.23	2.24 ± 1.46

Table 4.10 Seasonwise mean variation in the sediment organic matter and granulometry of marine sediments from the Indian subcontinent.

Parameters	Premonsoon	Monsoon	Postmonsoon	Total
Sand	76.23 ± 9.21	79.15 ± 6.34	81.50 ± 8.74	78.96 ± 7.75
Silt	21.28 ± 8.60	19.22 ± 6.17	15.66 ± 8.11	18.72 ± 7.37
Clay	2.49 ± 2.00	1.64 ± 0.41	2.84 ± 2.85	2.32 ± 1.91
Organic matter	1.68 ± 1.33	2.81 ± 1.38	2.23 ± 1.82	2.24 ± 1.46

a mean value of 0.79 ± 0.23 was observed. Seasonal mean values revealed that monsoon accounts for high OM content of 2.81 ± 1.38 whereas the low value of 1.68 ± 1.33 was observed during premonsoon. An investigation on the infaunal macrobenthos along the west coast of India revealed that the percentage of OM ranged from 0.24% to 6.71%. The study pointed out that higher OM is retained by finer sediments (Joydas & Damodaran, 2009).

4.5.3 Granulometry of marine sediments from Indian subcontinent

Marine sediments, due to their quality or their quantity, have an impact on the ecological quality of aquatic systems. The sediment composition and its resulting permeability regulate the magnitude of solute transport processes, the oxygen penetration depth, and the structure of biological communities (Ahmerkamp et al., 2020). In the material cycles of waters, sediments form a natural filter system and buffer. The physical characteristics of the sediments are greatly determined by the depositions from the environment and these are the important variables to understand geological events in marine environments. A wide range of colors is found in marine sediments. Along the southwest coast of India, the Arabian Sea surficial sediment color pattern was found to be greenish, greenish-gray, or gray color (Smitha, 2011), and also it showed variation slightly from greenish, greenish-black, black, or cream (Joydas, 2002; Sajan, 2003). The sediment colors change due to several reasons. The sediment associated with coarse-grained quartz or limestone deposits is usually lighter colored which may be white or pale tint. The darker sediments will have a predominance of mineral substances, such as iron oxide or manganese oxide Their colors may be red, blue, or green and are usually found in mud The black sediments are often found in undisturbed areas. The grain size of marine sediments varies from the finest clay to large boulders. The ocean floor in some areas contains deposits of either organic or inorganic origin predominately of siliceous or calcareous remains. Sediment texture is one of the key features of marine sediments and the study on the textural characteristics of sediments gives valuable information about the grain size which determines the transport and deposition in marine systems (Romano et al., 2017).

From the present study, the sediment textural analysis (Figs. 4.13–4.15; Tables 4.9 and 4.10) in marine systems from the Indian coast revealed that sand, silt, and clay showed spatial and temporal variations. Sediment textural composition in the marine sediments was predominated by sand followed by silt and clay. The percentage composition of sand ranged from 93% to 66%, silt ranged from 32.07% to 6.48%, and clay ranged from 7% to 0.52%. On investigating the sandy beach exposed coasts, it was found that they were the most dynamic of all sedimentary environments, being continuously exposed to breaking waves and currents varying with tidal range. Marine sediments found along linear, wave-dominated coasts, are commonly composed of well-sorted sand and gravel from which finer silt and clay size fractions have been winnowed by waves and currents. Fine sediment that escapes from the coastal zone is deposited on the adjacent continental shelf (Harris, 2020).

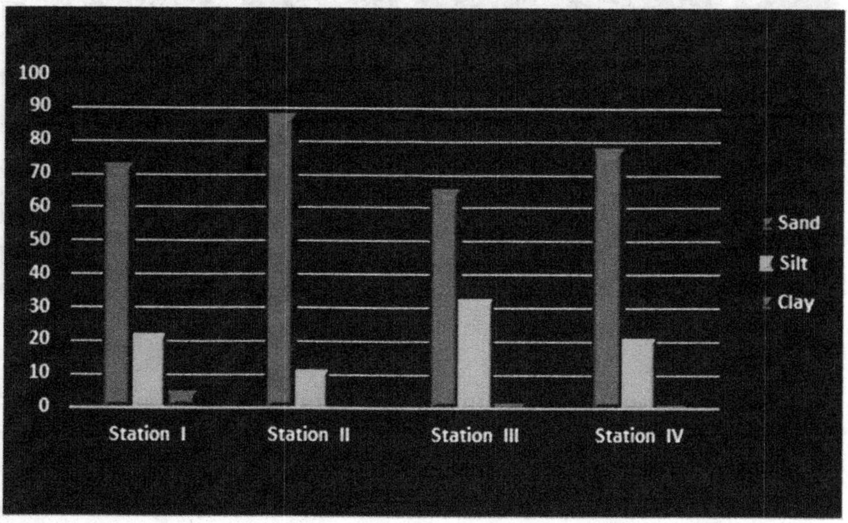

Figure 4.13 Premonsoon variation in the marine sediment textural composition.

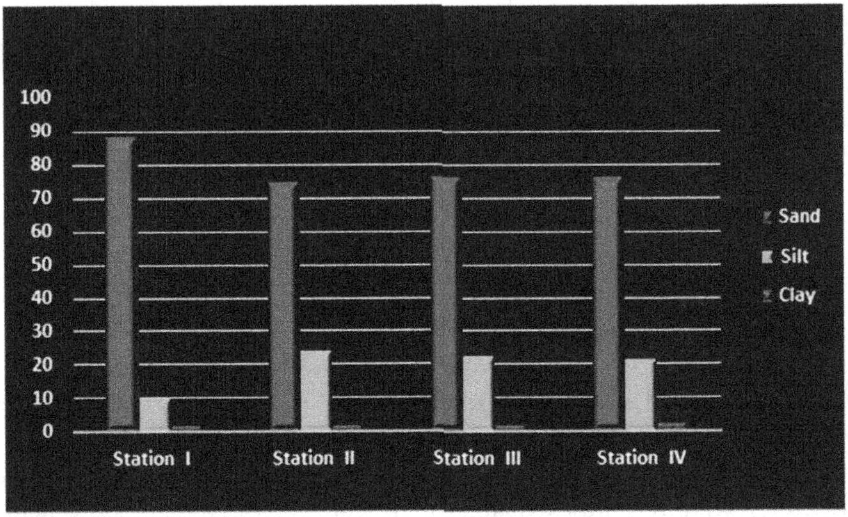

Figure 4.14 Monsoon variation in the marine sediment textural composition.

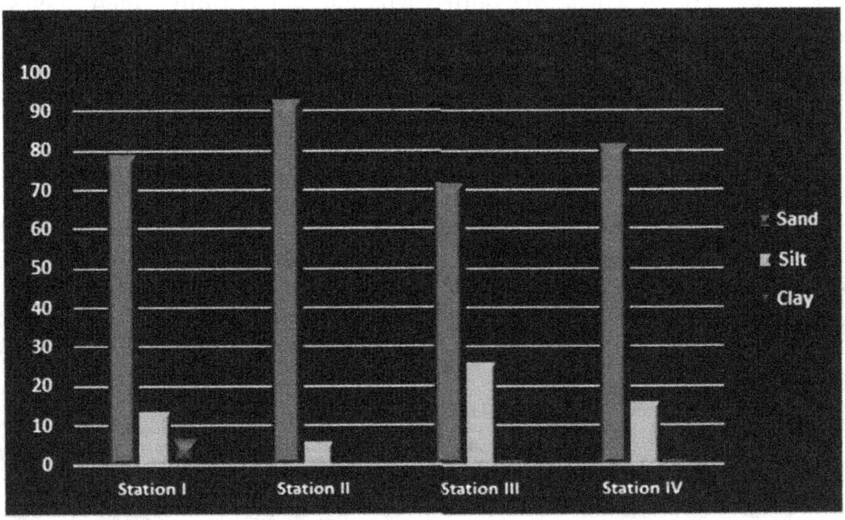

Figure 4.15 Postmonsoon variation in the marine sediment textural composition.

At Kochi (station I), the granulometric composition was predominated by sand. Sand in percentage was high during monsoon (88.6%) whereas low during premonsoon (73%) with a mean value of 80.2 ± 7.89, which was mainly due to influx of sand from nearby riverine source as part of monsoonal rainfall. Most sandy sediments that occur along wave-dominated regions, exhibit a wide range of geomorphic and sedimentological properties governed by fundamental processes that include wave energy, tidal range, sediment supply, and sea-level change (Masselink & Hughes, 2003). Silt in percentage at Kochi coast (station I), was high during premonsoon 21.55% whereas low during monsoon (10.08%) with a mean value of 15.21 ± 5.83. Clay in percentage was high during postmonsoon (7%) whereas the low value of 1.32% was observed during monsoon with a mean value of 4.59 ± 2.93. In a study carried out along the west coast of India, the textural composition in marine sediments in the nearshore regions was found to be of sandy texture with silt as the least-contributing fraction. With the increase in depth, this sand fraction was found to be replaced by the finer sediment fractions (Joydas, 2002).

At Alappuzha coastal region (station II), a high percentage of the sand value of 93% was recorded during postmonsoon and a low value of 75% during monsoon with a mean value of 85.3 ± 9.27. Alappuzha coast accounted for the highest percentage of sand content when compared to all other stations in the marine sediments. Alappuzha region is blessed with a network of rivers flowing into it and the Arabian Sea. This coastal belt is renowned for its high-quality black mineral-rich beach sands. The sandy fraction (high-quality black sand) in the sediments has the richest deposits of heavy minerals like ilmenite, rutile, leucosin (black sand with titanium), zircon, garnet, sillimanite, and monocyte (uranium and thorium). A review on the sand dunes of Kerala indicated that the heavy mineral content is estimated at 17 million tons out of a total raw sand reserve of 242 million tonnes and the land is naturally radioactive (Sekhar & Jayadev, 2003). Silt fraction in sediments of the coast recorded high value during monsoon (23.57%) and a low value of 6.48% during postmonsoon with a mean of 13.7 ± 8.84. Clay values in percentage were recorded high during monsoon (1.43%) with a mean of 1.00 ± 0.46. The difference in grain size distribution at different locations could be due to the variations in wave energy experienced by different sampling locations as well as the geomorphology (Dora et al., 2011).

A study on the sediment textural analysis along the continental shelf region of the southwest and southeast coasts of India revealed four types of substrata, viz., sand, sandy silt, clayey silt, and sand-silt-clay. Among these, sand was the predominant substrata along both the coasts. The study also pointed out that in the Kollam transect the sediment was clayey silt in nature and the percentage of sand is very low (Smitha, 2011). But in the present study, Kollam coast (station III) was predominated by sandy loam texture. Percentage of sand recorded a maximum value of 76.4% during monsoon and a minimum of 66% during premonsoon, with a mean value of 71.46 ± 5.22. Silt in percentage recorded a maximum value of 32.07% during premonsoon and a minimum of 22.04% during monsoon, with a mean value of 26.75 ± 5.042. Clay in percentage recorded a maximum value of 1.93% during premonsoon and a minimum of 1.56% during monsoon, with a mean value of 1.78 ± 0.19.

At Thiruvananthapuram coast (station IV), the percentage of sand recorded a maximum value of 82% during postmonsoon and a minimum of 76.6% during monsoon, with a mean value of 78.86 ± 2.80. Seasonal mean values revealed that postmonsoon accounts for high sand content of 81.50 ± 8.74 whereas the low value of 76.23 ± 9.21 was observed during premonsoon. In a previous survey, it was found that in Thiruvananthapuram coast, during the first survey, the percentage of sand was considerably high (84%) while during the second and third surveys the percentage decreased significantly (35% and 13%, respectively) and accordingly considerable variation in the silt content (4%, 44%, and 69% during the first, second, and third survey, respectively) (Jaleel, 2012). Silt in percentage was high during monsoon 21.17% whereas the low value of 16.02% was observed during postmonsoon with a mean value of 19.22 ± 2.79. Seasonal mean values revealed that premonsoon accounts for high silt content of 21.28 ± 8.60 whereas the low value of 15.66 ± 8.11 was observed during postmonsoon. Percentage of clay recorded a maximum value of 2.23% during monsoon and a minimum of 1.53% during premonsoon, with a mean value of 1.91 ± 0.35. Seasonal mean values revealed that postmonsoon accounts for high clay content of 2.84 ± 2.85 whereas the low value of 1.64 ± 0.41 was observed during monsoon. A study conducted in the surficial sediments of the Arabian Sea showed that a considerable spatial variation in their composition with the coarse sand, with biogenic materials like calcareous shells, occurred on the shelf edge especially off the southern tip of India. There were significant depth-related differences in the sand, silt, and clay content. In the deeper areas, the finer fractions, viz., silt and clay together constituted over 80% sediment texture whereas nearshore regions are with coarser fractions (Jaleel, 2012).

Marine sediments analyzed from the Indian coast revealed the occurrence of three textural classes of sediments (Fig. 4.16; Table 4.11), viz. sandy, loamy sand, and sandy loam. Stations I and II were characterized by relatively higher percentages of sand content and loam (sandy and loamy sand texture), whereas station III was characterized by a combination

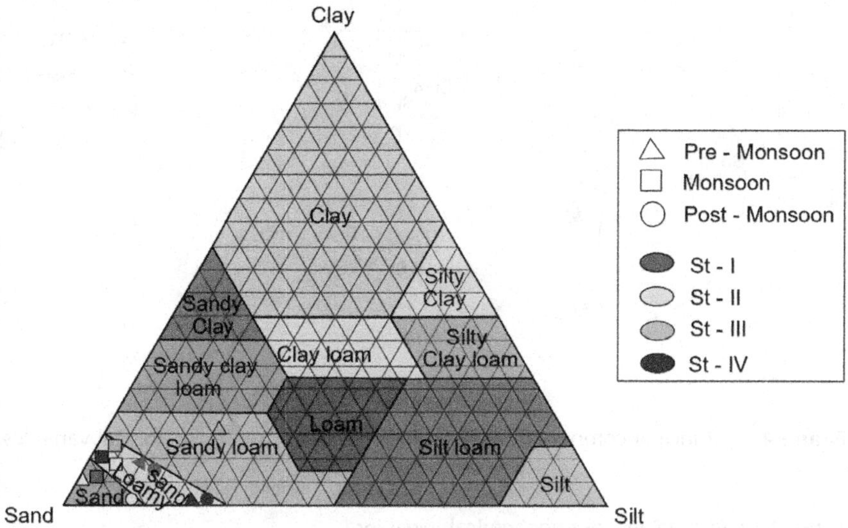

Figure 4.16 Trilinear plot for the seasonal and stationwise distribution of sediment textural classes in marine sediments.

Table 4.11 Seasonal and stationwise distribution of sediment textural classes.

Stations	Premonsoon	Monsoon	Postmonsoon
Station I	Loamy sand	Sandy	Loamy sand
Station II	Sandy	Loamy sand	Sandy
Station III	Sandy loam	Loamy sand	Sandy loam
Station IV	Loamy sand	Loamy sand	Loamy sand

of loamy sand and sandy loam texture. Station IV had comparatively higher percentages of loamy sand texture. Sediment textural analysis conducted along the continental shelf region of the southwest and southeast coast of India revealed altogether four types of textural classes, viz., sand, sandy silt, clayey silt, and sand-silt-clay (Smitha, 2011).

4.5.4 Principal component analysis for marine sedimentological variables

PCA analysis for sedimentological variables (Fig. 4.17; Tables 4.12 and 4.13) with a total of three canonical axes explains a cumulative variance of 99.2%. Of the total variables analyzed, PC1 and PC2 explain about 87.9% of the variance. Sand and silt, have a greater contribution to PC1 which accounted for 55.5% of the cumulative variance with an eigenvalue

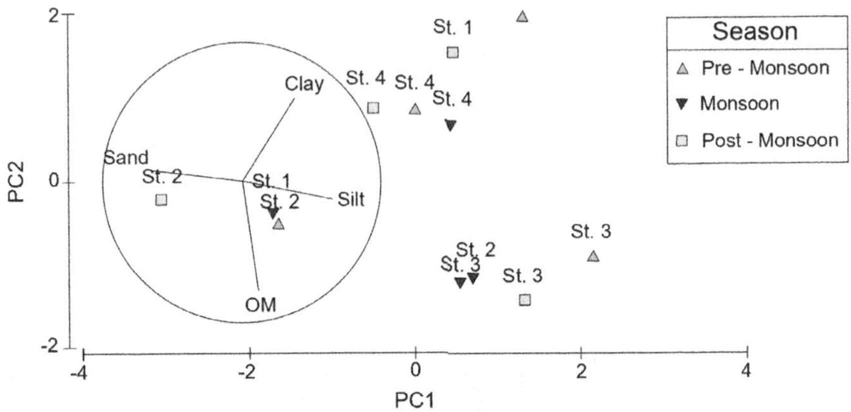

Figure 4.17 Principal component analysis ordinations for sedimentological variables.

Table 4.12 PCA values for geochemical variables.

PC	Eigenvalues	%Variation	Cum. %Variation
1	2.22	55.5	55.5
2	1.29	32.4	87.9
3	0.455	11.4	99.2

Table 4.13 Pattern matrix table for geochemical variables.

Variables	PC1	PC2	PC3
Sand	− 0.659	0.079	− 0.167
Silt	0.639	− 0.134	0.349
Clay	0.381	0.593	− 0.7
OM	0.112	− 0.79	− 0.601

of 2.22. Clay and OM have a greater contribution with PC2 with 87.9% of the cumulative variance with an eigenvalue of 1.29.

PCA plot illustrated the distribution of sand ordinated to the left to station II which was characterized by high sand content only during postmonsoon. Silt, clay, and OM were ordinated to the right to stations I, III, and IV indicating their greater concentration in those stations. From the PCA plot, it is clear that sand is negatively correlated to silt, clay, and OM as they are positioned oppositely in the PCA plot. The sedimentological variable having high loadings to PC1 such as sand and silt gives a clear picture of variations along with the stations.

4.5.5 Role of marine sediments in macrobenthic group composition and occurrence

Marine macrobenthos is metazoan bioresources (Stratmann et al., 2020) in the marine benthic system which has a distinct biological and operational identity. In marine systems, the seafloor is known to harbor high macrobenthic biodiversity extending from the wave splash zone on the beach to the sediments of the deepest oceanic trenches. They are considered a key group among the biota in all coastal aquatic ecosystems (Levin et al., 2000). Benthic macrofauna plays a central role in OM retention, remineralization, and benthic-pelagic exchange, both directly through ingestion, biomass production, respiration, and excretion (Ehrnsten et al., 2019). Due to their sedentary habit, marine macrobenthos are excellent indicators of environmental stress. They form critical links in the marine food web and are involved in energy transfer. Benthos form a connecting link between the nutrients present in the detritus and proteins that are found in the body parts of fishes and shell fishes and exist as a significant dietary component (Jędruch et al., 2019). They help in the energy transfer among other benthic communities by carrying out primary production (benthic algae) and secondary production as grazers and predators. So, they possess active participation in benthic-pelagic trophic relations (Boesch et al., 1977).

Marine macrobenthos makes up a large component of the benthic community, and mediate the ecosystem processes such as sediment decomposition. Much information on ecosystem health, stability, and persistence has been provided by studies examining the benthic marine community's diversity and structure, through time. Benthic infaunal monitoring is widely accepted as the fundamental step to most recent interdisciplinary studies on the contaminant effects on ecosystems. From the monitoring perspective, marine benthos offers mainly three positive attributes: (1) they are relatively sedentary and long-lived hence this infauna cannot avoid being affected by the sediments that are polluted or contaminated. (2) they are having an important position in the trophic level which is intermediate. (3) Infaunal communities are composed of a diverse array of species that respond differentially to varying environmental conditions, like high mortality of pollution-sensitive species and increasing abundance or frequency of pollution-tolerant species (Weisberg et al., 1993). Recent studies on benthic fauna in the Arabian Sea were carried out along the western continental margin (Gage et al., 2000). Various investigations were carried out on the influence of oxygen and OM with sediment characteristics on benthic fauna along the northern continental margins of the Arabian Sea (Gooday et al., 2009; Hughes et al., 2009; Lamont & Gage, 2000; Levin et al., 2009). Considerable research work along the southeast and west coast of India was carried out by various investigators (Damodaran, 1972; Govindankutty & Nair, 1966; Jaleel, 2012; Joydas, 2002; Priyalakshmi, 2008; Sajan & Damodaran, 2007; Sajan et al., 2010; Smitha, 2011). Due to their unique living patterns and sedentary mode of life, the health of macrobenthic communities is a good bioindicator to reflect ecosystem health and the long-term changes of the marine ecosystem. Marine macrofaunal community composition will alter in relation to the sediment organic carbon and composition. Their distribution highly depends on the physical nature of the substratum nutritive content, degree of stability, oxygen content, and level of hydrogen sulfide (Anbuchezhian et al., 2009).

Macrofaunal analysis from selected coastal zones from the Arabian Sea revealed that macrobenthic invertebrates (Figs. 4.18–4.21; Tables 4.14 and 4.15) consisted mainly of four major groups, viz., polychaetes, crustaceans, molluscs, and other groups. Among various faunal groups, polychaetes accounted for greater abundance on the Arabian Sea coast. The abundance of macrobenthic polychaetes in the Arabian Sea coastline will in turn boost the demersal fishery resources of the region. As macrobenthic polychaetes follow the detritus food chain, they take part in the breakdown of particulate organic material and export energy to higher trophic levels and can potentially support off-shore and pelagic communities too. The developmental stages of many macrobenthic

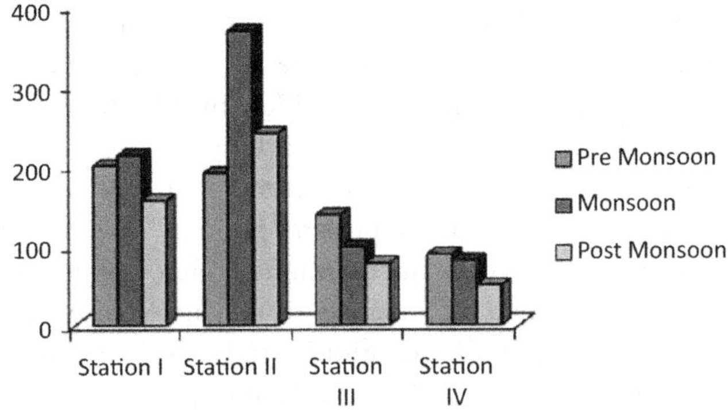

Figure 4.18 Spatiotemporal variation in the composition and distribution of macro benthic polychaetes (ind./m^2).

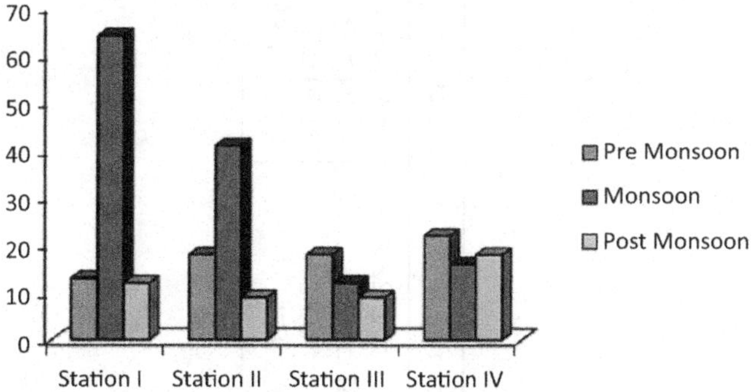

Figure 4.19 Spatiotemporal variation in the composition and distribution of macrobenthic crustaceans (ind./m^2).

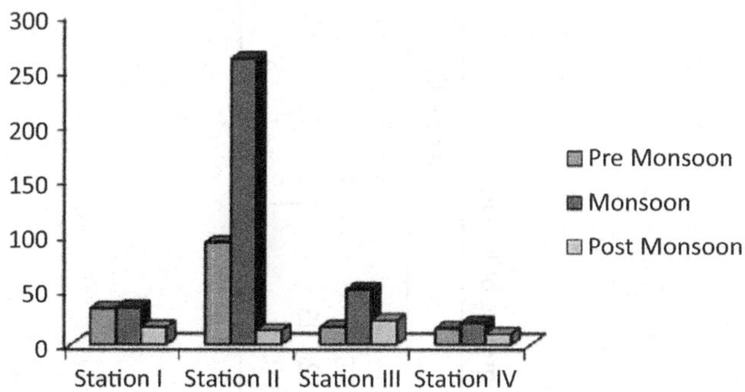

Figure 4.20 Spatiotemporal variation in the composition and distribution of macrobenthic molluscs (ind./m^2).

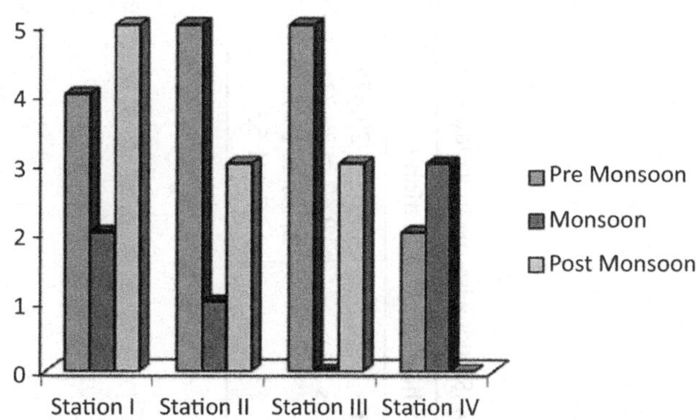

Figure 4.21 Spatiotemporal variation in the composition and distribution of macrobenthic other groups (ind./m^2).

Table 4.14 Stationwise mean variation in the macrobenthic groups of Indian seas.

Macrobenthos	Station I	Station II	Station III	Station IV	Total
Polychaetes	19.67 ± 29.87	27 ± 91.80	104.33 ± 30.99	73.00 ± 20.22	158 ± 90.3
Crustaceans	29.67 ± 29.74	22.67 ± 16.50	13.00 ± 4.58	18.67 ± 3.06	21.00 ± 16.00
Molluscs	27.67 ± 10.12	122.00 ± 126.03	29.33 ± 18.15	15.00 ± 5.00	48.50 ± 70.49
Others	3.67 ± 1.53	3.00 ± 2.00	2.67 ± 2.52	1.53 ± 2.75	2.75 ± 1.82

Table 4.15 Seasonwise mean variation in the macrobenthic groups of Indian seas.

Macrobenthos	Premonsoon	Monsoon	Postmonsoon	Total
Polychaetes	189.67 ± 29.87	266 ± 91.80	104.33 ± 30.99	158 ± 90.3
Crustaceans	29.67 ± 29.74	22.67 ± 16.50	13.00 ± 4.58	21.00 ± 16.00
Molluscs	27.67 ± 10.12	122.00 ± 126.03	29.33 ± 18.15	48.50 ± 70.49
Others	3.67 ± 1.53	3.00 ± 2.00	2.67 ± 2.52	2.75 ± 1.82

organisms are pelagic, forming important components of the plankton community, that in turn are consumed by fish and thus have a high influence on pelagic fisheries. Thus to assess the fishery production of a particular area, the estimation of benthic production is very useful. Moreover, the coastline of the Arabian Sea is densely populated with fisherman communities and fishing-dependent populations. Polychaetes from an important component in the marine food chain especially for bottom fish, as they form an important source of food for demersal fish (Herman et al., 2000) and are also being used for biomonitoring programs to check the health of the marine environment (Jayaraj et al., 2007).

Alappuzha coast, station II accounted for higher dominance (368 No./m^2) during monsoon and lower value at Thiruvananthapuram coast, station IV (50 No./m^2) during postmonsoon with a mean of 158 $\pm$ 90.3. High faunal abundance especially in polychaetes was due to the sandy texture of the coast with moderate OM content. Ansari et al. (1977) reported high density and biomass of polychaetes are associated with sandy substrate. It was noticed that the west coast of India is rich in polychaetes in terms of total abundance and diversity than the east coast (Musale & Desai, 2011). The west coast of the Arabian Sea is more productive than the Bay of Bengal due to geophysical processes, that is, southwest monsoon, wind-driven upwelling leading to nutrient enrichment along the west coast (Goes et al., 1992). Sediment composition is most important to marine benthic organisms (Ingole et al., 1998) which provides shelter and food in the form of OM (Gray, 1981). However, higher organic carbon is reported to cause a decline in species diversity, abundance, and biomass, possibly due to the oxygen depletion and buildup of toxic by-products such as ammonia and sulfide (Hyland et al., 2005). Kollam coast recorded high OM content and sandy loam texture with low crustacean abundance. Crustaceans recorded high abundance at Kochi coast, station I (64 No./m^2) during monsoon and low value (9 No./m^2) at both Alappuzha coast (station II) and Kollam coast (station III) during postmonsoon with a mean of 21.00 $\pm$ 16.00. Molluscs showed greater abundance in Alappuzha coast at station II (260 No./m^2) during monsoon, whereas lesser abundance in Thiruvananthapuram coast at station IV (10 No./m^2) during postmonsoon with a mean of 48.50 $\pm$ 70.49. Other groups recorded higher dominance at Kochi coast, the station I (5 No./m^2) during postmonsoon and nil value at Kollam coast, station III (0 No./m^2) during monsoon with a mean of 2.75 $\pm$ 1.82. Thiruvananthapuram coast was having a loamy sand texture which supported low faunal populations.

Due to the high availability of oxygen, the OM in coarse-grained sediments is readily mineralized, as oxygen is the most efficient electron acceptor which penetrates deeply in the sediment matrix. Such sediments have a relatively low amount of organic carbon but have relatively high oxygen consumption rates (Brenner et al., 2016).

4.6 Conclusion

From the present study, it can be safely concluded that both estuarine and marine macrobenthic group distributions have been affected by both sediment grain size and OM content. In both marine and estuarine ecosystems, sandy texture with moderate OM imparts the abundance of macrobenthic groups. However, recently, these valuable ecosystems have been subjected to multiple stressors due to various anthropogenic interventions such as local encroachment, pilgrim influx, tourist influx, urbanization, industrialization, aquaculture, dumping of plastics, residential and sewage waste disposal, dredging, overexploitation of abiotic and biotic resources, improperly treated industrial waste discharge, etc. In addition to the fishery resources, the estuarine and marine zones are renowned for their esthetic beauty, ecological significance, pilgrim centers, and historical beaches. Hence, biodiversity and quality of water and sediment will gradually deplete due to continuous human interferences. Aquatic sediments act as ultimate sink for natural and anthropogenic inputs and they can store contaminants for millennia. Therefore, if degraded, they can release large amounts back to the atmosphere also. So it is highly essential to protect these valuable aquatic ecosystems and also their biodiversity, by regular biomonitoring and sustainable utilization of biotic and abiotic resources in the pelagic-benthic zones of estuarine and coastal zones in the Indian subcontinent for preserving it for future generations and for maintaining the ecological balance of the entire biosphere.

Acknowledgment

I am extremely thankful to Fatima Mata National College (Autonomous), Kollam, for providing the necessary facilities to carry out this Research Project.

References

Ahmerkamp, S., Marchant, H. K., Peng, C., Probandt, D., Littmann, S., Kuypers, M. M. M., & Holtappels, M. (2020). The effect of sediment grain properties and porewater flow on microbial abundance and respiration in permeable sediments. *Scientific Reports*, *10*, 1–12. Available from https://doi.org/10.1038/s41598-020-60557-7.

Akita, L. G., Laudien, J., Akrong, M., Biney, C., Nyarko, E., & Addo, S. (2020). Physicochemical influence on the spatial distribution of fecal bacteria and polychaetes in the Densu estuary, Ghana. *African Journal of Environmental Science and Technology, 14* (6), 139–158. Available from https://doi.org/10.5897/AJEST2020.2830.

Al-Shiwafi, N., Rushdi, A. I., & Ba-Issa, A. (2005). Trace metals in surface seawaters and sediments from various habitats of the Red Sea Coast of Yemen. *Environmental Geology, 48*, 590–598.

Anbuchezhian, R. M., Rameshkumar, G., & Ravichandran, S. (2009). Macrobenthic composition and diversity in the coastal belt of Thondi, southeast coast of India. *Global Journal of Environmental Research, 3*(2), 68–75.

Ansari, Z. A., Parulekar, A. H., Harkantra, S. N., & Nair, A. (1977). Shallow water macrobenthic along the central west coast of India. *Mahasagar, 10*(3–4), 123–127.

Asha, C. V., Retina, I. C., Suson, P. S., & Bijoy Nandan, S. (2016). Ecosystem analysis of the degrading Vembanad wetland ecosystem, the largest Ramsar site on the south west coast of India—Measures for its sustainable management. *Regional Studies in Marine Science, 8*(3), 408–421.

Babu, K. N., Omana, P. K., & Mohan, M. (2009). Water and sediment quality of Ashtamudi estuary, a Ramsar site, southwest coast of India—A statistical appraisal. *Environmental Monitoring and Assessment, 165*(1–4), 307–319. Available from https://doi.org/10.1007/s10661-009-0947-0.

Barber, A., Brandes, J., Leri, A., Lalonde, K., Balind, K., Wang, J., & Gelinas, Y. (2017). Preservation of organic matter in marine sediments by inner-sphere interactions with reactive iron. *Scientific Reports*, 7, 1–10.

Baroni, I. R., Palastanga, V., & Slomp, C. P. (2020). Enhanced organic carbon burial in sediments of oxygen minimum zones upon ocean deoxygenation. *Frontiers in Marine Science, 6*, 839. Available from https://doi.org/10.3389/fmars.2019.0083.

Batvari, B. P. D., & Mariappan, V. E. N. (2016). Grain size and sediment characterization of lagoon sediments in Pulicat lake, north Chennai coastal region, southeast coast of India. *Rasayan Journal of Chemistry, 9*(3), 424–430.

Belan, T. A. (2003). Marine environmental quality assessment using polychaete taxonomic characteristics in Vancouver harbour. Marine *Environmental Research, 57*, 89-101.

Bhowmik, M., & Mandal, S. (2021). Do seasonal dynamics influence traits and composition of macrobenthic assemblages of Sundarbans Estuarine System, India? *Oceanologia, 63*, 80–98. Available from https://doi.org/10.1016/j.oceano.2020.10.002.

Bianchi, T. S. (2006). *Biogeochemistry of estuaries*. Oxford: Oxford University Press.

Boesch, D. F., Kraeuter, J. N., & Serafy, D. K. (1977). *Distribution and structure of communities of macrobenthos on the outer continental shelf of middle Atlantic Bight: 1975–1976 Investigations, Virginia Institute of Marine Science. Special Report in Applied Marine Science and Ocean Engineering* (p. 175) Virginia Institute of Marine Science.

Boggemann, M. (2005). *Revision of the Goniadidae. AbhandlungendesNaturwissenschaftlichen Vereins in Hamburg NF 39* (pp. 1–354). KelternWeiler: Goecke and Evers.

Bragadeeswaran, S., Rajasegar, M., Srinivasan, M., & Rajan, U. K. (2007). Sediment texture and nutrients of Arasalar estuary, Karaikkal, south-east coast of India. *Journal of Environmental Biology, 28*(2), 237–240.

Brenner, H., Braeckman, U., LeGuitton, M., & Meysman, F. J. R. (2016). The impact of sedimentary alkalinity release on the water column CO_2 system in the North Sea. *Biogeosciences, 13*, 841–863. Available from https://doi.org/10.5194/bg-13-841-2016.

Burdige, D. J. (2007). Preservation of organic matter in marine sediments: Controls, mechanisms, and an imbalance in sediment organic carbon budgets? *Chemical Reviews, 107*(2), 467–485. Available from https://doi.org/10.1021/cr050347q.

Canuel, E. A. (2001). Relations between river flow, primary production and fatty acid composition of particulate organic matter in San Francisco and Chesapeake Bays: A multivariate approach. *Organic Geochemistry, 32*, 563–583.

Canuel, E., & Hardison, A. (2016). Sources, ages, and alteration of organic matter in estuaries. *Annual Review of Marine Science, 8*(1), 409–434.

Choudhary, G. N. S., Kumar, N. A., & Khar, T. N. (2018). Source, processes and productivity from distribution of surface sediments, Prydz Bay, East Antarctica. *Polar Science, 18*, 63–71. Available from https://doi.org/10.1016/j.polar.2018.06.002.

Cividanes, S., Incera, M., & Lopez, J. (2002). Temporal variability in the biochemical composition of sedimentary organic matter in an intertidal flat of the Galician coast (NW Spain). *Oceanologica Acta, 25*, 1–12.

Costa, B. V. M. D., Moreira, M. Z., Bastos, R. F., & Yogui, G. T. (2021). Distribution and origin of sedimentary organic matter in a eutrophic estuary: Pina Sound—NE Brazil. *Anais da Academia Brasileira de Ciências, 93*(1), e20190638. Available from https://doi.org/10.1590/0001-3765202120190638.

Countway, R. E., Canuel, E. A., & Dickhut, R. M. (2007). Sources of particulate organic matter in surface waters of the York River, VA estuary. *Organic Geochemistry, 38*, 365–379.

Damodaran, R. (1972). Meiobenthos of the mud banks of the Kerala coast. *Proceedings of the Indian National Science Academy, 38*, 288–297.

Day, J. H. (1967). *A monograph on the Polychaeta of Southern Africa. Trustees of British Museum (Natural History).*

Dean, W. E., & Gorham, E. (1998). Magnitude and significance of carbon burial in lakes, reservoirs, and peatlands. *Geology, 26*, 535.

Denga, L., Bolsterlia, D., Kristensenb, E., Meilec, C., Sud, C., Bernasconie, S. M., Seidenkrantzf, M. S., Glombitzaa, C., Lagostinaa, L., Hana, X., Jorgenseng, B. B., Royg, H., & Levera, M. A. (2020). Macrofaunal control of microbial community structure in continental margin sediments. *Proceedings of the National Academy of Sciences of the United States of America, 117*(27), 15911–15922. Available from https://doi.org/10.1073/pnas.1917494117.

Dianto, A., Ridwansyah, I., & Subehi, L. (2019). Organic matter and organic carbon levels in sediments of Lake Maninjau, West Sumatra. *IOP Conference Series: Earth and Environmental Science, 535*, 012030. Available from https://doi.org/10.1088/1755-1315/535/1/012030.

Dora, G. U., Kumar, V. S., Philip, C. S., Johnson, G., Vinayaraj, P., & Gowthaman, R. (2011). Textural characteristics of foreshore sediments along Karnataka shoreline, west coast of India. *International Journal of Sediment Research, 26*(3), 364–377.

Ehrnsten, E. B., Norkkoa, A. B., Timmermannc, K., & Gustafssona, B. G. (2019). Benthic-pelagic coupling in coastal seas—Modelling macrofaunal biomass and carbon processing in response to organic matter supply. *Journal of Marine Systems, 196*, 36–47.

El-Serehy, H. A., Al-Misned, F. A., & Al-Rasheid, K. A. (2015). Population fluctuation and vertical distribution of meiofauna in the Red Sea interstitial environment. *Saudi Journal of Biological Sciences, 22*, 459–465, ISSN 0921-8181.

El-Sikaily, A., Ghoniem, D. G., Emam, M. A., & El-Nahrery, E. M. A. (2021). Biological indicators for environmental quality monitoring of marine sediment in Suez Gulf, Egypt. *Egyptian Journal of Aquatic Research, 47*(2), 125–132. Available from https://doi.org/10.1016/j.ejar.

El-Wakeel, S. K., & Riley, J. P. (1957). The determination of organic carbon in marine muds. *ICES Journal of Marine Science, 22*, 180–183.

FAO. (1998). *Corporate document repository site selection for mollusc culture*, NACA-SF/WP/88/8.

Fauvel, P. (1953). *The fauna of India including Pakistan, Ceylon, Burma and Malaya.* Allahabad: The Indian Press.

Gage, J. D., Levin, L. A., & Wolff, G. A. (2000). Benthic processes in the deep Arabian Sea: Introduction and overview. *Deep-Sea Research Part II*, *47*, 1–8.

Gandhi, S. M., Jisha, K., Jeshma, P., & Tarun, R. (2017). Foraminiferal and sediment geochemistry studies in and around Cochin backwaters, South West coast of India. *Indian Journal of Geo-Marine Sciences*, *46*(11), 2303–2313.

Giglioli, S., Colombo, L., Contestabile, P., Musco, L., Armiento, G., Somma, R., Vicinanza, D., & Azzellino, A. (2020). Source apportionment assessment of marine sediment contamination in a post-industrial area (Bagnoli, Naples). *Water*, *12*, 2181. Available from https://doi.org/10.3390/w12082181.

Goes, J. I., Gomes, H., Kumar, A., Gouveia, A. D., Devassy, V. P., Parulekar, A. H., & Rao, L. V. G. (1992). Satelrlite and ship studies of phytoplankton along the west coast of India. In B. N. Desai (Ed.), *Oceanography of the Indian Ocean* (pp. 67–80). Oxford and IBH.

Goni, M. A., Maria, J. T., & David, W. P. (2003). Sources and distribution of organic matter in a river-dominated estuary (Winyah Bay, SC, USA). *Estuarine, Coastal and Shelf Science*, *57*, 1023–1048.

Gooday, A. J., Levin, L. A., Aranda da Silva, A., Bett, B. J., Cowie, G. L., Dissard, D., Gage, J. D., Hughes, D. J., Jeffreys, R., Lamont, P. A., Larkin, K. E., Murty, S. J., Schumacher, S., Whitcraft, C., & Woulds, C. (2009). Faunal responses to oxygen gradients on the Pakistan margin: A comparison of foraminiferans, macrofauna, and megafauna. *Deep-Sea Research*, *56*, 488–502.

Gorringe, R., & Nicholls, J. (1982). *The flora and fauna of the Mediterranean Sea* (pp. 1–320). London: Hamlyn Publishing Group.

Govindankutty, A. G., & Nair, N. B. (1966). *Preliminary observations on the interstitial fauna of the southwest coast of India. Hydrobiologia* (28, pp. 101–122).

Gray, J. S. (1981). *The ecology of marine sediments: An introduction to the structure and function of benthic communities* (p. 185) Cambridge: Cambridge University Press.

Griffiths, J. R., Kadin, M., Nascimento, F. J. A., Tamelander, T., Tornroos, A., Bonaglia, S., Bonsdorff, E., Bruchert, V., Gard-mark, A., Jarnstrom, M., Kotta, J., Lindegren, M., Nord- strom, M. C., Norkko, A., Olsson, J., Weigel, B., Zydelis, R., Blenckner, T., Niiranen, S., & Winder, M. (2017). The importance of benthic-pelagic coupling for marine ecosystem functioning in a changing world. *Global Change Biology*, *23*(6), 2179–2196. Available from https://doi.org/10.1111/gcb.13642.

Harris, P. T. (2020). The fate of microplastic in marine sedimentary environments: A review and synthesis. *Marine Pollution Bulletin*, *158*, 111398.

Hashimi, N., Kidwai, R. M., & Nair, R. R. (1981). Comparative study of the topography and sediments of the western & eastern continental shelves around Cape Comorin *Indian Journal of Geo-Marine Sciences*, *10*, 45–50.

Herman, P. M. J., Middelburg, J. J., Widdows, J., Lucas, C. H., & Heip, C. H. R. (2000) Stable isotopes as trophic tracers: Combining field sampling and manipulative labeling of food resources for macrobenthos. *Marine Ecology Progress Series*, *204*, 79–92.

Hermand, R., Salen-Picard, C., Alliot, E., & Degiovanni, C. (2008). Macrofaunal density, biomass and composition of estuarine sediments and their relationship to the river plume of the Rhone River (NW Mediterranean). *Estuarine, Coastal and Shelf Science*, *79*, 367–376.

Hughes, D. J., Levin, L. A., Lamont, P. A., Packer, M., Feeley, K., & Gage, J. D. (2009) Macrofaunal communities and sediment structure across the Pakistan margin oxygen minimum zone, North-East Arabian Sea. *Deep-Sea Research Part II*, *56*, 434–448.

Hulse, D., Arndt, S., Wilson, J. D., Munhoven, G., & Ridgwell, A. (2017) Understanding the causes and consequences of past marine carbon cycling variability through models. *Earth-Science Reviews*, *171*, 349–382.

Hyland, J., Balthis, L., Karakassis, I., Magni, P., Petrov, A., Shine, J., Vestergaard, O., & Warwick, R. (2005). Organic carbon content of sediments as an indicator of stress in the marine bentho. *Marine Ecology Progress Series*, *295*, 91–103.

Ingole, B. S., Ansari, Z. A., & Parulekar, A. H. (1998). Spatial variation in meiofaunal abundance of some coralline beaches of Mauritius. *Tropical Ecology*, *39*(1), 103–108.

IPCC. (2007). Climate Change 2007: Synthesis Report. Contribution of Working Group I, II and III to the Fourth Assessment Report of the Intergovernmental Panel on Climate Change (pp. 104). Geneva, Switzerland: IPCC.

Jaleel, K. U. (2012). *Macrobenthos of the continental margin (200–1000 m) of south eastern Arabian Sea with special reference to polychaetes* (Ph.D. thesis) (p. 238). Cochin University of Science and Technology.

Janas, U., Burska, D., Kendzierska, H., Pryputniewicz-Flis, D., & Lukawska-Matuszewska, K. (2019). Importance of benthic macrofauna and coastal biotopes for ecosystem functioning—Oxygen and nutrient fluxes in the coastal zone. *Estuarine, Coastal and Shelf Science*, *225*, 106238. Available from https://doi.org/10.1016/j.ecss.2019.05.020.

Jayaraj, K. A., Jayalakshmi, K. V., & Saraladevi, K. (2007). Influence of environmental properties on macrobenthos in the northwest Indian shelf. *Environmental Monitoring and Assessment*, *127*(1–3), 459–475.

Jędruch, A., Bełdowska, M., & Ziółkowska, M. (2019). The role of benthic macrofauna in the trophic transfer of mercury in a low-diversity temperate coastal ecosystem (Puck Lagoon, southern Baltic Sea). *Environmental Monitoring and Assessment*, *191*(3), 137. Available from https://doi.org/10.1007/s10661-019-7257-y.

Jofre, R., Ågren, G. I., Gillon, D., & Bosatta, E. (2001). Organic matter quality in ecological studies: Theory meets experiment. *Oikos*, *93*, 451–458.

Jorgensen, B. (1996). Material fux in the sediment. In B. Jorgensen, & K. Richardson (Eds.), *Coastal and estuarine studies* (pp. 115–216). Washington, DC: American Geophysics Union.

Jorgensen, B. B., Findlay, A. J., & Pellerin, A. (2019). The biogeochemical sulfur cycle of marine sediments. *Frontiers in Microbiology*, *10*, 849. Available from https://doi.org/10.3389/fmicb.2019.00849.

Joydas, T. V. (2002). *Macrobenthos of the shelf waters of the west coast of India* (Ph.D. thesis) (p. 130). Cochin University of Science and Technology.

Joydas, T. V., & Damodaran, R. (2009). Infaunal macrobenthos along the shelf waters of the west coast of India, Arabian Sea. *Indian Journal of Geo-Marine Sciences*, *38*, 191–204.

Juez, C., Hassan, M. A., & Franca, M. (2018). The origin of fine sediment determines the observations of suspended sediment fluxes under unsteady flow conditions. *Water Resources Research*, *54*, 5654–5669.

Kara, G. T., Kara, M., Bayram, A., & Gunduz, O. (2017). Assessment of seasonal and spatial variations of physicochemical parameters and trace elements along a heavily polluted effluent-dominated stream. *Environmental Monitoring and Assessment*, *189*(11), 585. Available from https://doi.org/10.1007/s10661-017-6309-4.

Kim, G., Burnett, W. C., Dulaiova, H., Swarzenski, P. W., & Moore, W. S. (2001). Measurement of 224Ra and 226R activities in natural waters using a radon-in-air monitor. *Environmental Science & Technology*, *35*, 4680–4683.

Krishnakumar, A., & Das, R. (2019). Grain size variations and organic matter distribution in the surficial sediments of Ashtamudi estuary, Kerala, India: A study in Anthropocene perspectives. In *International conference on frontiers in Marine Science—Challenges and prospects*, Directorate of Public Relations and Publications, CUSAT, ISBN: 978-81-922264-3-9.

Kristensen, E., Delefosse, M., Quintana, C. O., Flindtand, M. R., & Valdemarsen, T. (2014). Influence of benthic macrofauna community shifts on ecosystem functioning in shallow estuaries. *Frontiers in Marine Science*, *1*(41). Available from https://doi.org/10.3389/fmars.2014.00041.

Krumbein, W. C., & Pettijohn, F. J. (1938). *Manual of sedimentary petrography* (p. 549) New York: Apple ton-Century Crafts.

Kumar, S. P., & Khan, A. B. (2012). Identification of mangrove water quality by multivariate statistical analysis methods in Pondicherry coast, India. *Environmental Monitoring and Assessment*, *184*, 3761–3774. Available from https://doi.org/10.1007/s10661-011-2222-4.

Lamont, P. A., & Gage, J. D. (2000). Morphological responses of macrobenthic polychaetes to low oxygen on the Oman continental slope, NW Arabian Sea. *Deep-Sea Research Part II*, *47*, 9–24.

Lessin, G., Artioli, Y., Almroth-Rosell, E., Blackford, J. C., Dale, A. W., Glud, R. N., Middelburg, J. J., Pastres, R., Queiros, A. M., Rabouille, C., Regnier, P., Soetaert, K., Solidoro, C., Stephens, N., & Yakushev, E. (2018). Modelling marine sediment biogeochemistry: Current knowledge gaps, challenges, and some methodological advice for advancement. *Frontiers in Marine Science*, *5*(19). Available from https://doi.org/10.3389/fmars.2018.00019.

Levin, L., Gage, J., Martin, C., & Lamont, P. (2000). Macrobenthic community structure within and beneath the oxygen minimum zone, NW Arabian Sea. *Deep-Sea Research Part II*, *47*, 189–226.

Levin, L. A., Whitcraft, C. R., Mendoza, G. F., Gonzalez, J. P., & Cowie, G. (2009). Oxygen and organic matter thresholds for benthic faunal activity on the Pakistan margin oxygen minimum zone (700–1100 m). *Deep-Sea Research Part II*. Available from https://doi.org/10.1016/j.dsr2.2008.05.032.

Liu, W., Ma, L., & Abuduwaili, J. (2020). Anthropogenic influences on environmental changes of Lake Bosten, the largest inland freshwater lake in China. *Sustainability*, *12*, 711. Available from https://doi.org/10.3390/su12020711.

Lone, A., Fousiya, A. A., Shah, R., & Achyuthan, H. (2018). Reconstruction of paleoclimate and environmental fluctuations since the early Holocene Period using organic matter and C:N proxy records: A review. *Journal Geological Society of India*, *91*, 209–214.

Lu, X., Zhou, F., Chen, F., Lao, Q., Zhu, Q., Meng, Y., & Chen, C. (2020). Spatial and seasonal variations of sedimentary organic matter in a subtropical bay: Implication for human interventions. *International Journal of Environmental Research and Public Health*, *17*, 1362. Available from https://doi.org/10.3390/ijerph17041362.

Manning, A. J., Langston, W. J., & Jonas, P. J. C. (2010). A review of sediment dynamics in the Severn estuary: Influence of flocculation. *Marine Pollution Bulletin*, *61*, 37–51.

Masselink, G., & Hughes, M. G. (2003). *Introduction to coastal processes and geomorphology* (p. 354) London: Hodder & Stoughton.

Mathew, J., Gopinath, A., & Martin, G. D. (2019). Spatial and temporal distribution in the biochemical composition of sedimentary organic matter in a tropical estuary along the west coast of India. *SN Applied Sciences*, *1*, 150. Available from https://doi.org/10.1007/s42452-018-0128-2.

Mathew, R., & Winterwerp, J. C. (2020). Sediment dynamics and transport regimes in a narrow microtidal estuary. *Ocean Dynamics*, *70*, 435–462. Available from https://doi.org/10.1007/s10236-020-01345-9.

McGlathery, K. J., Sundback, K., & Anderson, I. C. (2007). Eutrophication in shallow coastal bays and lagoons: The role of plants in the coastal filter. *Marine Ecology Progress Series*, *348*, 1–18. Available from https://doi.org/10.3354/meps07132.

Mestdagh, S., Bagaço, L., Braeckman, U., Ysebaert, T., De Smet, B., Moens, T., & Van Colen, C. (2018). Functional trait responses to sediment deposition reduce macrofauna-mediated ecosystem functioning in an estuarine mudflat. *Biogeosciences*, *15*, 2587–2599. Available from https://doi.org/10.5194/bg-15-2587.

Middelburg, J. J. (2018). Reviews and syntheses: To the bottom of carbon processing at the seafloor. *Biogeosciences*, *5*, 413–427.

Middelburg, J. J., & Herman, P. M. (2007). Organic matter processing in tidal estuaries. *Marine Chemistry*, *106*(1–2), 127–147. Available from https://doi.org/10.1016/j.marchem.2006.02.007.

Modayil, M. J. (2007). *A field guide to common marine molluscs of India* (pp. 1–79). Kochi: Central Marine Fisheries Research Institute.

Mohan, M., Chandran, M. S. S., Jayasooryan, K. K., & Ramasamy, E. V. (2014). Mercury in the sediments of Vembanad lake, the western coast of India. *Environmental Monitoring and Assessment, 186*, 3321–3336. Available from https://doi.org/10.1007/s10661-014-3620-1.

Murty, P. S. N., & Veerayya, M. (1981). Studies on the sediments of Vembanand lake. Kerala state part IV distribution of trace metals. *Indian Journal of Geo-Marine Sciences, 10*, 165–172.

Musale, A. S., & Desai, D. V. (2011). Distribution and abundance of macrobenthic polychaetes along the South Indian coast. *Environmental Monitoring and Assessment, 178* (1–4), 423–436.

Nair, N. B., AbdulAzis, P. K., Krishna Kumar, K., Dharmaraj, K., & Arunachalam, M. (1984). Ecology of Indian estuaries: Physico-chemical conditions in Kadinamkulam Backwater, south-west coast of India. *Indian Journal of Geo-Marine Sciences, 13*, 69–74.

Ouillon, S. (2018). Why and how do we study sediment transport? Focus on coastal zones and ongoing methods. *Water, 10*, 390. Available from https://doi.org/10.3390/w10040390.

Oyedotun, T. D. T. (2016). Sediment characterisation in an estuary-beach system. *Journal of Coastal Zone Management, 19*, 433. Available from https://doi.org/10.4172/2473-3350.1000433.

Parulekar, A. H., Dhargalkar, V. K., & Singbal, S. Y. S. (1980). Benthic studies in Goa estuaries: Part III—Annual cycle of macrofaunal distribution, production and trophic relations. *Indian Journal of Geo-Marine Sciences, 9*, 189–200.

Pearson, T. H., & Rosenberg, R. (1978). Macrobenthic succession in relation to organic enrichment and pollution of the marine environment. *Oceanography and Marine Biology, 16*, 229–311.

Priju, C. P., & Narayana, A. C. (2007). Heavy and trace metals in Vembanad lake sediments. *International Journal of Environmental Research, 1*(4), 280–289.

Priyalakshmi, G. (2008). *Studies on the taxonomy of interstitial fauna of some prominent beaches of Kerala* (Ph.D. thesis) (pp. 1–284). Cochin University of Science and Technology.

Pusceddu, A., Bianchelli, S., Gambi, C., & Danovaro, R. (2011). Assessment of benthic trophic status of marine coastal ecosystems: Significance of meiofaunal rare taxa. *Estuarine, Coastal and Shelf Science, 93*, 420–430.

Rajan, D. S. (2018). A glimpse on the textural analysis of an estuarine ecosystem. *International Journal of Research and Scientific Innovation (IJRSI), 5*(5), 85–96, ISSN 2321–2705.

Ralston, D. K., Geyer, W. R., Traykovski, P. A., & Nidzieko, N. J. (2013). Effects of estuarine and fluvial processes on sediment transport over deltaic tidal flats. *Continental Shelf Research, 60*, 40–57. Available from https://doi.org/10.1016/j.csr.2012.02.004.

Ramya, K. D., Jacob, J. C., Correya, N. S., Singh, B., & Philip, R. (2013). Biogeochemistry of the shelf sediments of south eastern Arabian Sea: Effect on benthic bacterial heterotrophs. *Advances in Applied Science Research, 4*(3), 315–328.

Rao, N. V. S. (2003). Indian seashells (Part 1) Polyplacoptera and Gastropoda. *Zoological Survey of India*, 426.

Reddy, K. R., & Laune, R. D. D. (2008). *Biogeochemistry of wetlands: Science and applications* (p. 770) CRC Press.

Renjith, K. R., Chandramohanakumar, N., & Joseph, M. M. (2011). Fractionation and bioavailability of phosphorus in a tropical estuary, southwest India. *Environmental Monitoring and Assessment, 174*, 299–312.

Romano, E., Giudici, G. D., Bergamin, L., Andreucci, S., Maggi, C., Pierfranceschi, G., Magno, M. C., & Ausili, A. (2017). The marine sedimentary record of natural and anthropogenic contribution from the Sulcis-Iglesiente mining district (Sardinia, Italy). *Marine Pollution Bulletin, 122*(1-2), 331–343. Available from https://doi.org/10.1016/j.marpolbul.2017.06.070.

Rowe, L. D. E., Arndtb, S., Bradleyc, J. A., Estese, E. R., Hoarfrostf, A., Langg, S. Q., Lloydh, K. G., Mahmoudii, N., Orsij, W. D., Shah Walterl, S. R., Steenh, A. D., & Zhaon, R. (2020). The fate of organic carbon in marine sediments—New insights from recent data and analysis. *Earth-Science Reviews, 204*, 103146. Available from https://doi.org/10.1016/j.earscirev.

Rullkotter, J. (2000). *Organic-matter: The driving force for early diagenesis. Marine geochemistry* (pp. 129–172). Berlin: Springer.

Sajan. (2003). *Meiobenthos of the shelf waters of the west coast of India with special reference to free-living marine Nematodes* (Ph.D. thesis) (p. 214). Cochin University of Science and Technology.

Sajan, S., & Damodaran, R. (2007). Faunal composition of meiobenthos from the shelf regions off the west coast of India. *Journal of Marine Biological Association of India, 49*(1), 119–126.

Sajan, S., Joydas, T. V., & Damodaran, R. (2010). Meiofauna of the western continental shelf of India, Arabian Sea. *Estuarine, Coastal and Shelf Science, 86*, 665–674.

Sanders, H. L. (1956). Oceanography of long island sound 1952–1954, biology of marine bottom communities. *Bulletin of Bingham Oceanographic Collection, 15*, 345–414.

Sekhar, L. K., & Jayadev, S. K. (2003). Karimanal (mineral beach-sand) mining in the Alappuzha coast of Kerala—A people's perspective. In *Proceedings of the third international conference on environment and health* (pp. 470–488). Chennai, India.

Shah, R. A., Achyuthan, H., Razi-Sadath, P. V., Derwaish, U., & Rafq, M. (2019). Sediment distribution pattern and environmental implications of physico-chemical characteristics of the Akkulam-Veli lake, South India. *Applied Water Science, 9*, 188. Available from https://doi.org/10.1007/s13201-019-1054-1.

Shepard, F. P. (1954). Nomenclature based on sand-silt-clay ratios. *Journal of Sedimentary Petrology, 24*(3), 151–158.

Sivaleela, G., & Venkataraman, K. (2013). *Diversity and distribution of polychaete from Tamil Nadu coast, India. Ecology and conversation of tropical marine faunal communities* (pp. 55–62). Springer. Available from https://doi.org/10.1007/978-3-642-38200-0.4.

Smitha, C. K. (2011). *Marine benthos of the continental shelf of southwest and southeast coast of India* (Ph.D. thesis) (p. 418). Cochin University of Science and Technology.

Stratmann, T., Vanoevelen, D., MartínezArbizu, P., Wei, C., Liao, J., Cusson, M., Scrosati, R. A., Archambault, P., Snelgrove, V. R., Ramey-Balci, P. A., Burd, B. J., Kenchington, E., Gilkinson, K., Belley, R., & Soetaert, K. (2020). The Ben Bio Den database, a global database for meio-, macro- and megabenthic biomass and densities. *Scientific Data, 7*, 206. Available from https://doi.org/10.1038/s41597-020-0551-2.

Takasu, H., Uchino, K., & Mori, K. (2020). Dissolved and particulate organic matter dynamics relative to sediment resuspension induced by the tidal cycle in macrotidal estuaries, Kyushu, Japan. *Water, 12*, 2561. Available from https://doi.org/10.3390/w12092561.

Testa, J. M., Murphy, R. R., Brady, D. C., & Kemp, W. M. (2018). Nutrient- and climate-induced shifts in the Phenology of linked biogeochemical cycles in a temperate estuary. *Frontiers in Marine Science, 5*, 114. Available from https://doi.org/10.3389/fmars.2018.00114.

Trask, P. D. (1939). *Recent marine sediments* (p. 736) Tulsa, OK: American Association of Petroleum Geologists Publication.

Turchyn, A. V., Bradbury, H. J., Walker, K., & Sun, X. (2021). Controls on the precipitation of carbonate minerals within marine sediments. *Frontiers in Earth Science, 9*, 618311. Available from https://doi.org/10.3389/feart.2021.618311.

Wang, M., Shan, Z., Liu, T., Xiong, Z., Yan, Q., & Wang, Y. (2019). Chemical-physical mechanism of marine sediments. *IOP Conference Series: Earth and Environmental Science*, *304*, 052052. Available from 10.1088/1755-1315/304/5/052052.

Weisberg, S. B., Frithsen, J. B., Holland, A. F., Paul, J. F., Scott, K. J., Summers, J. K., Wilson, H. T., Valente, R. M., Heimbuch, D. G., Gerritsen, G., Schimmel, S. C., & Latimer, R. W. (1993). *EMAP-Estuaries, Virginian Province. Demonstration Project Report. EPAl920/R-9~ /006.* Narragansett, RI: U.S. Environmental Protection Agency, Environmental Research Laboratory, Office of Research and Development.

Woulds, C., & Cowie, G. L. (2009). Sedimentary pigments on the Pakistan margin: Controlling factors and organic matter dynamics. *Deep-Sea Research Part II*, *56*, 347–357.

Yang, B., Gao, X. L., & Xing, Q. G. (2018). Geochemistry of organic carbon in surface sediments of a region in the coastal waters of northern Shandong Peninsula summer hypoxic. *Continental Shelf Research*, *171*, 113–125.

Yu, R., Zhang, W., Hu, G., Lin, C., & Yang, Q. (2016). Heavy metal pollution and Pb isotopic tracing in the intertidal surface sediments of Quanzhou Bay, southeast coast of China. *Marine Pollution Bulletin*, *105*(1), 416–421. Available from http://doi.org/10.1016/j.marpolbul.2016.01.047.

Zhang, Q., Li, J., Hu, G., & Zhang, Z. (2019). Bioturbation potential of a macrofaunal community in Bohai Bay, northern China. *Marine Pollution Bulletin*, *140*, 281–286. Available from https://Doi.Org/10.1016/J.Mar.PollBull.2019.01.063.

Zonneveld, K. A. F., Versteegh, G. J. M., Kasten, S., Eglinton, T. I., Emeis,K., C., Huguet, C., Koch, B. P., Lange, G. J., de Leeuw, J. W., Middelburg, J. J., Mollenhauer, G., Prahl, F. G., Rethemeyer, J., & Wakeham, S. G. (2010). Selective preservation of organic matter in marine environments; processes and impact on the sedimentary record. *Biogeosciences*, 7, 483–511.

CHAPTER 5

Anthropogenic impacts on the benthic habitat morphology

Prince S. Godson[1], N. Chandrasekar[2], S. Krishnakumar[3] and Salom Gnana Thanga Vincent[4]

[1]Centre for Marine Science and Technology, Manonmaniam Sundaranar University, Tamil Nadu, India
[2]Department of Geology, Francis Xavier Engineering College, Tirunelveli, Tamil Nadu, India
[3]Department of Geology, Malankara Catholic College, Mariyagiri, Tamil Nadu, India
[4]Department of Environmental Sciences, University of Kerala, Thiruvananthapuram, Kerala, India

5.1 Introduction

The improper and excessive use of natural resources has resulted in long-term environmental degradations for short-term economic benefits. Due to rapid urbanization, water and soil get polluted and cause a massive impact on the marine environment (Halpern et al., 2008; Hoegh-Guldberg et al., 2007). Specifically, human activities in the marine environment can lead to homogenization of ecosystems due to reductions in food-web complexity, diversity within functional groups, distribution range, biogenic habitat structure, and size of organisms (Dulvy et al., 2008). These trends are exacerbated by the growing human populations in coastal areas and the increasing need for marine resources.

Habitat destruction is considered the most pervasive threat to any coastal marine ecosystem's diversity, structure, and functioning, and the services they lend (Crain et al., 2008; Halpern et al., 2008). Destruction of marine habitat has occurred for the last fifteen decades. Direct or indirect human activities are now the primary cause of changes to marine biodiversity. Hence assessing the impacts of humans on marine habitats and protecting habitats from harmful effects are critical for the renewal of resources and the sustainable use of marine biodiversity. However, a complex suite of interacting factors limits the identification of conservation priorities, implementation of mitigation strategies, and accomplishment of effective restoration plans for marine habitats. Significant impediments are related to the general lack of baseline data on marine ecosystems before large-scale human impacts and substantial gaps in knowledge on the current distribution and classification of different habitat types (Fraschetti et al., 2008; Halpern et al., 2008). Because of these impediments, the effects of current and emerging threats to marine habitats are largely unknown.

Ecology and Biodiversity of Benthos
DOI: https://doi.org/10.1016/B978-0-12-821161-8.00011-8

Systematic assessments of habitats' vulnerability and sensitivity to stressors have been based on alternative means, such as expert opinion (Halpern et al., 2008). Although specialist opinion approaches can be used as a proxy for actual impacts on habitats, they are not as meaningful as quantitative assessments, which are grossly lacking. The need for more quantitative estimates of the human effects on habitat is particularly pressing in the coastal and marine areas due to its specific physiographic and biological characteristics and the high anthropogenic pressure on its habitats.

Some of the most prevalent human activities that directly impact the biodiversity of an ecosystem are coastal development activities including pollution, exploitation of living and nonliving resources, overfishing, and introduction of alien species. One of the factors that causes the most destruction to marine biodiversity at all levels of the organization is anthropogenic activity. Fragmentation of natural habitats is also one of the foremost causes of local species extinction. From this, it is evident that the future of the marine ecosystem has to be determined by man (Sala & Knowlton, 2006). Overexploitation by intensive and longtime harvesting has led to a severe decline in some benthic invertebrates, such as macro and meiofaunal species. Coastal development, habitat destruction, and sediment loading have notably reduced the extension of important habitats for marine diversity, such as intertidal ecosystems. Land-based eutrophication has been decreased in many coastal regions in recent years. However, fish and mussel farms in formerly oligotrophic regions bring about an increasing input of nutrients, which, in turn, affects the meiofauna, especially in the coastal sediments. Another prevalent anthropogenic environmental factor affecting the meiobenthic environment is the pollution by pesticides, heavy metals, and antifouling compounds. But these factors only have a local impact and hence are not considered to be general habitat factors.

The sandy beaches are extensively developed along the coasts by the processes of constructive wave action, caused by continuous deposition of littoral sediments from the adjacent areas. Sand is a dynamic substrate that provides habitats for animals in two ways. The first way is to provide the surface and upper layer of sand for the habitat. The second way is to offer the interstitial habitat formed by the spaces between the sand grains (McLachlan, 1983). Beach morphology, which includes ecosystem size and other physical qualities like substrate type, slope, etc. influence the broader community structure (Peterson et al., 2000). The above-specified habitat produces a set of intertidal to supra-tidal conditions, which explain the distribution of organisms within and between beach types and thus influences associated resource accessibility and availability to higher-order consumers

(Defeo & McLachlan, 2005). The beach morphological changes by frequent erosion and accretion processes along the coastal area within short-term periods are mainly due to unregulated coastal sand mining and natural hazards such as storm surge, tsunami, and cyclone. Some of the potentially toxic compounds like heavy metals and petroleum hydrocarbons occur naturally in the sea, and human introductions add to natural concentrations. Some externally introduced pollutants will decompose in time or will be attenuated by the enormous volume of the oceans so that their effect will not be noticeable. Some of the notable anthropogenic-induced pollutants, impact, and their source are summarized in Table 5.1.

Table 5.1 Major forms of marine pollution and their effects.

Pollutants	Sources	Effects
Heavy metals	Paint and other chemical industrial effluents	Bioaccumulation or biomagnification effects on marine organisms
Sewage	Human settlement runoff	Eutrophication and altered community structure. Formation of pathogens
Petroleum hydrocarbon	Fishing activity	Mass mortality of marine benthos and other flora.
Pesticides and related compounds	Local outfalls and agricultural runoff	Acute toxicity
Plastics	Beaches or floating debris	Esthetically disturbing the marine ecosystem. Entanglement of animals; ingestion by animals
Radioactive wastes	Power plants situated nearby the coast	Generally considered to be below harmful levels
Thermal effluents	Local power plants	Warming leads to altered community structure
Placer mining effluents	Local rare earth element extraction unit	Affects the diversity of marine benthos and alters geomorphology of a coast
Coir retting effluents	Coir industries	Alters the physicochemical property of both marine sediment and sea water
Shell bleaching effluents	Local small scale industries	Degradation of marine sediment and sea water.

5.2 Habitat loss

Studies on abiotic factors have dominated since the 1960s. Systematic studies on the impact of biotic factors on benthic communities have started increasing around 1975. Studies on biotic factors in recent years show that they can significantly influence the population structure of benthos especially meiofauna and the benthic habitats (Woodin & Jackson, 1979). The biotic factors in sediments range from dead organic matter and biogenic structures to mucus masses, free-floating plankton, bacteria, and living organisms. The organic component includes mainly detritus and also fecal matter, mucous agglutinations, and dissolved organic substances. These organic substances enhance bacterial growth, colloids, aggregate and biofilm production, and re-mineralization (Azam, 1998). In the broader sense, the activities of other fauna, such as disturbance, competition, and predation, also represent biotic habitat factors for meiofauna. Watling (1991) has pointed out that the classical methods of sediment analysis, derived from geology, give a misleading, denatured, and biologically rather an irrelevant picture of the sedimentary habitat. Living organisms also impact the meiobenthic habitats. Biotic factors interact in multiple ways which are difficult to assess, which is one reason for the fewer researches on the biotic parameters than the abiotic ones. For a better understanding of this biotic habitat interaction, the various components are presented separately below.

5.3 Development of coastal landforms

In both marine and lacustrine settings, depositional landforms contain various accretional and erosional elements formed by the interaction of wind as well as tidal, riverine, and other geological processes. These regions are geomorphologically characterized by dunes, beaches, deltas, barriers, strand plains, back-barrier marshes, lagoons, and tidal flats. From the beginning of the human era, anthropogenic-induced factors have attempted to slow or alter the dynamic coastal zone. The anthropogenic impacts on the coastal environments include the creation of inlets and dunes, dredging of waterways for shipping, placer mining for mineral exploration, and construction of jetties, groins, and seawalls. These changes to the coastal features severely affect the coastal processes and ecosystems. The proper understanding of coastal modification, including how human activities change coastal environments and resources is important to protect and conserve coastal areas.

5.4 Hard structures

Hard structures are generally kept in the coastal environment to prevent soil erosion in areas deficient in sediment, and to prevent accretion in areas rich in sediments. However, these anthropogenic activities bring about erosion in nearby downdrift areas, increasing the need for additional hard structures. The creation of new hard structures is currently banned in many countries or strongly discouraged as a coastal management practice. Groins are built perpendicular to the coast to control longshore sediment transport. Permeable groins are becoming popular and may negate some of the adverse effects of impermeable groins. Another type of shore-perpendicular hard structure is jetty, which is usually placed adjacent to tidal inlets to control inlet migration and to minimize sediment deposition within the inlet.

Hard structures that are built parallel to the coastline include seawalls, bulkheads, and revetments which protect coastal property. Though seawalls are built to prevent erosion, in many places they actually accelerate erosion. Sometimes wave energy is reflected off the seawall thereby increasing erosion in areas opposite to it. They decrease the sediment supply near the seawalls, increasing erosion in nearby beaches. Evidences state that beaches have even disappeared due to the presence of seawalls. Other anthropogenic structures including breakwaters, headlands, sills, and reefs alter tidal action thereby preventing coastline erosion.

5.5 Rocky intertidal environment

Rocky shores and intertidal regions are dynamic and distributed worldwide and are one of the most widely studied ecosystems for several decades (Menge, 1976). Hence it is suitable for investigating community structure and functioning (Hawkins et al., 2008) and studying the effects of single and multiple anthropogenic driver(s) on the rocky shore. The biogeographic ranges of the species present in the rocky intertidal ecosystem depend on several terrestrial and marine factors of this ecosystem. The several potential drivers have their impact on the rocky shore species. Hence the environmental impact on these species is very severe compared to other coastal ecosystems (Branch et al., 2008). In a diverse ecosystem, functioning can be maintained when many functional group species are present because species eradication can be compensated (Yachi & Loreau, 1999). This functional redundancy is a major ecological concept and it is a major reason for the high tolerance level of rocky shores to natural drivers (Naeem, 2008). Intertidal rocky shore environments have developed amidst changing environmental factors. This environment has survived in spite of being very close to its tolerance limit (Deutsch et al., 2008). This tolerance limit might be surpassed if conditions become more extreme. For example, the threshold heat tolerance can be surpassed under increasing sea surface temperatures caused by climate change. Hence natural impacts including extreme weather and seasonal changes affect the ability of the rocky coastline to withstand changes. This ultimately alters the ecosystem structure as well as functioning (Hughes et al., 2019).

Intertidal rocky shores are highly affected by climate change and other anthropogenic impacts (Doney et al., 2012). The intertidal and rocky shore species undergo anthropogenically driven changes. The negative impact of these stress, such as increased atmospheric CO_2 levels and changes made to the coastal geomorphology enhance the temperature of the air and sea surface (Hoegh-Guldberg et al., 2018; Mieszkowska, 2016). Temperature changes lead to modifications in the number and distribution of intertidal species. Research has revealed that warm water species are increasing in abundance, whereas cold water species are either moving poleward (Wernberg et al., 2011) or contracting their ranges (Smale et al., 2017).

5.6 Impact of ocean acidification

Climate change also brings about ocean acidification. This happens as a result of increased atmospheric CO_2 dissolving into the ocean, lowering the pH of seawater (Harley et al., 2006). This impacts calcifying species and macroalgae, particularly in their earlier stages of life (Przeslawski et al., 2015). Calcifying species such as foraminifera, bivalves, gastropods, barnacles, limpets, and top shells are most vulnerable to ocean acidification and have been affected significantly during the second half of the century (Mieszkowska, 2016).

Marine gastropods are supposed to be more affected in their predatory behavior and are more vulnerable to predation themselves under raising ocean acidification (Queirós et al., 2015). Juvenile stages of marine living organisms are shown to be most sensitive to changing environments (Przeslawski et al., 2015). Mieszkowska (2016) has showed that ocean acidification altered behavior and reduced metabolism and growth rates in a juvenile marine gastropod (*Concholepas concholepas*). Furthermore, aquatic larvae and eggs are assumed to be the worst affected and, in some scenarios, will not reach the calcifying stage (Byrne, 2011), which would have immense consequences for subsequent generations and the distribution range of calcifying organisms.

The impact of acidification of the ocean on macroalgae will be more complex due to the production of photosynthesis with improved atmospheric CO_2, but brings an unknown effect of raised pH (Cornwall et al., 2017; Zweng et al., 2018). These studies have shown a positive correlation between their growth and their capacity to use HCO_3 as source material. This capability is more strongly encouraged in fleshy algae, which are not obligate calcifiers. However, the overall situation is more complex: macroalgae are likely to be more vulnerable to acidification during earlier life stages. Furthermore, ocean acidification impacts the benthic organisms and community structure on rocky shores and the overall environmental conditions (Kroeker et al., 2017). The toxicity levels of heavy metals are increased with respect to raising the levels of partial pressure of carbon dioxide (pCO_2) (Millero et al., 2009). Overall, studies evident that multiple driver impacts on single organisms and communities will further increase with ongoing climate change (Kroeker et al., 2017; Mieszkowska, 2016).

5.7 Coralligenous environment

The Coralligenous environment is created due to the edification process as a result of precipitation, by organisms like algae of the carbonated calcium present in seawater. This edification process is dependent on the season as well as the location. It takes thousands of years for the edification process to be over. In terms of biodiversity, the coralligenous environment is considered as the second benthic ecosystem in the coastal habitat. It includes around 1700 species, including 1300 algae, 1200 invertebrate species, in particular, the well-known Mediterranean red coral (*Corallium rubrum*) and more than a hundred fish species.

Research shows that this ecosystem is vulnerable to anthropogenic activities like trawling or the exploitation of the red coral by prohibited gears. It is also subjected to the indirect impacts of climate change. These affect the stability of the ecosystem and strongly mortgage its future maintenance.

5.8 Seagrass environment

Seagrasses are common components of the shallow sublittoral sheltered sandy shores in tropical, subtropical, and temperate regions. Seagrasses consist of blades or fronds, rhizomes or horizontal stems, and roots. Rhizomes and roots serve for anchorage and absorption of nutrients. Seagrasses may be found even in the lower intertidal regions of sandy beaches. These seagrasses build an exclusive habitat by binding the bottom sediments, purifying the water, and thereby providing a habitat for plants and animals. These habitats can be especially important for fish and prawns, particularly commercially important species, which may use them as nursery areas. Seahorses, pipefishes, and gobies are very typical of seagrass habitats. Most fish feed on invertebrates in the seagrass, especially crustaceans.

Seagrass meadows are naturally fragmented by extreme weather, waves, and currents. However, anthropogenic factors like eutrophication, sewage disposal, recreational activities, and coastal constructions which need dredging and infilling also increase fragmentation. Fragmentation in seagrass affects faunal assemblages, especially the exclusive species which are more vulnerable to changes in the habitat. This impact of fragmentation can even change the ecological connectivity between different coastal habitats bringing about change in ecosystem functioning. Macrofauna also play a substantial role in nutrient cycling. Macrofauna including crustaceans, gastropods, bivalves, polychaetes, and echinoderms have been implicated in: (1) structuring the benthic environment through processes such as bioturbation and burrow construction, (2) aeration of the sediment fostering aerobic bacterial activity, and (3) acting as food sources for higher trophic levels including many ecologically and commercially important fish species. The recovery of seagrass is often very slow, to the verge of nonexistence (Holmquist, 1997), partially because of the naturally slow recovery of many species, and that seagrasses to such an extent create their living conditions (Duarte, 1995; Van der Heide et al., 2007).

5.9 Marine biodiversity indicators

The status of indicator species (presence, absence, or abundance of species) reveals a particular environmental condition. It can indicate a change in the biological condition of an ecosystem and help in investigating the health of an ecosystem. However, these indicator species need to be used carefully. These species serve as research and monitoring tools. They assist as an "early warning system" to conservation managers and biologists. Macrofauna and meiofauna have been used as sentinels for pollution indicators of environmental quality. Pollution brings about an increase of individuals belonging to certain opportunistic species (Murray, 1973). These opportunistic species living in variable environments react fast to favorable conditions by quick rates of colonization and reproduction. Macrofauna represent a significant component that influences the productivity of the habitat to a large extent. Certain infaunal taxa are believed to be environmental indicators because of their opportunistic lifestyle.

The benthic environment is a fundamental compartment of any aquatic ecosystem. The serious effects of environmental stressors on benthic organisms are changes in behavior, growth, and reproduction. These changes in individual organisms, known as biological indicators (bioindicators), can eventually result in large-scale changes in biological communities and even ecological systems. All these measurable changes at any scale from cells to ecosystems may serve as bioindicators of pollutant stress. The meiobenthic nematodes are bound to the sediment throughout their life history and therefore are sensitive to many toxicants (Ingole et al., 2009).

The regional level and local baseline studies are important to find the significant regional-wise changes. The drastic changes along the coast due to human impact will inevitably increase and diversify anthropogenic pressures in oceans. The success of assessment and management is partially dependent on the availability of scientific tools to managers. A common and detailed practice on marine habitat change specifically focuses on sensitive habitats and endangered organisms. An ecological indicator is intended to highlight the status of the system and they are classified as distinct types of indicators depending on their applications: descriptive indicators, performance indicators, efficiency indicators, and total welfare indicators (Smeets & Weterings, 1999).

5.10 Anthropogenic impact

Sandy beaches are one of the common habitats along continental margins and bordering 70% of all marine coasts (Schlacher et al., 2015). These habitats provide an interesting model system to examine general food web dynamics of coupled aquatic-terrestrial systems (Porri et al., 2011). Sandy beach livelihoods are affected by various environmental factors such as the tidal system and sediment source. These factors interact with one another to bring about changes in beach morphology (Jaramillo et al., 2001). Correspondingly, as a proxy for ecosystem size, it has been anticipated that beach width with other important physical traits may influence the availability of resources in sandy beach ecosystems (Dugan et al., 2013; McLachlan & Dorvlo, 2005). Beach morphology also has an impact on the broader community structure (Peterson et al., 2000). In sand mining, habitat dislocation and destruction can affect the biota of marine benthos and other microbial biota. Overexploitation of sand mining has also affected the available resources in light of current and future demand. The mining of beach sand has its direct impact on the intertidal area, and the effect of this activity has conventionally targeted bottom substrate and the associated fauna (Desprez, 2000; Van Dalfsen et al., 2000).

Some of the coastal habitats have acted as spawning and nursery ground for commercially important fishes, mollusks, crustaceans, and other species that constitute the inshore fisheries. Some human activities have physically altered estuarine shorelines and habitats, impacting biotic communities. The manmade structures including bay shore housing, lagoons, bulkheads, and other shore protection features, harbor and marine development, dredging and dredged material disposal, wetland reclamation, and channel and inlet stabilization generate severe impact on marine biota. In addition, the upland modifications (e.g., dams and reservoirs, deforestation and habitat fragmentation, and channelization) can significantly exacerbate other anthropogenic stresses leading to greater pressures in estuarine systems.

The marine/coastal livelihoods are dramatically affected by nutrient enrichment and contaminants from anthropogenic sources. These contaminants have a tremendous influence on the meiobenthos around the world (Nixon, 1995). Physical adsorption and chemical bonds largely supplement the nutrients and the pollutants in the sediment when compared to that of the water. The geochemical process like absorption of inorganic and organic nutrients is significantly high in carbonate sediments (Rasheed et al., 2003a). Once contaminants are bound in the sediment, these substances are only slowly released into the water column. Retarded microbial decomposition in the oxygen-deficient deeper layers of many bottoms combined with high absorptive forces can cause long-lasting or chronic negative effects on the benthos. Measurement of the amount of nitrogen and total phosphorus is not sufficient for judging the actual condition. Eutrophication and pollution result in a patchy distribution of meiofauna (Lambshead & Hodda, 1994). Terrigenous eutrophication has been decreased in many areas in recent years. However, impacts from fish and mussel farms bring increased nutrient input, which affects meiofauna, especially in sheltered coastal sediments. Other widespread man-induced environmental factors in the meiobenthic habitat include sediment accumulated heavy metals, antifouling compounds, and pesticides.

5.11 Placer mineral mining

Placer mining has become a serious environmental issue all over the world. Placer mining often causes land degradation, changes in beach morphology, loss of agricultural lands, marine and coastal biodiversity changes, and turbidity in the near-shore regions. Similarly, the mining associated activities can be responsible for considerable environmental damage such as increasing turbidity, low pH, high electrical conductivity (EC) and raising of salinity, level of toxic heavy metals, low dissolved oxygen (DO), and high biological oxygen demand (BOD). To understand the ecological significance of the problem of the geographical distribution of the benthic fauna, it is relevant to consider certain aspects of their biology in brief. The placer mineral mining affects the geomorphologic settings of the coast. The improper and unsustainable sand mining results in environmental imbalance and leads to coastal hazards like habitat destruction, loss of biodiversity, seawater intrusion, scarcity of food, etc. The habitat morphology of the marine benthos has been seriously affected in soft bodied animals such as polychaetes and nematodes. Physical disturbances, including mining, may be more damaging to the benthic community, whereas, natural land barriers like sand dunes, beach ridges, and cliffs reduce the effect of natural processes like littoral current, wave energy, longshore drift, and anthropogenic activities including habitat degradation and unregulated placer mining.

Most of the research on the effect of physical disturbances on marine benthos is focused on the reduction and complete defaunation of benthic organisms (Sivadas et al., 2012). The recurrent erosion and accretion along the coastal stretch within short-term periods are mainly due to unregulated coastal mining and natural hazards such as storm surge, tsunami, and cyclone. A case study on unsystematic beach garnet sand mining has disturbed the beach morphology, particularly the littoral zone of the coast (Chandrasekar & Rajamanickam, 2001).

Beach placer deposits worldwide are known for ilmenite, rutile, zircon, monazite, sillimanite, and garnet (Fig. 5.1). Ilmenite is found in the largest quantity in the Indian beaches, followed by garnet and sillimanite (Rajamanickam et al., 2004; Rao & Misra, 2009). The placer mineral reserves in India are reported by Atomic Mineral Division as follows: 278 MT of Ilmenite, 86 MT of garnet, 84 MT of Sillimanite, 13.49 MT of Rutile, 18 MT of Zircon, and 7 MT of Monazite (all the data in million tons; AMD, 2000). Placer minerals can be concentrated during the transport and deposition of sediments in certain areas due to their high-density nature. The extraction of beach sand disturbs the stability of the beach environment (Angusamy & Rajamanickam, 2006). The geological and environmental factors like geology of the area, coastal geomorphology, hinterland, subtropical to the tropical climate, and drainage system, along with waves and currents favor these placer mineral reformations (Chandrasekaran et al., 1997; Loveson & Rajamanickam, 1988).

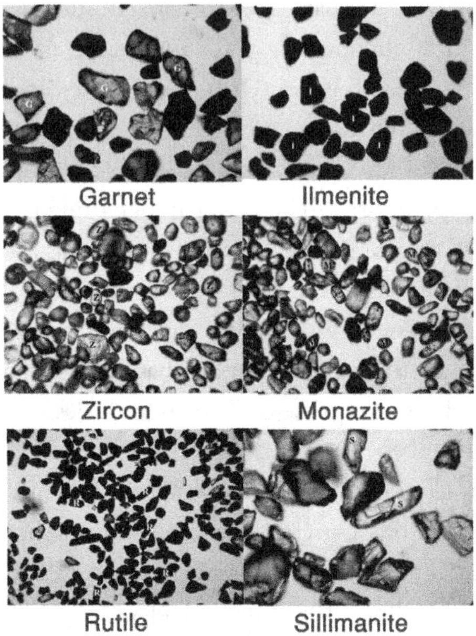

Figure 5.1 Heavy mineral assemblages in the study area.

The heavy mineralogical assemblages in the coast are dependent on the source rock of the area, regional climatic conditions, agent of sediment transport, and hydraulic conditions (Borreswar, 1957). Additionally, natural activities like tectonics, isostatic and eustatic sea-level changes, drainage patterns, and rainfall of the study region play a vital role in the deposition of placer deposits. Pettijohn et al. (1972) have pointed out that the diversity in heavy mineral assemblages is higher in young sediments. These placer minerals are highly sensitive and they help to find the very minute changes in the depositional environments (Mange & Maurer, 1992). These transitional ecosystems and their near-shore environment serve as habitats for preserving biodiversity and genetic resource, nesting grounds for birds, and prey grounds for terrestrial wildlife, thus establishing species diversity, biomass, and community structure on the beach environment (Brown & McLachlan, 1990).

5.12 Impact of habitat changes in faunal diversity

Anthropogenic changes could potentially alter habitat features and macrofaunal community structure, resulting in biodiversity loss (Defeo et al., 2009; Reyes-Martínez et al., 2015), loss of ecosystem services, and difficulties in facing a climate crisis. Benthic macrofaunal communities play a pivotal role in particles reworking (Aller, 1988; Kristensen et al., 2012), nutrient cycling (D'Andrea & DeWitt, 2009), and serving as food for other organisms (Carvalho et al., 2018; Constable, 1999). The ecological status of submerged sandy beaches can be assessed by analyzing the composition and abundance of macrofauna and through the elaboration of biotic indices (Borja et al., 2000; Muxika et al., 2007) however, these indices put in evidence the sensitivity to organic enrichment and they do not consider effects of other forms of impact.

Macrofauna living in the intertidal zone can be particularly vulnerable to beach activities (Bessa et al., 2014; Reis & Rizzo, 2019). In the recent decades, there have been some studies based on understanding the response of benthic communities and populations toward physical disturbances (Veloso et al., 2010). The community structure of macro-, meio-, and microfauna hosted in beach environments is influenced by the interaction of several physicochemical factors such as sand granulometry, mineralogy, and tides or beach exposure (Barboza & Defeo, 2015; Defeo et al., 2009). These influential factors suggest that alteration of natural beach dynamics due to human pressure could affect the ecological traits of these organisms and the overall functioning of the beach ecosystem (Thrush et al., 2017). Some of the coastal habitats have acted as spawning and nursery ground for commercially important fishes, mollusks, crustaceans, and other species that constitute the inshore fisheries. The physical features and number of related factors including seasonal variations, habitat dislocation and destruction (placer mining) influence coastal productivity and species diversity. The beach placer mining and mined sediments naturally induce the contamination level in the coastal environment. The grain-size alterations and contamination during mining operations may exceed flora and fauna tolerances, negatively impacting native ecosystem functions. The unregulated mining and overexploitation of sand has also deteriorated the benthic diversity and affected the available resources in the light of current and future demand.

5.13 Impact of mining on marine benthos: a case study of mining and nonmining beaches in the southwest coast of India

Marine benthic macrofauna have a major role in ecosystem processes like pollutant metabolism, nutrient cycling, secondary production, and dispersion and burial (Snelgrove, 1998). The investigation of these benthic communities serves as a tool for evaluating the functioning of the ecosystems (Gray, 1981). Coastal sediments and water samples have been collected seasonally to assess the status of marine benthos at mining and nonmining sites of the southwest (SW) coast of Tamil Nadu. The sampling period is classified into four seasons namely, postmonsoon (January–March), summer (April–June), SW monsoon (July–September), and NE monsoon (October–December).

5.13.1 Study area and sampling methods

The study area is located in the southernmost part of the peninsular region of India (Fig. 5.2). Some major taxonomic groups have been selected to identify their species-level in macro and meiofaunal communities. The selected taxonomic groups for macrofauna analysis are polychaetes, bivalves, gastropods, decapods, amphipods, and isopods. The meiofaunal communities chosen include polychaeta, nemetoda, foraminifera, copepoda, oligochaeta, ostracoda, halacaroidea, cladocera, and amphipoda. The sampling points are classified as mining (Manakudi, Kadiapattinam, Manavalakurichi, and Midalam) and nonmining areas (Kanyakumari, Sanguthurai, Rajakkamangalam Thurai, Colachel, and Inayam) based on coastal placer mining operations in the areas. The geomorphological structures of the study area are depicted in Fig. 5.3.

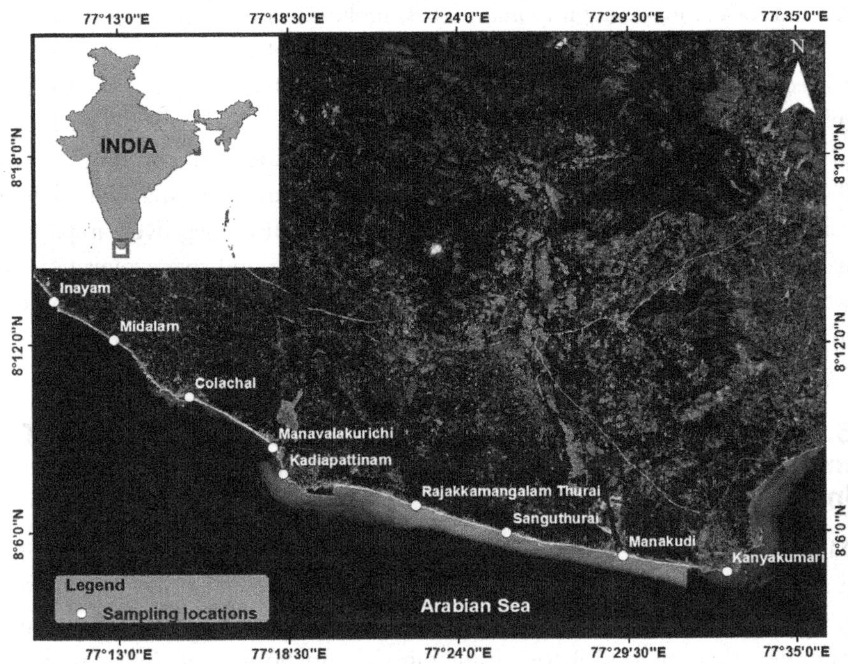

Figure 5.2 Location map of the study area.

Figure 5.3 Geomorphological structures of sampling locations and heavy mineral distribution. (A) Rocky coast seen in Kanyakumari. (B) Kanyakumari beach carpeted with rich concentrations of black sands (Ilmenite) and garnet. (C) Rocky armour constructed near Gandhi Mandabam in Kanyakumari. (D) Heavy erosion during NE monsoon in Manakudi. (E) Ilmenite placer seen in Manakudi. (F) Placer mineral beds in Sanguthurai. (G) Sand dune and coconut plantations in Sanguthurai. (H) Small lagoon like structure noticed in Rajakkamangalam Thurai. (I) Coconut plantation seen near the estuary in Rajakkamangalam Thurai. (J) Estuary mouth seen in Rajakkamangalam Thurai. (K) Rocky intertidal area in Kadiapattinam. (L) Placer mineral bed noticed in Kadiapattinam. (M) The weathered placer mineral source rock near Kadiapattinam. (N) Casuarina plantations in Colachel. (O) Heavy wave energy condition seen in Colachel. (P) Rocky armor noticed in Colachel. (Q) Mechanized placer mining activity noticed in Manavalakurichi. (R) Mining area in Manavalakurichi. (S) Placer mineral deposits in Midalam. (T) Rocky armor constructed in Inayam.

Figure 5.3 (Continued).

Figure 5.3 (Continued).

5.13.1.1 Kanyakumari

The coast of Kanyakumari is rocky, and consists of coarse to medium sands with a huge amount of shelly fragments. The beach has a narrow backshore surrounded by rocks. Rocks in the foreshore region are sometimes encrusted with fossils and seaweeds. Headland projection acts as a barrier to the beach and the rocky cliffs face the foreshore. The width of the beach varies from 17 to 25 m. During the monsoon, the beach faces seasonal erosion and accretion. The coastal area is highly polluted due to the impact of tourist activities. The Kanyakumari coast is protected from mining activities including beach placer mining.

5.13.1.2 Manakudi

Manakudi beach has a wider backshore and casuarina trees are planted along the beach ridges. The near-shore region is well protected by rock armor. The beach has a width of 19–52 m. River Pazhayar with its origin in the Western Ghats flows into it. The estuary is connected to the sea during the rainy season and remains separated for the rest of the time by sand bars. Small-scale industries such as the coconut husk retting industry, the lime shell dredging industry, and the salt pans are located near the estuary. The estuary receives heavy surface runoff from the nearby areas and coconut plantations. The mudflats contain heavy minerals and shell fragments. One of the tidal flats of the estuary has a mangrove. Mining is done near the estuary and also a few other places.

5.13.1.3 Sanguthurai

Sanguthurai beach consists of a significantly wide backshore area flourished by sand dunes. It is rich in vegetation including natural vegetation as well as artificially planted coconut trees. The steep near-shore region has scattered placer minerals. Medium to fine sand particles are found on the shore. The beach has a width of 40–50 m. The area is protected from mining and is flourished with coconut plantations in the landward region.

5.13.1.4 Rajakkamangalam Thurai

Rajakkamangalam Thurai beach is located on the west coast of Tamil Nadu, 15 km west of Kanyakumari. The width of the beach is around 50–70 m. Pantri estuary and Ayiramkal Mandapam creeks flow into the beach. Erosion is very high in Rajakkamangalam Thurai due to the groins in the east. The foreshore region is steep and the backshore region has large stabilized sand dunes.

5.13.1.5 Kadiapattinam

Kadiapattinam is a narrow curved beach in the north of Muttom headland. The width of the beach is less than 10 m. Medium to fine sands are present. The beach is protected by rocks and seawalls. Human settlements are found very near the beach resulting in the mixing of domestic sewage into the sea. There is also the chance of mixing of water from the Valliyar

River, which flows north of this beach, and from small streams adjoining the Valliyar, bringing Teri sand from Muttom hinterland. Heavy mineral deposits are found in this beach. The beach width ranges from 9 to 70 m. The backshore region is covered with beach vegetation and a gentle sloping foreshore. The beach is prevented from erosion by the construction of rock armor. The near-shore region is severely affected by coastal mining.

5.13.1.6 Manavalakurichi

Manavalakurichi is an open coast enriched with beach placers. The Indian Rare Earth Limited (IREL) for exploring placer deposits is located here. Coconut plantations are seen all over the coastal region. The region is noted for its beach placer mining activities. The Indian Rare Earths Limited (IREL), a public sector undertaking of the Government of India under the directorial control of the Department of Atomic Energy, mines and processes ilmenite along with other heavy minerals like garnet, monazite, rutile, sillimanite, and zircon from this area. Manavalakurichi is situated in the southern tip of the Indian peninsula.

5.13.1.7 Colachel

This beach is long and wide and is located north of Manavalakurichi. Jetty and local fishing harbor construction are done in the area by private government-approved organizations. The beach is free from structures like seawalls and groins. The width of this beach is up to 100 m from the benchmark. The area is covered with rich heavy mineral deposits. Pambar estuary joins in the western part and a small creek adjoins the eastern part. The creek is connected to the sea during the northeast (NE) monsoon season and closed by sand spit during the rest of the year. A small river Pampoori on the north, and a small creek on the south adjoin the Colachal beach. Rocks are present in the offshore region. The slope of the beach is highly steep in the low-tide region. High-energy condition prevails all through the coast.

5.13.1.8 Midalam

Midalam is a long and wide sandy beach. It is a medium-barrier beach with seawalls as protective structures. Three small creeks adjoin the sea near Midalam. The beach is highly inclined with a width of around 100 m. Black sands with garnet and other heavy minerals are found. Thick heavy mineral accumulation is seen along the backshore region. The annual heavy mineral-rich accretion trend supports the continuous heavy mineral mining activities at Midalam. The width of the beach ranges from 27 to 70 m. Coconut plantations are found throughout the backshore region. Midalam is a mining area on the west coast of India.

5.13.1.9 Inayam

Inayam is an enclosed beach with medium to fine sands. Tidal flats and erosional features are found on the beach. Black sands and heavy minerals are also found. Rubble-Mound Seawalls are constructed on the beach. Shell fragments, broken brachiopods, and gastropods are also observed in the sand. There are many rivers near this beach. The beach width varies from 7 to 70 m from the benchmark. Coconut plantations are seen parallel to the coast on the backshore region. The beach is free from mining.

5.14 Transect and quadrat sampling

Transect sampling is a technique for sampling both flora and fauna in which sediment sampling is done in two transects along the study area to represent the features of the entire beach. The structural characteristics of the intertidal fauna are investigated by this method (Creese & Kingsford, 1998; Littler & Littler, 1985). The species diversity along the different intertidal regions is not constant. Therefore the sampling transects are laid perpendicular to the coast to collect samples from all the spatial scales of the intertidal areas.

Quadrates are used for ecological sampling (Greig-Smith, 1983). Quadrate frequency is defined based on the total length of the intertidal area and the sampling site. Macrofaunal sampling is done from a depth of 25 cm using a core pipe of 10.5 cm diameter and meiofaunal sampling is done at 8 cm depth (Palacin et al., 1991). Five cores for macrofauna and four cores for meiofauna are collected at high and mid-tide regions, while 3 cores for macrofauna and 2 for meiofauna are collected from the low-tide region (Fig. 5.4). The samples are collected from the mining sites, that is, Manakudi, Kadiapattinam. Manavalakurichi and Midalam and nonmining areas like Kanyakumari, Sanguthurai, Rajakkamangalam Thurai, Colachal, and Inayam of the study area.

5.15 Subsampling of cores

The macro and meiofaunal studies are done in a small core to measure faunal distribution. The core samples are subsampled for grain size analysis,

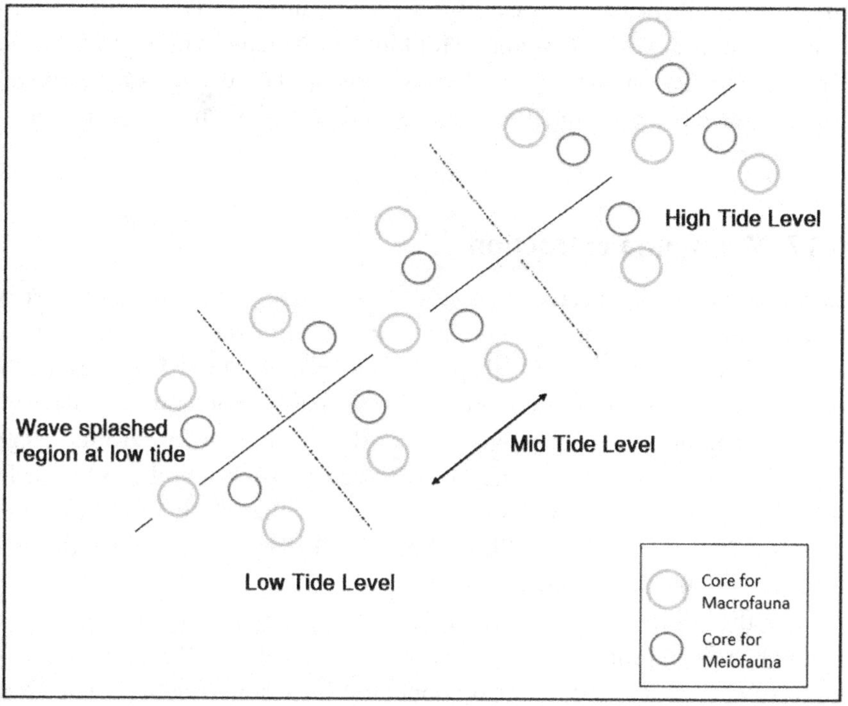

Figure 5.4 Sampling design for a single transect.

total organic carbon (TOC), and macro and meiofaunal analysis. The core samplers are cylindrical in shape and are made from tubing or piping of any firm material. The diameter of the core is based on the depth and volume of the sample required (core inner diameter ~5 cm). The core samples were collected at 20 cm depth for biological samples, whereas in sandy beaches, the animals are collected at 10 cm due to the maximum diversity of species at top layers (Brazeiro, 2001; McLachlan, 1983).

5.16 Macrofauna extraction

To sample macrofauna, a hand coring device is used, typically a plastic corer with an inner diameter of 10.5 cm (two corers were used for every replicate). The corer is inserted into the sediment (in a vertical position) to a depth of 20 cm (if sediments are that deep). Sandy-beach macrobenthos is usually collected by excavating quadrates or taking cores in transects across the intertidal zone and passing the sand through sieves of the finest mesh to let the sand through but trap the fauna. After sieving, the samples are placed in a sample bag or jar and preserved with 7% formalin buffered with seawater (Hopkins & Mudge, 2004) and "Rose Bengal" (1 g/L) for staining the live organisms. The density of the organism is given by counting the actual number and then converting it to number/m^2.

5.17 Meiofauna extraction

In the laboratory, collected sediment samples are washed through a set of 0.5and 0.063 mm sieves. The sediment retained in the 0.063 mm sieve is decanted to extract meiofaunal species (Higgins & Thiel, 1988). Isolation of meiofauna from the sediment is carried out by flotation and decantation technique. The effectiveness of this technique is around 95% (Danovaro et al., 2004; Giere, 2009). Sieved specimens and residual sediments are transferred into plastic bottles and fixed at 5% neutral formaldehyde or 70% alcohol with "Rose Bengal" (1 g/L) to determine the live organisms using the staining method.

For the separation of the foraminifera, the washed samples are passed through a 0.063 mm sieve and then dried (Walton, 1952). Later, one gram of sediment is taken out by cone and quartering method. The foraminifera species are handpicked and the residues are again separated by the CCl_4 method (Nigam, 1984). The picked out specimens are mounted and then counted for species-wise distribution. The identification and counting is done under a stereoscopic microscope—Novex model P-20 Euromex (Holland). All intertidal macrofauna and meiofauna observed are recorded properly and later systematic classification is carried out at their Phyla level. All the meiofauna samples are mounted onto glass slides using the formalin-ethanol-glycerol technique by Vincx (1996) and photographs are taken by using Olympus CX 41 Phase contrast microscope. The numerical abundance of the meiofauna is expressed in individuals/m^2 (Giere, 2009).

5.18 Geochemical analysis of seawater and marine sediments

Seawater samples are collected using a Niskin water sampler at neap high tide and neap low tide. Marine sediment and seawater temperatures are recorded using a digital standard Celsius thermometer. pH, EC, salinity, and DO of the water samples are recorded using a multiprobe water quality analyzer

The concentration of chloride is found using the argentometric titration method. The sulfate concentration is measured by the nephelometric method (AFNOR, 1997). The cations such as Ca, Mg, Na, and K are determined using Flame AAS. Biochemical oxygen demand is measured by the method recommended by the APHA (1995). Orthophosphate analysis is done using the spectrophotometer-ascorbic acid method and nitrate is determined using the cadmium column reduction method.

In marine sediments calcium carbonate ($CaCO_3$) content is analyzed as suggested by Loring and Rantala (1992). Organic carbon (OC) concentration is determined by exothermic heating and oxidation with potassium dichromate and concentrated H_2SO_4. Excess dichromate is titrated with 0.5 N ferrous ammonium sulfate solution (Gaudette et al., 1974). Carbohydrate (CHO) content is determined using the phenol-sulfuric acid method (Dubois et al., 1956; Kochert, 1978). The amino acid is estimated by the Ninhydrin method. The protein is analyzed by Lowry's method (Herbert et al., 1971). The lipids were extracted and determined from freeze-dried sediments (Bligh & Dyer, 1959; Parsons et al., 1984).

Sieve analysis is done in ASTM (American Standard Tested Mesh) sieves at 0.25 φ intervals. The sieves are stacked in descending order based on their size and are shaken continuously for around 20 min using a Ro-tap sieve shaker. The sieved materials are collected and the dry weight of individual fractions are measured and tabulated for granulometric studies. The graphical and moment methods are applied for granulometric studies. Textural parameters including mean, standard deviation, skewness, and kurtosis are calculated for all the processed samples (Friedman, 1967). Detailed granulometric studies are done using GRADISTAT 4.0 software package.

5.19 Ternary diagram

One of the most common graphical techniques used by sedimentologists is to plot the basic gravel, sand, silt, and clay percentage on equilateral triangular diagrams. This is a simple method that helps in the rapid classification and comparison of samples. Sediment classification has also been done by plotting the percentage of sand, silt, and clay in a triangular diagram proposed by Folk and Ward, 1957; Folk, 1974. The nature of sand, silt, clay percentage with respective seasons is shown in Fig. 5.5. During postmonsoon, most of the samples (74%) fall in the sandy triangle region and the rest of the samples (26%) fall in the silty sand region. In the SW

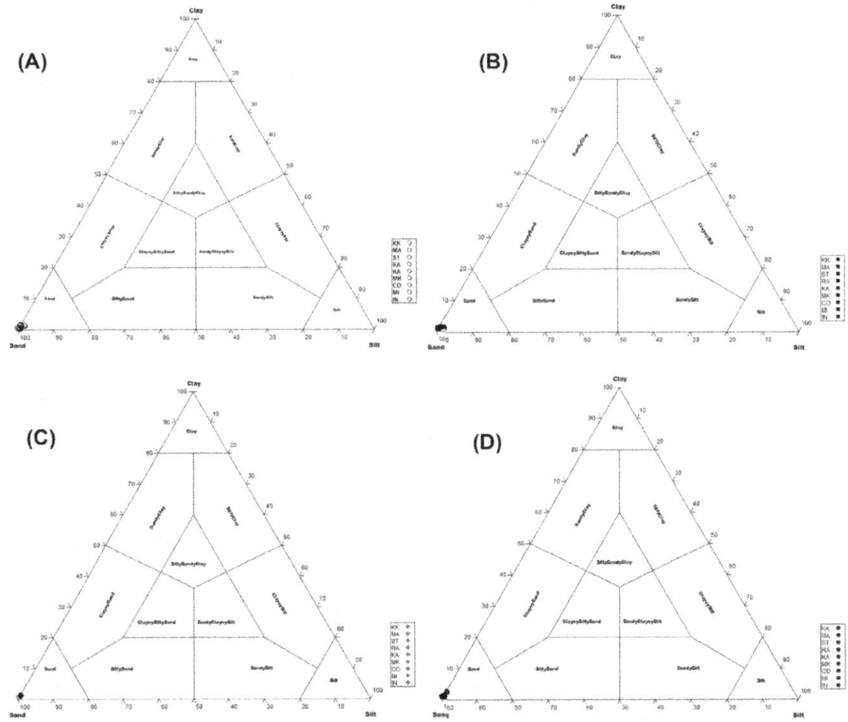

Figure 5.5 Ternary diagrams illustrating the nature of sand silt and clay percentage in the study region with respective seasons (A, Postmonsoon; B, Summer; C, SW Monsoon; D, NE Monsoon).

monsoon, 63% of samples fall in the sandy region and 37% of samples fall in the silty sand region. During summer, 74% of samples fall in the sandy region and 26% of samples fall in the silty sand region. In the NE monsoon, 78% of samples fall in the sandy region and the rest 22% of samples fall in the silty sand region. This indicates that during postmonsoon and summer seasons, sediment distribution is more or less the same. During SW and NE monsoon seasons, prominent changes have been observed in the textural characteristics.

The study shows that the concentration of heavy minerals is lesser in coarser fractions. The coarse heavy mineral fractions are separated by sieve analysis and then by density separation techniques using bromoform solution (sp.gr.2.89). The separated fractions are washed with distilled water and then dried in a hot air oven (60°C) for complete moisture removal. The dried particles are weighed and their weights are tabulated. The heavier particles are then classified based on their magnetic properties using an isodynamic separator. The separated magnetic fractions are weighed and kept aside for further analysis.

5.20 Physicochemical properties of seawater and marine sediments

The diversity of the marine benthic community depends upon the physicochemical properties of sea water and marine sediments.

5.20.1 Seawater

The tropical aquatic living organisms flourish well at water temperatures ranging between 25°C and 30°C. Heat fluctuation can lead to the death of marine organisms. Rainfall directly impacts the temperatures of the surrounding water and the salinity of the water. It also has a severe influence on the coastal aquatic ecosystem. The amount of freshwater from river discharge into the estuarine and marine environments bring a broad variation in salinity, and the water temperature is decided by rainfall. Further rainfall influences the water level of the estuarine ecosystem. It also controls the inorganic and organic nutrient contents in the marine ecosystem. The salinity of the coastal ecosystem and offshore environment is severely affected by various environmental parameters like river runoff, upwelling, precipitation, evaporation, and surface current patterns. The DO in seawater reflects the condition of the marine environment. Concentrations below 5 mg/L may lead to the death of many fish and other species (Bruckner & Burrows, 2005). The nutrient enhancement of the marine ecosystem is determined by their levels of concentration in the riverine and coastal water sources and their properties in the mixture. Inorganic phosphate content is a valuable index of the state of eutrophication of water bodies. Water receiving raw or untreated sewage, agricultural runoff, and unavoidable industrial waste usually contains a significant phosphate concentration. The nitrate source to the study area is primarily controlled by natural runoff, domestic sewage, agricultural waste, and industrial wastes.

During summer, environmental parameters like temperature, pH, EC, salinity, TDS, BOD, Cl, Na, Mg, and SO_4 are found in higher concentrations, and DO, K, Ca, PO_4, NO_2, and NO_3 are present in lower concentrations. During SW monsoon K alone is found in higher concentrations and temperature and SO_4 are in lower concentrations. During NE monsoon, DO, Ca, PO_4, NO_2, and NO_3 are found in higher concentrations and pH, EC, salinity, TDS, BOD, Cl, Na, and Mg are in a lower concentration range (Supplementary Table 5.S1). The high concentration level of nitrate in the present study may be due to heavy monsoon rains which bring in a large number of organic nutrients into the marine ecosystem through rivers such as Pazhayar, Panniyar, Valliyar, and Pampuar. The temperature fluctuation is one of the important factors in the swamp and estuarine ecosystem, which may affect the physicochemical properties and influence the overall distribution and abundance of flora and fauna (Soundarapandian et al., 2009). The availability of oxygen in the water is a very important parameter in water analysis. It serves as an important tool for the physical, chemical, and biological activities of the aquatic ecosystem (Manikannan et al., 2011). Two major sources of DO in the aquatic ecosystem are by diffusion of oxygen from the air and from photosynthetic activity. Oxygen is normally considered to be the primary limiting factor in water bodies with a high concentration of organic materials. In general, temperature and salinity affect the availability of oxygen (Vijayakumar et al., 2000). This may be due to the mixing of freshwater runoff in the monsoon season. There is a slight deviation between mining and nonmining sites in the study area. Significant spatial and seasonal variations in all the physicochemical properties of seawater and marine sediments are observed during the study period.

5.20.2 Marine Sediments

With regard to the marine sediment, during postmonsoon CHO, lipid, and sand are recorded in the highest level and clay is recorded in low concentration. During summer temperature, pH, TOC, amino acid, protein, and silt are found in higher concentrations. During SW monsoon $CaCO_3$ is in higher concentration and temperature, TOC, amino acid, and lipid are found in a lower concentration. During NE monsoon, clay is in higher value but the other parameters like pH, CHO, protein, sand, and silt are in low concentrations (Supplementary Table 5.S2). In this study, various multivariate statistical methods are utilized to evaluate seasonal variations in surface water quality and sediment quality in the mining and nonmining areas. The chemical composition of surface water and sediment in each season is strongly influenced by external forces and their interactions, dissolution, dilution, and anthropogenic inputs. The overall distribution of the geochemical data of seawater and marine sediment in the beaches of the SW coast indicates that there is a significant variation among the mining and nonmining beaches in the study area.

5.21 Statistical analysis

Specific statistical approaches are carried out for the explanation of the gathered biological data. Univariate and multivariate analyses are performed using the PRIMER version 6.1.10 (Plymouth Routines In Multivariate Ecological Research) software package developed at the Plymouth Marine Laboratory (Clarke & Gorley, 2001; Clarke & Warwick, 1994). The univariate methods include a set of species diversity indices carried out to estimate the community structure of macro and meiofauna along the SW coast of India. The Margalef's species richness index (Margalef, 1958), Pielou's evenness index (Pielou, 1975), and Shannon–Wiener index (Shannon, 1949) are used to study species diversity. Multivariate statistical analysis is conducted to analyze similarity coefficients between studied parameters. Cluster analysis is carried out using the Bray–Curtis coefficient with square root transformation to determine the similarities between groups (Bray & Curtis, 1957). The multiple k-dominance plot facilitates discrimination of polychaetes based on species-relative contribution to abundance. The biological and environmental (BIO-ENV) analysis is done to measure the rank correlations of fauna with essential environmental variables such as depth, temperature, salinity, DO, organic matter %, and textural characteristics.

5.22 Macrofaunal distribution

The availability and abundance of group-wise and season-wise macrofaunal assemblages in the intertidal sediments along the study area are tabulated in Tables 5.2 and 5.3. Among the 88 species of macrofauna, polychaetes constitute the major diversity with 56 individual species. Gastropods are the second dominant group with 12 individual species, bivalves rank the third with six species, decapods and isopods rank fourth with three species each, and amphipods are in the last position with two species. All the macrofauna are identified up to their species level, and a detailed list of macrofaunal composition is presented in Table 5.4.

Polychaetes dominate the macrofaunal communities in species composition and numerical abundance in all seasons. The abundance of polychaetes during the postmonsoon season is high in the Colachel coast with

Table 5.2 Availability of group-wise and season-wise macrofaunal assemblages in the intertidal sediments along the study area.

POST MONSOON

Macrofauna	KK			MA			ST			RA			KA			MK			CO			MI			IN		
	HT	MT	LT	HT	MT	LT	HT	MT	LT	HT	MT	LT	HT	MT	LT	HT	MT	LT	HT	MT	LT	HT	MT	LT	HT	MT	LT
Polychaeta	*	*	*		*		*	*		*		*		*	*		*	*	*	*	*	*	*	*		*	
Bivalvia	*				*		*	*	*	*	*		*	*				*	*	*	*		*		*	*	
Gastropoda	*	*				*	*	*	*	*					*	*			*		*				*	*	
Decapoda		*	*		*		*	*					*	*					*	*	*		*			*	*
Amphipoda																											
Isopoda		*	*										*	*			*										

SUMMER

Macrofauna	KK			MA			ST			RA			KA			MK			CO			MI			IN		
	HT	MT	LT	HT	MT	LT	HT	MT	LT	HT	MT	LT	HT	MT	LT	HT	MT	LT	HT	MT	LT	HT	MT	LT	HT	MT	LT
Polychaeta	*	*	*	*	*	*	*	*	*	*		*	*	*	*	*	*		*	*	*	*	*	*	*	*	*
Bivalvia	*					*		*	*		*										*						
Gastropoda		*	*					*			*														*		
Decapoda															*		*									*	
Amphipoda	*																										
Isopoda	*	*	*	*			*			*			*	*										*	*		

SW MONSOON

Macrofauna	KK			MA			ST			RA			KA			MK			CO			MI			IN		
	HT	MT	LT	HT	MT	LT	HT	MT	LT	HT	MT	LT	HT	MT	LT	HT	MT	LT	HT	MT	LT	HT	MT	LT	HT	MT	LT
Polychaeta	*	*	*	*	*	*			*	*	*	*	*	*	*	*	*		*	*	*	*	*	*	*	*	*
Bivalvia											*			*			*			*	*			*		*	*
Gastropoda									*			*				*	*							*			
Decapoda																*	*							*			
Amphipoda															*											*	
Isopoda	*	*	*			*				*				*	*											*	*

NE MONSOON

Macrofauna	KK			MA			ST			RA			KA			MK			CO			MI			IN		
	HT	MT	LT	HT	MT	LT	HT	MT	LT	HT	MT	LT	HT	MT	LT	HT	MT	LT	HT	MT	LT	HT	MT	LT	HT	MT	LT
Polychaeta	*		*	*		*		*	*	*	*	*	*	*	*	*			*	*	*	*	*	*			*
Bivalvia										*																	
Gastropoda		*								*					*			*									
Decapoda																	*										
Amphipoda																											
Isopoda			*	*				*				*	*	*	*		*										

* Species availability.
KK, Kanyakumari; *MA*, Manakudi; *ST*, Sanguthurai; *RA*, Rajakkamangalam Thurai; *KA*, Kadiapattinam; *MK*, Manavalakurichi; *CO*, Colachel; *MI*, Midalam; *IN*, Inayam; *HT*, high tide; *MT*, mid tide; *LT*, low tide.

Table 5.3 Numerical abundance of macrofaunal assemblages in the intertidal sediments along the study area (SD ± average).

POST MONSOON

Macrofauna	KK			MA			ST			RA			KA		
	HT	MT	LT	HT	MT	LT	HT	MT	LT	HT	MT	LT	HT	MT	LT
Polychaeta	2.55±13.24	1.79±11.39		0	1.91±6.12		6.38±26.78	5.56±17.69	0	1.86±9.11	0	0.09±0.67	0	1.14±3.45	0.84±2.69
Bivalvia	8.83±8.95	0	1±4.53	0	1.83±3.25	0	1.67±2.88	1.17±2.04	0.67±1.03	2±4.90	1.17±2.86	0	0.67±1.63	0.50±1.22	0
Gastropoda	5.11±8.71	3.39±6.40	0	0	0	0	0.33±1.41	0.28±1.18	0.11±0.47	2.22±4.41	0	0	0	0	0.44±1.04
Decapoda	0	8.33±9.07	0	0	0.33±0.58	0.33±0.84	8±13.86	6±10.39	0	0	0	0	15.67±11.93	5.67±4.93	0
Amphipoda	0	0	7.33±8.08	0	0	0	0	0	0	0	0	0	0	0	0
Isopoda	0	4±6.93	0	0	0	0	0	0	0	0	0	0	0.67±1.15	2±3.46	0

Macrofauna	MK			CO			MI			IN		
	HT	MT	LT	HT	MT	LT	HT	MT	LT	HT	MT	LT
Polychaeta	0	0.66±2.96	0.16±0.97	2.26±12.76	10.63±34.48	2.27±9.29	1.07±2.63	0.95±3.34	0.14±0.67	0	8.09±22.33	0
Bivalvia	0	0	0.33±0.82	1.50±2.07	1±1.26	0.33±0.82	0	0.33±0.82	0	1.83±2.99	1.33±2.16	0
Gastropoda	0.50±1.04	0	0	2.33±3.55	0	0.78±1.93	0	0	0	0.50±1.04	0.56±1.76	0
Decapoda	0	0	0	2.67±4.62	4±6.93	3.33±2.89	0	1±1	0	7.33±3.79	8±8	0
Amphipoda	0	0	0	0	0	0	0	0	0	0	0	0
Isopoda	0	1±1	0	0	0	0	0	0	0	0	0	0

SUMMER

Macrofauna	KK			MA			ST			RA			KA		
	HT	MT	LT	HT	MT	LT	HT	MT	LT	HT	MT	LT	HT	MT	LT
Polychaeta	4.04±13.90	1.34±10.02	0.46±2.42	1.05±3.54	0.73±2.77	0.09±0.48	4.48±9.18	5.48±12.15	0.54±2.37	3±13	0	0.39±2.94	0.39±1.53	1.07±3.96	0.20±0.82
Bivalvia	4.50±5.96	0	0	0	0	0.50±1.22	0	4.33±3.44	2±1.41	0	2.50±6.12	0	0	0	0
Gastropoda	0	10.44±12.09	1.83±3.28	0	0	0	0	4.17±9.48	0	0	0.67±1.78	0	0	0	0
Decapoda	0	0	0	0	0	0	0	0	0	0	0	0	0	0	10±1.73
Amphipoda	3±4.24	0	0	0	0	0	0	0	0	0	0	0	0	0	0
Isopoda	8±2	14.33±6.81	5±2	1±1.73	0	0	14±8.19	0	0	11.33±3.06	0	0	13±2.65	16.67±7.77	0

Macrofauna	MK			CO			MI			IN		
	HT	MT	LT	HT	MT	LT	HT	MT	LT	HT	MT	LT
Polychaeta	0.46±1.84	0.57±2.30	0	1.89±9.97	4.25±14.11	0.16±0.60	0.53±2.78	1.71±4.90	1.81	4.70±15.24	3.71±19.29	2.07±7.06
Bivalvia	0	0	0	0	0	1.33±1.21	0	0	0	0	0	0
Gastropoda	0	0	0	0	0	0	0	0	0	4.44±8.79	0	0
Decapoda	0	10±1.73	0	0	0	0	0	0	0	0	0.67±1.15	0
Amphipoda	0	0	0	0	0	0	0	0	0	0	0	0
Isopoda	0	0	0	0	0	0	0	0	6±5.29	1±1.73	0	0

SW MONSOON

Macrofauna	KK			MA			ST			RA			KA		
	HT	MT	LT	HT	MT	LT	HT	MT	LT	HT	MT	LT	HT	MT	LT
Polychaeta	0.34±1.49	3.05±7.95	1.13±2.86	0.32±2.41	1.89±6.28	0.41±2.16	0	0	0.77±2.43	2.95±7.16	1.55±5.86	0.82±3.13	0.20±1.47	1.23±4.53	0.50±3.74
Bivalvia	0	0	0	0	0	0	0	0	0	0	1.67±2.58	0	0	0.33±0.52	0
Gastropoda	0	0	0	0	0	0	0	0	1.17±2.50	0	0	0.78±2.29	0	0	0
Decapoda	0	0	0	0	0	0	0	0	0	0	0	0	0	0	0
Amphipoda	0	0	0	0	0	0	0	0	0	0	0	0	0	0	0.50±0.71
Isopoda	4.00±6.93	1.00±1.73	0.67±1.15	0	0	1.33±1.53	0	0	0	0.33±0.58	0	0	0	0.67±1.15	1.67±2.89

Macrofauna	MK			CO			MI			IN		
	HT	MT	LT	HT	MT	LT	HT	MT	LT	HT	MT	LT
Polychaeta	0.80±4.30	0.43±3.21	0	0.71±3.76	1.00±5.25	0.43±1.33	0.98±3.79	2.21±7.07	0.25±1.21	1.21±3.73	3.48±8.16	0.80±1.89
Bivalvia	0	1.00±1.55	0	0	3.00±4.65	0.67±1.03	0	0	1.00±1.55	0	1.67±2.58	0.67±1.03
Gastropoda	2.67±5.63	4.11±7.37	0	0	0	0	0	0	1.28±2.63	0	0	0
Decapoda	0.33±0.58	0.33±0.58	0	0	0	0	0	0	0.33±0.58	0	0	0
Amphipoda	0	0	0	0	0	0	0	0	0	0	0.50±0.71	0
Isopoda	0	0	0	0	0	0	0	0	0	0	2.00±3.46	1.67±1.53

(Continued)

Table 5.3 (Continued)

Macrofauna	NE MONSOON														
	KK			MA			ST			RA			KA		
	HT	MT	LT	HT	MT	LT	HT	MT	LT	HT	MT	LT	HT	MT	LT
Polychaeta	0.04±0.27	0	0.09±0.48	0.27±1.41	0	0.13±0.66	0	0.14±1.07	0.05±0.40	0.13±0.94	0.48±2.54	0.11±0.59	0.13±0.94	0.20±1.47	0.25±0.96
Bivalvia	0	0	0	0	0	0	0	0	0	5.83±8.35	0	0	0	0	1.50±1.76
Gastropoda	0	2.11±3.18	0	0	0	0	0	0	0	1.56±2.48	0	0	0	0	1.11±2.22
Decapoda	0	0	0	0	0	0	0	0	0	0	0	0	0	0	0
Amphipoda	0	0	0	0	0	0	0	0	0	0	0	0	0	0	0
Isopoda	0	0	1.33±1.53	2.67±3.06	0	0	0	2±2.65	0	0	0	1±1.73	0.67±1.15	2.50±3.54	1.33±1.53

MK			CO			MI			IN		
HT	MT	LT	HT	MT	LT	HT	MT	LT	HT	MT	LT
0.23±1.10	0	0	0.64±2.73	1.05±4.87	0.29±1.09	0.14±1.07	0.79±2.89	0.20±0.92	0	0	0.27±0.96
0	0	0	0	0	0	0	0	0	0	0	0
0	0	0.06±0.24	0	0	0	0	0	0	0	0	0
0	3±3	0	0	0	0	0	0	0	0	0	0
0	0	0	0	0	0	0	0	0	0	0	0
0	2.33±2.08	0	0	0	0	0	0	0	0	0	0

KK, Kanyakumari; *MA*, Manakudi; *ST*, Sanguthurai; *RA*, Rajakkamangalam Thurai; *KA*, Kadiapattinam; *MK*, Manavalakurichi; *CO*, Colachel; *MI*, Midalam; *IN*, Inayam; *HT*, high tide; *MT*, mid tide; *LT*, low tide.

an average and SD of (10.63 ± 34.48) and low in the Manavalakurichi coast (0.16 ± 0.97). This dominancy of polychaetes is due to their high adaptability to a wide range of environmental factors including beach placer mining. The second dominant macrofaunal group is the gastropods. The highest abundance of gastropods is observed in the Kanyakumari coast (5.11 ± 8.71) and the lowest abundance in the mining sites, that is, Manakudi and Midalam coasts (0.50 ± 1.04). Polychaetes also show a similar kind of abundance in both mining and nonmining areas. The third dominant macrofaunal group is the Decapoda. Decapoda is high in abundance in the Kadiapattinam coast (15.67 ± 11.93) and low in the Midalam coast (1 ± 1) during the postmonsoon season. The fourth dominant macrofaunal group is the Bivalvia. Bivalvia is abundant in the nonmining areas during the postmonsoon season (8.83 ± 8.95), and the mining areas have a low population (0.33 ± 0.82) except in the Kadiapattinam coast. The abundance of Bivalvia (8.83 ± 8.95) is noticeably high in the nonmining sites of the Kanyakumari coast and low in the Manavalakuriach and Midalam coasts (0.33 ± 0.82). Isopods are very low in number in all the stations except the Kanyakumari coast (4 ± 6.93). The distribution of Amphipoda in the postmonsoon season shows the natural absence of habitat in all stations. This shows that the biodiversity of benthos is directly

Table 5.4 Macrofauna recorded in the study area.

I	Polychaeta
1	*Aonides oxycephala*
2	*Capitella capitata*
3	*Cirratulus* sp.
4	*Cossura coasta*
5	*Cossura longocirrata*
6	*Diplocirrus* sp.
7	*Dispio* sp.
8	*Eteone longa*
9	*Eunice indica*
10	*Glycera convoluta*
11	*Glyceridae* sp.
12	*Glycinde nordmanni*
13	*Hesiospina vestimentifera*
14	*Heteromastus filiformis*
15	*Heteromastus similis*
16	*Laeonereis culveri*
17	*Levensteiniella kincaidi*
18	*Levinsenia gracilis*
19	*Magelona* sp.
20	*Malacoceros indicus*
21	*Mediomastus* sp.
22	*Minuspio cirrifera*
23	*Nephtys polybranchia*
24	*Notomastus aberrans*
25	*Paraonis gracilis*
26	*Parougia wolfi*
27	*Phalacrophorus pictus*
28	*Phyllodoce* sp.
29	*Pisione remota*
30	*Pisione* sp. 1
31	*Pisione* sp. 2
32	*Pisione* sp. 3
33	*Pisionidens* sp.
34	*Polydora ciliata*
35	*Polygordius appendiculatus*
36	*Polygordius neapolitanus*
37	*Prionospio pinnata*
38	*Prionospio* sp.
39	*Protodrilus rubropharyngeus*
40	*Rhodine gracilior*
41	*Saccocirrus sonomacus*
42	*Saccoclrrus* sp.

(*Continued*)

Table 5.4 (Continued)

43	*Sphaerosyllis levantina*
44	*Spio filicornis*
45	*Spionidae* sp.
46	*Spiophanes bombyx*
47	*Syllid* sp. 1
48	*Syllid* sp. 2
49	*Syllides longocirrata*
50	*Syllidia armata*
51	*Syllis cornuta*
52	*Syllis gracilis*
53	*Syllis* sp. 3
54	*Syllis* sp. 4
55	*Tharyx* sp.
56	*Typosyllis nipponica*
II	**Bivalvia**
1	*Donax deltoides*
2	*Donax incarnatus*
3	*Donax spiculum*
4	*Meretrix meretrix*
5	*Cardium setosum*
6	*Anadara granosa*
III	**Gastropoda**
1	*Admete gracilior*
2	*Caecum glabrum*
3	*Cerithedia cingulata*
4	*Daphniola exigua*
5	*Turritella attenuata*
6	*Eulima glabra*
7	*Hydrobia ulvae*
8	*Iravadia quadrasi*
9	*Leucorhynchia* sp.
10	*Natica albula*
11	*Natica didyma*
12	*Odostomella doliolum*
13	*Rissoina crassi*
14	*Setia levantina*
15	*Setia subvaricosa*
16	*Stenothyra messageri*
17	*Turbonilla conradi*
18	*Umbonium vestiarium*

(Continued)

Table 5.4 (Continued)

IV	**Decapoda**
1 2 3	*Emerita asiatica* *Emerita holthuisi* *Gastrosaccus spinifer*
V	**Amphipoda**
1 2	*Apohyale prevostii* *Leptocheirus plumulosus*
VI	**Isopoda**
1 2 3	*Microcerberus abbott* *Angeliera phreaticola* *Paramunnidae vanhoffen*

proportional to the accumulation of heavy mineral sand during postmonsoon and indirectly proportional to the mining areas (Godson, 2016).

The numerical abundance of polychaetes during summer is found to be high in the Sanguthurai and Inayam coasts with an average and SD of 5.48 ± 12.15 and low in the Manavalakurichi coast (0.57 ± 2.30). High numerical abundance is noticed in nonmining areas such as Kanyakumari, Sanguthurai, Rajakkamangalam Thurai, Colachel, and Inayam coasts. The mining areas such as Manakudi, Kadiapattinam, Manavalakurichi, and Midalam coasts have been noticed with a low abundance of polychaetes compared with the nonmining areas (Table 5.4). The high abundance of bivalves is noticed in the Sanguthurai coast (4.33 ± 3.44) and low in the Kadiapattinam, Manavalakurichi, Midalam, and Inayam coasts (0 No./sq.m) as these coasts are engaged in mining activity, except Inayam where other climatic and natural factors may have affected their distribution in the summer season. The abundance of gastropods is noticed to be high in the Kanyakumari coast (10.44 ± 12.09) followed by Inayam, Sanguthurai, and Rajakkamangalam Thurai coasts having an abundance of 4.17 ± 9.48, 0.67 ± 1.78, and 4.44 ± 8.79, respectively. Low abundance is noticed in the mining areas of Manakudi, Kadiapattinam, Manavalakurichi, and Midalam coasts (0.33 ± 0.82). The low abundance of Amphipoda group reveals the absence of natural habitat in the summer season in all the stations except the Kadiapattinam and Kanyakumari coasts (7.33 ± 8.08), whereas, the abundance of isopods is found to be low in the Manavalakurichi and Colachel coasts where the numerical abundance is (0). The reason for such varied abundance in the mining and nonmining areas is due to the effect of postplacer deposition during the summer season, where mining activity takes place in the placer rich zones, and the habitat is already damaged due to intense mining in the postmonsoon season, thus further affecting the species abundance. The temporal distribution of benthos shows high density with a high diversity of faunal assemblages during the postmonsoon and summer seasons. The decrease in the diversity of benthos in the monsoon seasons may be due to low temperature and a decrease in salinity as a result of the mixing of freshwater from fluvial sources.

The numerical abundance of polychaetes during SW monsoon is comparatively low in the postmonsoon and summer seasons. The high abundance is found in the Inayam and Rajakkamangalam Thurai coasts (3.48 ± 8.16) and low in the Sanguthurai coast (0.77 ± 2.43). High numerical abundance is noticed in the nonmining areas like Kanyakumari, Rajakkamangalam Thurai, and Colachel coasts. However, significant abundance is noticed in the mining areas of Manakudi, Kadiapattinam, Manavalakurichi, and Midalam coasts. This highlights the replenishment of nutrients and habitat restoration during the SW monsoon season due to low anthropogenic impact. Mining activities are suspended during the monsoons due to unavailability of workforce, unfavorable climatic conditions, and occurrence of beach erosion. The high abundance of bivalves is noticed in the Colachel coast (3.00 ± 4.65) and low value in the Kanyakumari, Manakudi, and Sanguthurai coasts (0) as these coasts are situated in the eastern part of the study area, where significant wave action influences the sediment distribution pattern during the SW monsoon. The meteorological and oceanographic parameters during SW monsoon control the distribution of bivalves in the study area. The impact of mining is low in these regions except Manakudi during the SW monsoon and the impact of Manakudi estuary also plays a significant role in the sediment distribution and nutrient dynamics. The abundance of gastropods is noticed to be high in the Manavalakurichi coast (4.11 ± 7.37) followed by Midalam, Sanguthurai, and Rajakkamangalam Thurai coasts having an abundance of 1.28 ± 2.63, 1.17 ± 2.50, and 0.78 ± 2.29, respectively. The low abundance is noticed in Manakudi, Kanyakumari, Kadiapattinam, Colachel, and Inayam coasts (0). Moreover, no significant relationship is noticed in species abundance between mining or nonmining areas. This might be due to the overall disturbance through natural as well as anthropogenic activities. The abundance of decapods is noticed in the mining regions of Midalam and Manavalakurichi (0.33 ± 0.58), but this number is negligible to take into consideration. All other stations are found with a zero value which indicates their natural distribution in the disturbed environment during the monsoons. The low abundance of amphipod group reveals the absence of natural habitat in all stations except the Kadiapattinam and Inayam coasts (0.50 ± 0.71), whereas, the abundance of isopods is found to be high in the Kanyakumari and Inayam coasts where the numerical abundance is 4.00 ± 6.93. The low abundance (0) is noticed in the coasts of Sanguthurai, Manavalakurichi, Colachel, and Midalam. The effect of the SW monsoon has influenced the distribution of macrofaunal assemblages in the study area and mining does not have any significant role in this season due to halting of mining activities in the season.

The numerical abundance of polychaetes during the NE monsoon is low when compared to other seasons. High abundance is found in the Colachel coast (1.05 ± 4.87) and it is low in the Kanyakumari coast (0.04 ± 0.27). The relationship between mining and species distribution is not very noteworthy during the NE monsoon. This may be due to the disturbed climatic condition during the monsoon season which is similar to that of the condition during the SW monsoon. The abundance of bivalves is noticed to be high in the Rajakkamangalam Thurai coast (5.83 ± 8.35) and low in the Kanyakumari, Manakudi, Sanguthurai, Manavalakurichi, Colachel, Midalam, and Inayam coasts (0). The abundance of gastropods is high in the Kanyakumari coast (2.11 ± 3.18) followed by Rajakkamangalam Thurai, Kadiapattinam, and Manavalakurichi coasts having an abundance of 1.56 ± 2.48, 1.11 ± 2.22, and 0.06 ± 0.24, respectively. Low abundance is noticed in Manakudi, Sanguthurai, Colachel, Midalam, and Inayam coasts (0). Such variations are due to the disturbances from natural as well as anthropogenic activities. The abundance of Decapoda is noticed only in the mining regions of Manavalakurichi (3 ± 3), but this number is negligible to take into account. All other stations are found to have a zero value, which points out their natural distribution in the disturbed environment during the NE monsoon. The low abundance of amphipods group reveals the absence of natural habitat in all the stations. The numerical abundance of isopods is found to be high in the Kadiapattinam coast where it is 2.50 ± 3.54. Low abundance (0) is noticed in the coasts of Colachel, Midalam, and Inayam. The NE monsoon has influenced the distribution of macrofaunal assemblages in the study area, and mining does not have any significant impact during this season due to halting of mining activities, which is unfavorable in the season.

5.22.1 Univariate analysis

The macro and meiofaunal assemblages are analyzed using various indices: Univariate measures such as Margalef's species richness (d), Shannon–Wiener diversity (H′ log) and Pielou's evenness (J′), graphical tools like k-dominance curve, and multivariate tools such as Bray–Curtis similarity, after suitable transformation of sample abundance data, classification (hierarchical agglomerative clustering using group-average linking), and ordination (multidimensional scaling, MDS), are used to interpret the data using PRIMER 6.1. software. The univariate methods include a set of species diversity indices. The below-mentioned diversity indices have been performed for the estimation of macrofaunal community structure.

5.22.1.1 Species richness (Margalef's index)

Margalef's index shows variation in range among the species in the study area. High species richness is seen in Manavalakurichi (1.479) during the postmonsoon season and low species richness is found in Kanyakumari (0.4) and Colachel (0.38) during SW monsoon and summer seasons, respectively. The high richness in the postmonsoon season is due to good habitat conditions like nutrient availability, high oxygen level, and fresh placer deposits. Species richness is either absent or very low in the Colachel, Midalam, and Inayam coasts during the NE monsoon season (Fig. 5.6). This may have resulted from the impact of climatic disturbances on sensitive species. It is also found that species richness is varied along the study area forming an oscillation curve showing the effect of littoral movement of sediments, transporting the sand and nutrients.

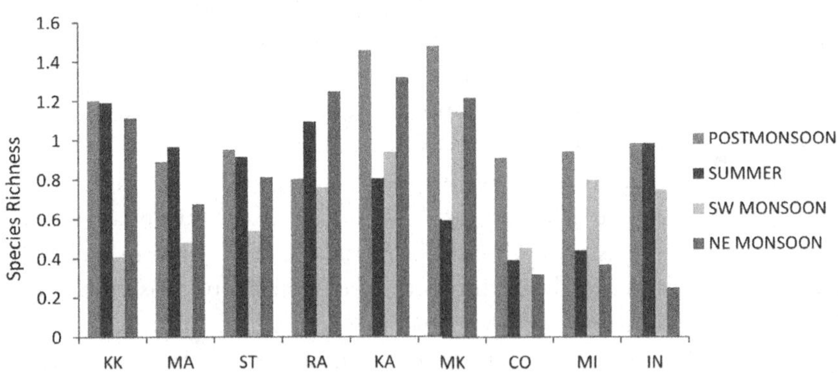

Figure 5.6 Species richness along the study area (Macrofauna).

5.22.1.2 Evenness index (Pielou's index)

The evenness index value is high in Manavalakurichi during the NE monsoon and low in Colachel, Midalam, and Inayam (Fig. 5.7). An increasing trend in evenness index is observed in Sanguthurai, which ranges from 0.7 to 0.98, from postmonsoon to NE monsoon. Consistent values are seen in the stations from Manakudi to Manavalakurichi during the postmonsoon season, which points out similar environmental conditions in these coasts. Other stations show an irregular trend of evenness distribution in all the seasons. This shows that the species are selective towards their habitat, and variation in the habitat through mining or climatic changes impact their distribution.

5.22.1.3 Species diversity index (Shannon index)

The species diversity index shows a wavy pattern. High Shannon index is seen in Kanyakumari during postmonsoon and summer, and a low value is observed in the coasts of Colachel, Midalam, and Inayam during NE monsoon. Beach disturbance is one major factor controlling species diversity. The drastic lowering of the Shannon index during the SW monsoon highlights the effect of climate in disturbing the species diversity in the Kanyakumari coast. Manakudi is also disturbed during the SW monsoon. The impact of pollution through the Manakudi estuary is one major reason for the low Shannon index. Manakudi is a mining coast, which ranges between 1.3 and 0.5. The Sanguthurai coast shows the influence of the monsoon, with low values during both the monsoons and high values during the postmonsoon and summer seasons. This high value shows the

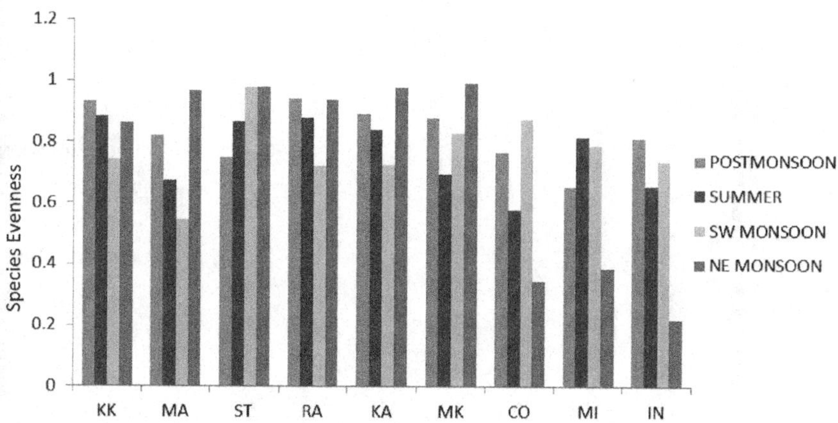

Figure 5.7 Species evenness along the study area (Macrofauna).

effect of controlled conditions as mining activities are very less in the area. The Rajakkamangalam Thurai coast also has a similar trend due to similar environmental conditions and lack of mining. The effect of the SW monsoon has lowered the species diversity in the area due to erosion. The Kadiapattinam and Manavalakurichi coasts show low species diversity in summer. This is due to the impact of mining during summer. The Colachel, Midalam, and Inayam coasts experience high erosion during the monsoons, which alters the diversity of macrofaunal assemblages, especially during the NE monsoon season (Fig. 5.8). The macrofaunal diversity brings to light specific benthos, sensitive to climatic changes and the impact of mining.

5.22.2 Multivariate analysis

Multivariate analysis is done to find out if biological assemblages respond to disturbances by small, but consistent changes in the relative abundance of species. These are characterized by comparisons of two or more samples on the extent to which these samples share particular species, at comparable levels of abundance. Cluster analysis including MDS and similarity index, dominance plot, and BIO-ENV have also been done. All these techniques are found on similarity coefficients to facilitate clustering and ordination.

5.22.2.1 Cluster analysis (Bray–Curtis similarity)

The similarities between various sampling stations concerning macrofaunal assemblages in all the seasons are calculated to support the

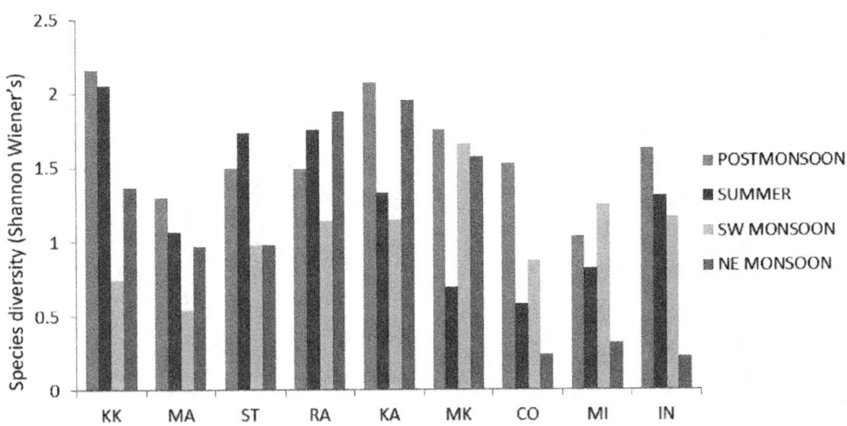

Figure 5.8 Species diversity (Shannon index) along the study area (Macrofauna).

similarity index along with the MDS plot. Figs. 5.9 and 5.10 highlight the results of hierarchical clustering and MDS ordination on macrofaunal abundance in the nine stations during the four seasons. Cluster analysis is carried out to find the similarities between different groups. The common clustering method is the hierarchical agglomerative

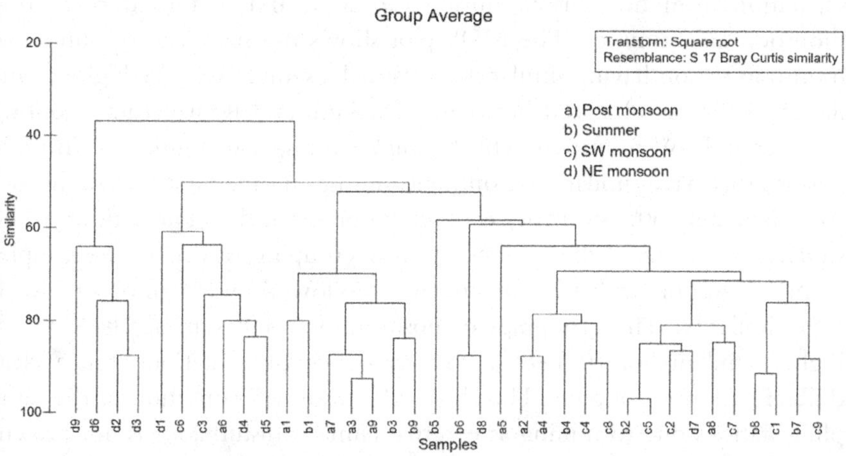

Figure 5.9 Bray curtis similarity in all the stations (Macrofauna).

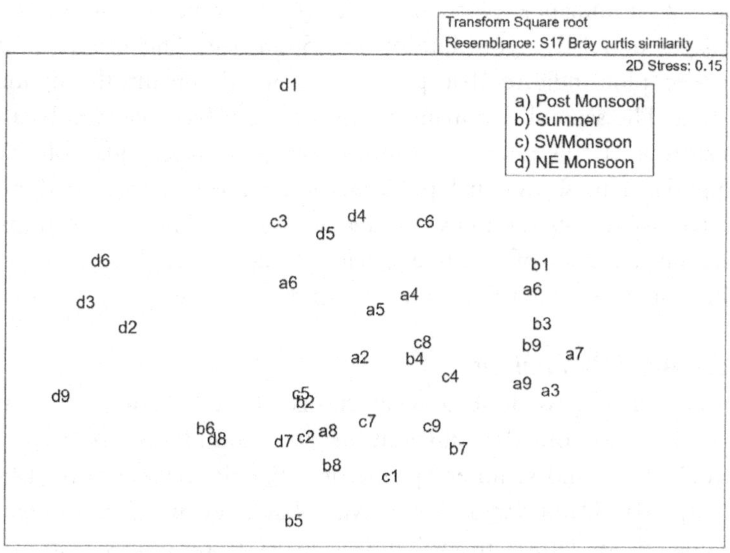

Figure 5.10 MDS plot describing the similarity among stations (Macrofauna).

clustering method, the results of which are denoted in a tree diagram or dendrogram. The X-axis of the dendrogram denotes the full set of samples and the Y-axis represents the similarity level at which the samples are fused. The Bray–Curtis coefficient (Bray & Curtis, 1957) with square root transformation has been used to produce the dendrogram.

The Bray–Curtis similarity and MDS plot show that there is a distinct separation from the stations during monsoon and stations during postmonsoon and summer. The MDS plot shows the significant grouping of monsoon season having similar macrofaunal assemblages. The highest similarity of 98% is observed between Manakudi and Kadiapattinam during summer and SW monsoon. This highlights the species abundance in these stations that have similar environments during summer and SW monsoon. Approximately 60% similarity is observed between different stations in the study area during all the seasons. A clear grouping is seen in the samples of NE monsoon with 62% of similarity, followed by SW monsoon with 60% similarity. The groupings of postmonsoon and summer have much higher similarity of 70%–80% concerning sampling stations, but a clear differentiation is not seen. This shows the disturbed condition of the sampling stations due to mining as external climatic disturbance is not present during postmonsoon and summer.

5.22.2.2 K-dominance

Multiple K-dominance plots are used to differentiate the macrofaunal assemblages to relative contribution to abundance. The results of the K-dominance plot indicate that polychaetes are dominant throughout the study area. The second dominant group is the bivalves followed by Isopoda and Gastropoda. The next dominant group is Decapoda followed by Amphipoda. This shows that polychaetes are more diverse as they adapt themselves to climatic and environmental changes. The groups amphipoda and decapoda consist of sensitive species. A slight change in either climate or habitat alters their distribution in the study area (Fig. 5.11).

5.22.2.3 BIO-ENV analysis

BIO-ENV analysis is done to characterize the effect of more than one ecological factor on the distribution and abundance of macrofauna. Physicochemical and sediment properties alter the distribution of benthic species (Joydas Damodaran, 2009; Manoharan et al., 2011). The BIO-ENV analysis of faunal macrobenthos in postmonsoon season indicates that total organic matter (TOM) and distribution of placer minerals affect

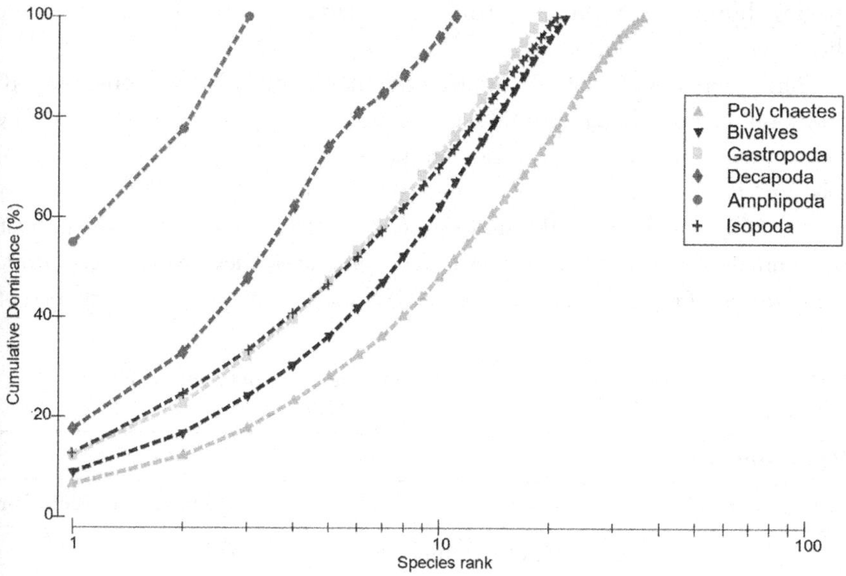

Figure 5.11 K-dominance plot discriminating the macrofaunal assemblages.

the abundance of the macrofauna ($r = 0.951$). This is also influenced by temperature ($r = 0.948$) and pH ($r = 0.944$). The influence of organic matter, pH, and temperature forms the biogenic support and the influence of placer minerals supports the habitat of the macrofauna. However, mining harms the abundance of the fauna.

In summer, total organic matter and sand have the most influence ($r = 0.715$), followed by temperature and placer minerals ($r = 0.704$), DO and placer minerals ($r = 0.695$), and pH and placer minerals ($r = 0.649$). This shows the similar influencing factors of the postmonsoon season. Sand and placer minerals support their habitat, and organic matter, temperature, and pH offer biogenic support. DO also alters their abundance as oxygen is reduced in summer. In aquatic environments, the temperature is inversely proportional to DO. Hence increased DO sustains their habitat during summer.

During the SW monsoon season, runoff from various sources affects faunal abundance. Nutrients from the land reach the coast through rivers and streams and this also affects faunal diversity. The major factors controlling their abundance include DO, TOM, sand, and silt ($r = 0.629$) followed by TOM, sand, silt, and clay ($r = 0.611$), then salinity, TOM, sand, and silt ($r = 0.602$), and DO, TOM, sand, and silt ($r = 0.591$). The impact of biogenic factors like DO, salinity, and TOC also determines the species abundance. Sediments from the estuaries and streams also influence the diversity of the fauna.

Similar environmental conditions exist in the study area during the NE monsoon season. Therefore similar results are observed in this season also. The factors influencing the most include temperature, TOM, and clay ($r = 0.594$) followed by TOM, silt, and clay ($r = 0.592$), then salinity, TOM, and clay ($r = 0.588$), and DO, TOM, and clay ($r = 0.588$). However, placer mineral concentration does not affect monsoonal sediments as they are very less during these seasons (Table 5.5). The BIO-ENV results show that the macrofauna

Table 5.5 Bio-Env analysis of macrofauna along the study area with respective seasons.

Postmonsoon		
No. of variables	**Parameters**	**Spearman correlation**
2	TOM, placer minerals	0.951
1	Temperature	0.948
1	Placer minerals	0.945
2	pH, placer minerals	0.944
Summer		
No. of variables	**Parameters**	**Spearman correlation**
2	TOM, sand	0.715
2	Temperature, placer minerals	0.704
2	Placer minerals, DO	0.695
2	pH, placer minerals	0.694
SW mosoon		
No. of variables	**Parameters**	**Spearman correlation**
3	Temperature, TOM, clay	0.594
3	TOM, silt, clay	0.592
3	Salinity, TOM, clay	0.588
3	DO, TOM, clay	0.588
NE monsoon		
No. of variables	**Parameters**	**Spearman correlation**
4	DO, TOM, sand, silt	0.629
4	TOM, sand, silt, clay	0.611
4	Salinity, TOM, sand, silt	0.602
4	DO, TOM, sand, clay	0.591

are much influenced by biogenic factors like DO and TOM, followed by physical factors like temperature, salinity, pH, and habitat bases like placer minerals, sand, silt, and clay. Any changes in these factors have adverse effects on the diversity and abundance of the macrofaunal community in the study area.

5.23 Meiofaunal distribution

The availability and abundance of group-wise and season-wise meiofaunal assemblages in the intertidal sediments along the study area are tabulated in Tables 5.6 and 5.7. The current study highlights the benthic community structure of the meiofauna and its habitat response toward the mining of placer minerals along the beaches of the study area. The numerical abundance of polychaetes during the postmonsoon season is high in Kanyakumari with an average and SD of 11.41 ± 41.51 and low in the Inayam coast (0 No/sq.m). But, their abundance is noticed more significantly in the nonmining areas such as Kanyakumari, Sanguthurai, Rajakkamangalam Thurai, and Colachel coasts. During summer, they are high in the Kadiapattinam coast (12.04 ± 20.62) and low in the Manavalakurichi coast (0.93 ± 4.78). Polychaetes are dominant in summer. Mining activity in the Manavalakurichi coast might have affected the distribution of polychaetes in the area. Their abundance during the SW monsoon is high in the Inayam coast (13.37 ± 19.23) and low in the Sanguthurai coast (0.83 ± 6.12). Their abundance during the NE monsoon is high in the Midalam coast (504 No/sq.m) and low in the Kanyakumari coast (7.94 ± 23.02). However, their abundance does not match significantly in the mining areas. The increasing trend of species diversity may be due to the impact of monsoonal disturbances. The abundance of oligochaetes is high in the Colachel coast (42.69 ± 38.97) and low in the Kadiapattinam coast during the postmonsoon season (1.69 ± 3.03). During summer, the abundance of oligochaetes is high in the Kanyakumari coast (39.50 ± 37.92) and low in the Manavalakurichi coast (0.75 ± 3). In SW monsoon, the abundance of oligochaetes is high in the Inayam coast (110.38 ± 56.50) and low in the Manavalakurichi coast (1.13 ± 4.50), and during NE monsoon, the abundance of oligochaetes is high in the Inayam coast (60.13 ± 63.58) and low in the Manavalakurichi coast (3.69 ± 10.09) as these coasts are engaged in mining. This shows a similar trend with the polychaetes in other mining and nonmining areas.

Table 5.6 Availability of meiofaunal assemblages in the intertidal sediments along the study area.

Meiofauna	KK			MA			ST			RA			KA			MK			CO			MI			IN		
POST MONSOON																											
	HT	MT	LT	HT	MT	LT	HT	MT	LT	HT	MT	LT	HT	MT	LT	HT	MT	LT	HT	MT	LT	HT	MT	LT	HT	MT	LT
Polychaeta	*		*		*		*	*		*		*		*	*	*		*	*	*	*	*	*	*			
Oligochaetes	*	*	*	*	*	*	*	*	*	*	*	*	*	*	*		*		*	*	*	*	*	*	*	*	
Foraminifera	*	*	*	*	*	*	*	*	*	*	*	*	*	*	*	*	*	*	*	*	*	*	*	*	*	*	*
Nemetoda	*	*	*	*	*	*	*	*	*	*	*	*	*	*	*		*	*	*	*	*	*	*	*	*	*	
Copepoda	*	*	*	*	*	*	*	*	*	*	*	*	*	*	*	*	*		*	*	*	*	*	*	*	*	
Halacaroidea		*	*	*			*				*	*	*		*			*			*	*	*	*		*	*
Ostracoda	*		*		*		*			*		*	*											*			
Cladocera	*																										
Amphipoda	*															*											
SUMMER																											
Meiofauna	KK			MA			ST			RA			KA			MK			CO			MI			IN		
	HT	MT	LT	HT	MT	LT	HT	MT	LT	HT	MT	LT	HT	MT	LT	HT	MT	LT	HT	MT	LT	HT	MT	LT	HT	MT	LT
Polychaeta	*	*	*	*	*	*	*	*	*	*		*	*	*	*	*	*		*	*	*	*	*	*	*	*	*
Oligochaetes	*	*	*	*	*	*	*	*	*	*	*	*	*	*	*	*			*	*	*	*	*	*	*	*	*
Foraminifera	*	*	*	*	*	*	*	*	*	*	*	*	*	*	*	*	*	*	*	*	*	*	*	*	*	*	*
Nemetoda	*	*	*	*	*	*	*	*	*	*	*	*	*	*	*		*		*	*	*	*	*	*	*	*	*
Copepoda	*	*	*	*		*	*		*				*	*	*	*	*		*	*	*	*	*	*	*	*	*
Halacaroidea	*	*	*	*		*		*	*		*	*	*	*		*	*		*	*	*	*	*	*	*	*	*
Ostracoda			*						*		*			*			*					*					
Cladocera	*									*			*							*		*	*	*	*		*
Amphipoda																											
SW MONSOON																											
Meiofauna	KK			MA			ST			RA			KA			MK			CO			MI			IN		
	HT	MT	LT	HT	MT	LT	HT	MT	LT	HT	MT	LT	HT	MT	LT	HT	MT	LT	HT	MT	LT	HT	MT	LT	HT	MT	LT
Polychaeta	*	*	*	*	*	*		*	*	*	*	*	*	*	*	*		*	*	*	*	*	*	*	*	*	*
Oligochaetes	*	*	*	*	*	*		*	*	*	*	*	*	*	*		*	*	*	*	*	*	*	*	*	*	*
Foraminifera								*		*	*	*		*	*	*			*	*	*	*	*	*	*	*	*
Nemetoda	*	*	*	*	*		*	*	*	*	*	*	*	*		*	*		*	*	*	*	*	*	*	*	*
Copepoda	*	*	*		*	*			*	*	*	*		*			*		*	*	*	*	*	*	*	*	*
Halacaroidea	*		*		*	*			*	*	*		*	*				*	*	*		*	*	*			
Ostracoda	*	*	*	*	*				*	*	*	*			*		*	*			*	*					
Cladocera		*	*		*				*								*	*									
Amphipoda																*											
NE MONSOON																											
Meiofauna	KK			MA			ST			RA			KA			MK			CO			MI			IN		
	HT	MT	LT	HT	MT	LT	HT	MT	LT	HT	MT	LT	HT	MT	LT	HT	MT	LT	HT	MT	LT	HT	MT	LT	HT	MT	LT
Polychaeta	*		*	*		*		*	*	*	*	*	*	*	*	*		*	*	*	*	*	*	*			*
Oligochaetes	*			*	*	*	*	*	*	*	*	*	*	*	*		*		*	*	*	*	*	*		*	*
Foraminifera	*	*	*			*	*			*																	
Nemetoda	*	*		*	*	*	*		*	*	*	*	*	*	*		*		*	*	*	*	*	*	*	*	*
Copepoda	*	*	*	*	*	*	*		*	*	*	*	*	*	*	*	*	*	*	*	*	*	*	*	*	*	*
Halacaroidea	*	*	*	*	*	*		*	*	*		*	*	*	*		*		*	*	*	*		*	*	*	*
Ostracoda													*			*					*						*
Cladocera																*		*									
Amphipoda																											

* Species availability.

KK, Kanayakumari; *MA*, Manakudi; *ST*, Sanguthurai; *RA*, Rajakkamangalam Thurai; *KA*, Kadiapattinam; *MK*, Manavalakurichi; *CO*, Colachel; *MI*, Midalam; *IN*, Inayam; *HT*, high tide; *MT*, mid tide; *LT*, low tide.

The abundance of foraminifera is high in the Kanyakumari coast (66.09 ± 73.23) and low in the Inayam coast during the postmonsoon season (2.73 ± 10.72). Foraminifera is the most dominant group during the postmonsoon season and shows a similar trend as that of polychaetes regarding mining and nonmining areas. During summer, the abundance is

Table 5.7 Numerical abundance of meiofaunal assemblages in the intertidal sediments along the study area (SD ± average).

POST MONSOON															
Macrofauna	KK			MA			ST			RA			KA		
	HT	MT	LT	HT	MT	LT	HT	MT	LT	HT	MT	LT	HT	MT	LT
Polychaeta	11.41±41.51	0	1.02±5.24	0	1.69±5.34	0	4.56±14.40	4.24±15.14	0.94±4.89	3.56±12.70	1.11±8.16	0.54±3.95	0	2.63±7.58	1.54±3.50
Oligochaeta	14.69±40.23	18.63±51.77	26.94±14.84	2.50±5.40	7.38±9.21	1.81±3.90	24.63±22.63	17±26.11	8.06±12.57	23.31±24.20	16.63±19.66	8.81±8.40	28.94±52.12	3.94±8.57	1.69±3.03
Foraminifera	66.09±73.23	13.19±29.20	28.38±16.54	3.42±5.79	33.69±8.51	1.44±3.38	35.25±18.09	16.66±23.88	2.67±4.22	22.30±22.58	22.53±27.52	4.48±4.56	15.59±31.17	6.86±10.30	4.55±4.49
Nemetoda	11±36.48	29.55±65.97	14.73±32.77	2.27±5.18	8.45±14.49	2.45±4.23	7.73±17.19	2.91±9.65	2.55±4.37	11.36±19.47	8.55±19.35	1.73±3.85	11.73±38.89	1.82±6.03	4.18±4.83
Copepoda	10.57±39.55	11.57±29.42	34.43±18.96	3.57±5.92	26.79±24.34	1.93±3.83	15.43±21.57	13.21±21.76	0.64±2.41	3±11.22	12.36±17.36	4.43±4.64	9.93±19.75	5.29±10.55	0.64±2.41
Halacaroidea	0	4.56±19.33	22.22±25.82	2.89±5.59	0	0	4.94±14.42	0	0	0	6.28±18.28	0.50±2.12	2.89±12.26	0	5.11±14.89
Ostracoda	15.63±44.19	0	5.88±16.62	0	13.75±25.51	0	6±16.97	0	0	6±16.97	0	1.13±3.18	5±14.14	0	0
Cladocera	9.50±13.44	0	0	0	0	0	0	0	0	0	0	0	0	0	0
Amphipoda	33.33±29.96	0	0	0	0	0	0	0	0	0	0	0	0	0	0

MK			CO			MI			IN		
HT	MT	LT	HT	MT	LT	HT	MT	LT	HT	MT	LT
0.06±0.30	0	0.02±0.14	5.57±28.69	11.13±39.72	1.41±10.34	5.78±13.11	3.81±12.33	0.46±1.95	0	7.09±18.62	2.72±8.55
0	9.44±20.30	0	23.06±35.40	42.69±38.97	3.69±4.48	19.25±22.72	26.19±30.97	3.88±4.13	4.44±17.75	6.94±14.93	0
8.28±17.41	22.20±37.22	0.61±2.38	25.97±32.31	56.50±39.87	5.36±4.65	30.44±21.57	24.52±22.95	4.45±4.65	2.73±10.72	17.92±23.04	6.19±10.98
0	19.18±26.64	1.82±4.05	15.82±28.48	15±33.39	2.55±4.37	8.27±18.47	7±15.58	1.55±3.45	5±16.58	25.64±29.51	0
2.71±10.16	7.14±18.17	0	10.57±21.03	17.93±35.66	0.64±2.41	18.36±22.18	27.14±28.26	5.14±4.66	10.57±21.03	7.86±20.01	0
0	0	2.89±12.26	0	0	0.50±2.12	9.78±18.86	12.17±23.45	0.50±2.12	0	2.67±11.31	0.50±2.12
0	0	0	0	0	0	0	0	1.13±3.18	0	0	0
0	0	0	0	0	0	0	0	0	0	0	0
54±46.78	0	0	0	0	0	0	0	0	0	0	0

SUMMER															
Macrofauna	KK			MA			ST			RA			KA		
	HT	MT	LT	HT	MT	LT	HT	MT	LT	HT	MT	LT	HT	MT	LT
Polychaeta	11.50±24.85	0.94±6.94	0.87±4.51	6.98±15.89	3.28±7.99	0.61±2.19	11.31±19.41	4.56±9.15	0.98±2.84	8.06±25.51	0	1.07±5.53	12.04±20.62	7.59±17.46	9.59±21.74
Oligochaeta	39.50±37.92	22.81±36.71	14.44±20.39	27.75±29.51	23.25±31.72	15±12.76	16.38±21.54	21.81±27.33	3.31±7.72	12.44±21.45	26.19±28.74	4.69±13.27	15.50±15.45	27.94±23.83	8.31±6.70
Foraminifera	5.89±20.85	7.55±14.82	4.84±9.36	9.25±19.70	6.08±13.99	1.67±5.50	8.42±14.34	7.88±16.27	2.94±6.55	11.25±20.95	7.08±15.88	3.31±9.28	8.22±18.93	4.45±14.42	3.13±6.15
Nemetoda	9.18±20.70	7±16.40	7.27±12.59	16.27±23.16	11.09±19.09	9.82±15.07	11.45±16.69	14.55±17.82	2±6.63	7.45±16.68	2.82±9.35	2.64±5.90	16.09±24.68	10.64±18.50	5.55±6.59
Copepoda	50.71±23.50	36.50±17.53	18.93±9.70	6.36±16.26	0	3.71±5.04	28.21±22.52	0	5.07±8.94	0	0	0	33±25.61	10.43±22.31	2.07±7.75
Halacaroidea	18.33±22.59	17.89±19.12	12.44±12.09	2±8.49	0	3.11±6.34	0	4.44±13.10	0.56±2.36	0	2.39±10.14	1.22±5.19	12.17±20.42	1.89±8.01	0
Ostracoda	0	0	1.75±4.95	0	0	0	0	0	2±5.66	0	3.38±9.55	0	0	4.25±8.50	0
Cladocera	17±24.04	0	0	0	0	0	0	0	0	22±31.11	0	0	21.50±30.41	0	0
Amphipoda	0	0	0	0	0	0	0	0	0	0	0	0	0	0	0

MK			CO			MI			IN		
HT	MT	LT	HT	MT	LT	HT	MT	LT	HT	MT	LT
0.93±4.78	1.50±5.51	0	6.39±26.63	5.44±17.48	1.70±3.64	1.67±8.60	16.56±35.24	5.07±10.85	10.15±29.15	1.85±9.77	1.91±6.84
0.75±3	0	0	18.50±20.37	24.38±27.34	14.69±15.25	16.19±25.39	27.56±23	12.31±9	18.38±14.15	8.75±15.97	7.38±9.01
2.91±9.18	2.73±6.06	0.95±3.76	8.27±18.54	0.61±2.04	0.78±2.16	0.94±2.43	0.39±1.38	1.05±2.10	0.89±2.08	0.58±1.73	0.33±1.31
0	3.09±5.52	0	18.36±26.58	5.36±5.68	1.55±1.92	1.91±3.30	2±2.24	1.09±1.58	1.91±2.34	0.36±1.21	0.82±1.83
1.71±6.41	3.79±7.75	0	21.21±24.10	0.36±0.50	0.50±0.52	0.50±0.52	0.29±0.47	0.57±0.51	0.36±0.50	0.14±0.36	0.07±0.27
3.83±11.38	4.50±7.49	0	6.94±17	0.11±0.32	0.28±0.46	0.11±0.32	0.11±0.32	0.06±0.24	0.06±0.24	0.06±0.24	0.11±0.32
0	4.13±7.68	0	0	0	0	0.13±0.35	0	0	0	0	0
0	0	0	0	0.50±0.71	0	0.50±0.71	0.50±0.71	0.50±0.71	0.50±0.71	0	0.50±0.71
0	0	0	0	0	0	0	0	0	0	0	0

SW MONSOON															
Macrofauna	KK			MA			ST			RA			KA		
	HT	MT	LT	HT	MT	LT	HT	MT	LT	HT	MT	LT	HT	MT	LT
Polychaeta	24.59±52.14	5.48±14.23	5.33±10.16	0.63±4.63	5.37±15.59	0.41±1.72	0	0.83±6.12	4.69±10.00	9.02±18.05	1.91±9.84	2.83±8.14	2.63±13.55	4.11±13.36	0.80±4.11
Oligochaeta	14.63±19.61	88.44±63.52	10.38±10.89	4.56±12.57	61.19±30.51	3.69±3.09	0	6.94±15.15	15.81±8.40	36.63±3.86	20.56±18.19	12.56±7.75	20.44±28.43	29.44±19.24	2.00±5.51
Foraminifera	0	0	0	0	0	0	0	5.86±14.87		8.38±16.89	12.92±20.14	3.92±7.95	0	6.61±14.18	3.81±7.35
Nemetoda	7.18±16.02	36.45±62.59	5.45±9.42	2.45±8.14	27.36±38.01	0	2.64±8.74	11.64±20.01	8.91±10.58	11.09±19.03	4.27±14.17	1.64±5.43	10.64±23.69	10.55±18.13	0
Copepoda	10.36±17.18	82.71±64.63	13.14±10.57		5.07±18.98	0.71±2.67	0	0	3.00±7.67	7.93±15.77	2.57±9.62	8.36±8.86	0	6.50±16.55	0
Halacaroidea	2.17±9.19		2.28±6.65		12.67±29.19	0.39±1.65	0	0	3.11±7.16	2.44±10.37	2.28±9.66		3.06±12.96	4.72±14.12	0
Ostracoda	5.25±14.85	15.50±43.84	8.63±11.99	4.00±11.31	9.63±27.22	0	0	0	5.63±10.42	10.00±18.52	4.88±13.79	1.88±5.30	0	0	6.25±8.70
Cladocera	0	66.00±93.34	12.00±16.97		36.50±51.62	0	0	0	23.50±2.12	0	0	0	0	0	0
Amphipoda	0	0	0	0	0	0	0	0	0	0	0	0	0	0	0

MK			CO			MI			IN		
HT	MT	LT	HT	MT	LT	HT	MT	LT	HT	MT	LT
3.31±11.86	0	0.48±3.54	2.59±10.80	1.59±5.69	3.43±9.02	3.37±12.05	11.11±29.08	3.33±8.78	6.78±19.55	47.52±77.60	13.37±19.23
0	8.75±16.01	1.13±4.50	33.06±26.93	19.69±5.80	26.69±1.92	21.75±22.56	51.50±41.32	3.31±9.05	36.00±25.82	110.38±56.50	30.19±15.94
1.92±8.74	0	0	7.98±17.73	6.98±11.36	9.80±12.98	13.23±21.58	20.31±35.57	8.06±12.68	13.11±25.71	43.70±71.07	12.84±20.35
7.09±15.80	17.82±11.85	0	15.55±21.62	5.91±10.14	8.73±12.22	8.45±18.84	23.82±40.81	8.55±11.98	26.64±31.08	83.73±80.81	24.55±24.05
0	4.93±12.71	0	11.29±22.55	10.64±12.91	1.50±5.61	18.43±22.23	5.79±21.65	3.43±8.75	13.71±27.43	20.64±52.49	5.71±14.55
0	0	1.17±4.95	2.39±10.14	1.33±5.66	0	5.06±14.72	18.72±36.08	9.72±12.69	0	0	0
0	3.63±10.25	4.00±7.43	0	0	3.63±10.25	5.88±16.62	0	0	0	0	0
0	35.50±3.54	18.00±1.41	0	0	0	0	0	0	0	0	0
13.00±22.52	0	0	0	0	0	0	0	0	0	0	0

(Continued)

Table 5.7 (Continued)

NE MONSOON															
Macrofauna	KK			MA			ST			RA			KA		
	HT	MT	LT	HT	MT	LT	HT	MT	LT	HT	MT	LT	HT	MT	LT
Polychaeta	0.70±5.17	0	0.72±3.05	0.39±2.86	0	0.48±2.04	0	1.76±7.40	0.57±2.99	1.80±7.57	1.43±7.39	0.69±3.56	1.70±8.82	0.93±4.91	1.72±5.08
Oligochaeta	7±15.19	0	0	7.63±10.50	19.44±30.10	0.31±1.25	1.38±5.50	1.94±7.75	3.44±7.47	3.25±8.95	15.63±18.74	2.56±5.59	16.38±22.47	8.13±12.78	8.06±7.82
Foraminifera	0.70±5.63	0.91±7.25	0.28±2.25	0	0	0.55±1.76	0.33±2.63	0	0	2.77±9.69	0	0	0	0	0
Nemetoda	8.91±29.55	12.18±27.28	0	5.18±11.58	15.91±27.31	1.55±2.21	2.91±9.65	0	3.64±8.10	6.18±13.82	3.18±10.55	1.±3.32	16.64±23.79	14.64±20.89	7±8.31
Copepoda	2±4.57	9.14±17.50	0.93±1.90	1.50±3.08	4.57±10.16	0.43±1.09	1.50±4.38	0	0.86±2.48	0.14±0.36	1.50±3.86	1.±2.57	3.14±7	3.07±7.19	1.14±2.18
Halacaroidea	2.44±4.84	6.22±10.42	0.67±1.57	1.17±2.96	4.11±9.98	0.44±1.10	0	1.33±3.56	0.61±1.46	0.11±0.47	0	0.61±1.85	2.39±5.54	2.50±5.82	0.72±1.87
Ostracoda	0	0	0	0	0	0	0	0	0	0	0	0	50±6.25	0	0
Cladocera	0	0	0	0	0	0	0	0	0	0	0	0	0	0	0
Amphipoda	0	0	0	0	0	0	0	0	0	0	0	0	0	0	0

MK			CO			MI			IN		
HT	MT	LT	HT	MT	LT	HT	MT	LT	HT	MT	LT
1.44±7.44	0	0.30±2.18	2.35±9.85	2.35±6.73	2.13±6.82	0.76±5.58	7.94±23.02	1.89±6.07	0	0	4.76±12.64
0	3.69±10.09	0	4.25±11.66	4.25±7.66	9.69±11.57	11.63±15.84	30.25±35.65	7±11.11	0	60.13±63.58	23.13±16.92
0	0	0	0	0	0	0	0	0	0	0	0
0	5.45±12.21	0	5.73±12.79	6.45±9.06	4.36±9.80	14.82±20.61	30.45±35.10	1.91±6.33	8.91±19.84	24.64±54.85	17.36±20.09
3.14±5.86	2.36±6.25	0.79±2.08	2.64±5.68	1.07±2.27	1.93±4.50	3.14±7.46	4.64±8.94	1.64±4.22	5.07±8.97	9±14.52	2.29±4.51
0	1.28±3.72	0	2.39±5.05	1.17±2.77	1.22±2.92	2.33±5.76	0	1.22±2.46	3.39±7.18	6.28±12.32	2.44±6.49
34±4.25	0	0	0	0	21±2.63	0	0	0	0	0	31±3.88
17.50±2.12	0	5.50±4.95	0	0	0	0	0	0	0	0	0
0	0	0	0	0	0	0	0	0	0	0	0

KK, Kanyakumari; *MA*, Manakudi; *ST*, Sanguthurai; *RA*, Rajakkamangalam Thurai; *KA*, Kadiapattin *MK*, Manavalakurichi; *CO*, Colachel; *MI*, Midalam; *IN*, Inayam; *HT*, high tide; *MT*, mid tide; *LT*, low tide.

high in the Kanyakumari coast (7.55 ± 14.82) and low in the Colachel coast (0.61 ± 2.04). In SW monsoon, the abundance is high in the Inayam coast (43.70 ± 71.07) and low in the Manakudi coast (0). During NE monsoon, the abundance is high in the Rajakkamangalam Thurai coast (2.77 ± 9.69) and low in the Manavalakurichi and Midalam coasts (0 No/sq.m) as these coasts are engaged in mining during postmonsoon and summer seasons followed by the monsoonal disturbance in SW and NE monsoon seasons.

The abundance of Nematoda is high in the Kanyakumari coast (29.55 ± 65.97) and low in the Sanguthurai coast (0.94 ± 4.89) during the postmonsoon season, which is due to increased human intervention. During summer, the abundance is high in the Manakudi coast (6.36 ± 16.26) and low in the Inayam coast (0.36 ± 1.21), which is due to erosion as a result of high wave dynamics. In the SW monsoon, the abundance is high in the Inayam coast (83.73 ± 80.81) and low in the Rajakkamangalam Thurai coast (1.64 ± 5.43). There is no major relationship between Nematoda and the mining areas during the SW monsoon season. During the NE monsoon, the abundance is high in the Midalam coast (30.45 ± 35.10) and low in the Sanguthurai coast (2.91 ± 9.65). This trend might be due to the high human intervention.

The abundance of Copepoda is high in the Midalam coast (27.14 ± 28.26) and low in the Manavalakurichi coast (2.71 ± 10.16) during the postmonsoon season. During summer, the abundance is high in the Kanyakumari coast (50.71 ± 23.50) and low abundance (0 No/sq.m) is noticed in the Rajakamangalam Thurai coast followed by Inayam, Midalam, and Manavalakurichi. In the SW monsoon, the abundance is high in the Kanyakumari coast (82.71 ± 64.63) whereas the low abundance (0.71 ± 2.67) is seen in the Manakudi coast. During the NE monsoon, abundance is high in the Inayam coast (9 ± 14.52) and low in the Sanguthurai coast (0.86 ± 2.48). This may be due to the impact of intense mining, monsoonal disturbance, and the disturbance of natural habitat in the area.

The abundance of Halacaroidea is high in the Midalam coast (12.17 ± 23.45) and low in the Colachel coast (0.50 ± 2.12) during the postmonsoon season. Most of the mining areas have low abundance which may be due to beach placer mining in the specific mining stations. During summer, the abundance is high in the Kanyakumari coast (18.33 ± 22.59) and low in the Inayam coast (0.06 ± 0.24). There is no significant relationship between species occurrence and mining activities. In SW monsoon, the abundance is high in the Midalam coast (18.72 ± 36.08) and low in the Inayam coast (0 No/sq.m). The Manavalakurichi coast has low abundance which might be due to the after-effects of mining during summer. During NE monsoon, the abundance is high in the Inayam coast (6.28 ± 12.32) whereas a low abundance of 0.11 ± 0.47 is observed in the Rajakkamangalam Thurai coast. There is no significant relationship between Halacaroidea and mining. The varied distribution is entirely dependent on climatic disturbances.

The abundance of Ostracoda is high in the Kanyakumari coast (15.63 ± 44.19) and low in the Manavalakurichi, Colachel, and Inayam coasts (0 No/sq.m) during the postmonsoon season. During summer, the abundance is high in the Kadiapattinam coast (4.25 ± 8.50) and low in the Manakudi, Colachel, and Inayam coasts (0 No/sq.m). In the SW monsoon, the abundance is high in the Kanyakumari coast (15.50 ± 43.84) and low in the Inayam coast (0 No/sq.m). During the NE monsoon, the abundance is high in the Manavalakurichi coast (34 ± 4.25) whereas the low abundance of 0 No/sq.m is found in the Kanyakumari, Manakudi, Sanguthurai, Rajakkamangalam Thurai, and Midalam coasts. Hence, there is no direct relationship between the abundance of Ostarcoda and the mining areas. This shows that Ostracoda is more sensitive to the location of the stations as low values are grouped on the western coast of the study area.

The abundance of Cladocera is high in the Kanyakumari coast (9.50 ± 13.44) whereas a low abundance of 0 No/sq.m is seen in all other stations during the postmonsoon season. During summer, the abundance is high in the Rajakamangalam Thurai coast (22 ± 31.11) and low in the Manakudi, Sanguthurai, and Midalam coasts (0.13 ± 0.35). In SW monsoon, the abundance is high in the Kanyakumari coast (66.00 ± 93.34) and low in Rajakkamangalam Thurai, Kadiapattinam, Colachel, Midalam, and Inayam coasts (0 No/sq.m). During NE monsoon, the abundance is high only in the Manavalakurichi coast (17.50 ± 2.12) whereas a low abundance of 0 No/sq.m is noticed in all other stations. This indicates the lack of good community structure to establish or the absence of natural habitat in the study area. *Daphnia pulex and Graptoleberis testudinaria* species are found during the entire period of study.

The distribution of Amphipoda indicates the absence of a natural habitat during the postmonsoon season in all the stations except Kanyakumari (33.33 ± 29.96) and Manavalakurichi (54 ± 46.78). The distribution of Amphipoda is nil in all the other stations which shows their sensitivity to changing climate or the natural absence of habitat in summer in all the stations. The distribution of Amphipoda in the SW monsoon season shows the natural absence of habitat in all stations except Manavalakurichi (13.00 ± 22.52). Amphipoda is also absent in the study area during the NE monsoon. The reason for their varied abundance is their natural habitat distribution in the selective beaches. Total meiofaunal diversity at their species level in the study area is recorded in (Table 5.8).

5.23.1 Univariate analysis

The univariate methods include a set of species diversity indices. The below-mentioned diversity indices have been performed for the estimation of the meiofaunal community structure.

5.23.1.1 Species richness (Margalef's index)

Margalef's index shows huge variation among the species in selective stations of the study area. High species richness is seen in Manavalakurichi (1.8) during the SW monsoon and a low value is seen in Inayam (0.74) during the SW monsoon season. The high species richness during SW monsoon is due to the availability of good habitat conditions like nutrient availability, sediment cycling, well-sorted grains replenishing the oxygen

Table 5.8 Total meiofauna recorded in the study area.

I	Polychaeta
1	*Aktedrilus* sp.
2	*Autolytus* sp.
3	*Cossura longocirrata*
4	*Dendronereis aesfuarina*
5	*Euclymene* sp.
6	*Eunice tubifex*
7	*Exogone clavator*
8	*Exogone* sp.
9	*Glycera tesselata*
10	*Glycera unicornis*
11	*Hesionides arenarius*
12	*Hipponoa* sp.
13	*Hydroides elegans*
14	*Lumbrineris heteropoda*
15	*Lumbrineris polydesma*
16	*Maldane sarsi*
17	*Marionina* sp.
18	*Mesonerilla intermedia*
19	*Microphthalmus aberrans*
20	*Myrianida* sp.
21	*Namalycastis indica*
22	*Neanthes chingrighattensis*
23	*Nerilla antennata*
24	*Ophryotrocha gracilis*
25	*Ophryotrocha puerilis*
26	*Paranerilla* sp.
27	*Paranerilla limicola*
28	*Pedinosoma* sp.
29	*Pelagobia longicirrata*
30	*Phalacrophorus* sp.
31	*Phyllochaetopterus* sp.
32	*Phylo* sp.
33	*Pionosyllis* sp.
34	*Platynereis dumerilii*
35	*Platyneries* sp.
36	*Polydora capensis*
37	*Polydora kempi*
38	*Polydora* sp.
39	*Polyophthalmus* sp.
40	*Protodrilus helgolandicus*
41	*Protodrilus rubropharyngeus*
42	*Pygospio elegans*

(*Continued*)

Table 5.8 (Continued)

43	*Rhodine* sp.
44	*Saccocirrus minor*
45	*Scoloplos* spp.
46	*Sphaerodorum gracile*
47	*Sphaerosyllis* sp.
48	*Spionidae* sp.
49	*Syllides longocirrata*
50	*Syllis gracilis*
51	*Syllis* sp.
52	*Thalassodrilus firmus*
53	*Trilobodrilus* sp.
54	*Trypanosyllis* sp.
II	**Oligochaeta**
1	*Aeolosoma quaternarium*
2	*Aulophorus furcatus*
3	*Aulophorus vagus*
4	*Chaetogaster* sp.
5	*Dero digitata*
6	*Grania acanthochaeta*
7	*Lumbricillus* sp.
8	*Marionina* sp.
9	*Nais barbata*
10	*Paranais litoralis*
11	*Pristina leidyi*
12	*Pristina longiseta*
13	*Pristina proboscidea*
14	*Tubifex Lamarck*
15	*Tubifex tubifex*
16	*Tubificidae* sp.
III	**Foraminifera**
1	*Ammonia beccarii*
2	*Ammonia tepida*
3	*Amphistegina lessonii*
4	*Arilina ammonoides*
5	*Asterigerina carinata*
6	*Bolivina kuriani*
7	*Bolivina limbata*
8	*Bolivina ordinaria*
9	*Borelis schlumbergeri*
10	*Buccella frigida*
11	*Calcarina* sp.

(*Continued*)

Table 5.8 (Continued)

12	*Charophyta* sp.
13	*Chryalidinella dimorpha*
14	*Cibicidoides wuellerstorfi*
15	*Cribothalammina alba*
16	*Cribroelphidium gerthi*
17	*Cymbaloporetta squammosa*
18	*Discorbinella complanata*
19	*Rosalina flondana*
20	*Discorbinella* sp.
21	*Elphidium advenum*
22	*Elphidium craticulatum*
23	*Elphidium crispum*
24	*Elphidium discoidale*
25	*Elphidium fichtellianum*
26	*Elphidium hispidulum*
27	*Elphidium incertum*
28	*Elphidium macellum*
29	*Elphidium senseni*
30	*Elphidium texanum*
31	*Epistominella exigua*
32	*Fissurina lacunata*
33	*Globigerinoides sacculifer*
34	*Heterostegina depressa*
35	*Miliammina fusca*
36	*Nonion commune*
37	*Nonion elongatum*
38	*Nonionella hantk*
39	*Operculina ammonoides*
40	*Ordorsalis umbonatus*
41	*Osangularia* sp.
42	*Planorbulina distoma*
43	*Planorbulinella* sp.
44	*Polymorphina* sp.
45	*Quinqueloculina bosciana*
46	*Quinqueloculina cuvieriana*
47	*Quinqueloculina laevigata*
48	*Quinqueloculina lamarckiana*
49	*Quinqueloculina parkeri*
50	*Quinqueloculina pseudoreticulata*
51	*Quinqueloculina rhodiensis*
52	*Quinqueloculina schlumberger*
53	*Quinqueloculina Seminulum*
54	*Quinqueloculina tropicalis*

(*Continued*)

Table 5.8 (Continued)

55	*Robulus stephensoni*
56	*Schlumbergerina alveoliniformis*
57	*Spiroculina* sp.
58	*Spiroloculina excavata*
59	*Spirophthalmidium acutimargo*
60	*Triloculina oblonga*
61	*Triloculina terquemiana*
62	*Triloculina tricarinata*
63	*Unknown* sp. 1
64	*Unknown* sp. 2
IV	**Nemetoda**
1	*Daptonema oxycerca*
2	*Desmodora pontica*
3	*Echinotheristus* sp.
4	*Enoplus* sp.
5	*Eurystomina terricola*
6	*Helicotylenchus pseudorobustus*
7	*Metachromadora remanei*
8	*Metoncholaimus* sp.
9	*Paramonohystera* sp.
10	*Sabatieria lucia*
11	*Terschellingia longicaudata*
V	**Copepoda**
1	*Ameiropsis angulifera*
2	*Attheyella illinoisensis*
3	*Caligus macarovi*
4	*Canthocamptus staphylinus*
5	*Clausocalanus lividus*
6	*Harpacticoid copepod*
7	*Malacopsyllus fragilis*
8	*Metridia longa*
9	*Microsetella norvegica*
10	*Notodelphyid copepod*
11	*Orthocyclops modestus*
12	*Osphranticum labronectum*
13	*Parastenocaris* sp.
14	*Pseudocalanus mimus*
VI	**Halacaroidea**
1	*Acanthohalacarus*
2	*Acarochelopodia*

(*Continued*)

Table 5.8 (Continued)

3	*Ameronothrus oribatidae*
4	*Ameronothrus* sp.
5	*Copidognathus biodomus*
6	*Copidognathus lamellosus*
7	*Enterohalacarus*
8	*Halacarellus*
9	*Halacarellus procerus*
10	*Halacarellus* sp.
11	*Hydrachnoidea*
12	*Limnohalacarus* sp.
13	*Limnohalacarus* sp.
14	*Porolohmannella violacea*
15	*Porolohmannella violacea*
16	*Rhodacarus roseus*
17	*Rhombognathus papuensis*
18	*Soldanellonyx monardi*
VII	**Ostracoda**
1	*Bythopussella Warne*
2	*Caudites javana*
3	*Leptocythere pellucida*
4	*Macrocyprididae* sp.
5	*Paracypris wrothamensis*
6	*Polycope orbicularis*
7	*Tanella gracilis*
8	*Xylocythere vanharteni*
VIII	**Cladocera**
1	*Daphnia pulex*
2	*Graptoleberis testudinaria*
IX	**Amphipoda**
1	*Ampelisca* sp.
2	*Grandidierella* sp.
3	*Themisto gaudichaudii*

level, and improved sediment texture. Species richness is low in the Colachel, Midalam, and Inayam coasts during postmonsoon, NE, and SW monsoon seasons, respectively (Fig. 5.12). This might be due to climatic disturbances and high wave energy conditions on the coasts. Mining resulting in habitat disturbance in the postmonsoon season and erosion in

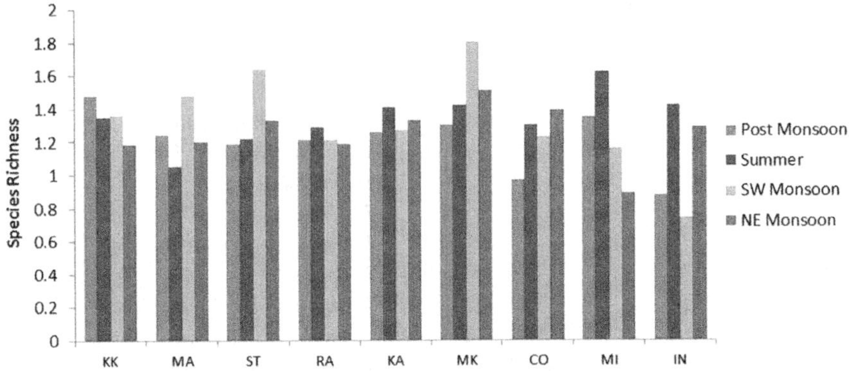

Figure 5.12 Species richness along the study area (Meiofauna).

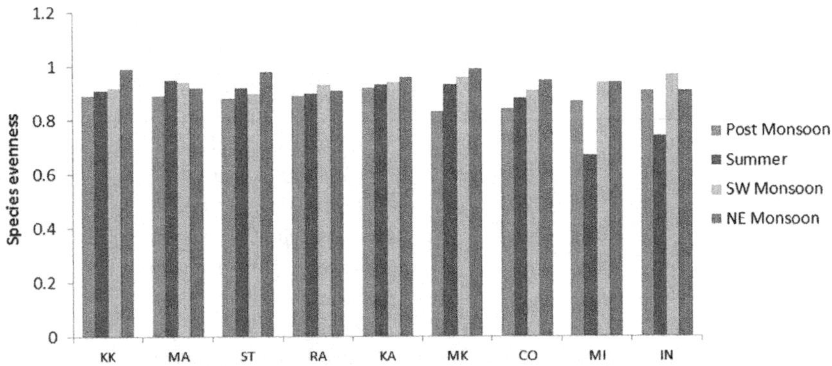

Figure 5.13 Species evenness along the study area (Meiofauna).

the monsoon in the Midalam coast may be the reason behind the low species richness. The erosional hazard is also high in the Inayam coast which has a low richness index during SW monsoon.

5.23.1.2 Evenness index (Pielou's index)

The evenness value of the study area is high in the coastal stretch from Kanyakumari to Kadiapattinam and has a similar evenness index in all the seasons ranging from 0.8 to 0.99 (Fig. 5.13). This indicates similar environmental conditions on these coasts. The Manavalakurichi coast shows an increasing trend from postmonsoon to the NE monsoon and a similar evenness value is observed in the Colachel coast. This indicates the effect of monsoon enriching the beach sediment through sediment cycling and nutrient recycling. A huge reduction in the evenness index is seen in the Midalam and Inayam coasts during summer, with values of 0.67 and 0.74, respectively. However, the condition comes back to normalcy in the monsoon season.

5.23.1.3 Species diversity index (Shannon index)

The species diversity index indicates an almost similar pattern ranging from 2 to 3 in all the seasons. High Shannon index is seen in the Manavalakurichi coast during SW and NE monsoons. A low index is observed in the coasts of Midalam and Inayam in the summer season. Similar values of the Shannon index during all the seasons show low disturbance. It might also mean that the meiofaunal community is not much affected by climatic conditions. An overall high index is observed during the SW monsoon season, revealing favorable habitat establishment as a result of high nutrient load and low anthropogenic influence like mining and other activities. The Kanyakumari and Manakudi coasts show a low-diversity index during the NE monsoon. This reveals the impact of erosion on the texture of the sediment as clay load will be high due to the overload from the Manakudi estuary. The Rajakkamangalam Thurai coast also shows a similar trend due to similar environmental conditions and the presence of estuary on this coast may influence the sediment texture during NE monsoon. The Kadiapattinam and Manavalakurichi coasts show low species diversity during the postmonsoon season due to the impact of mining during the season. This imbalance is retained during the SW and NE monsoon seasons through sediment cycling. A similar trend is detected in the Colachel coast due to sediment cycling. The high erosion in the Midalam and Inayam coasts affects the diversity of meiofauna especially during summer (Fig. 5.14).

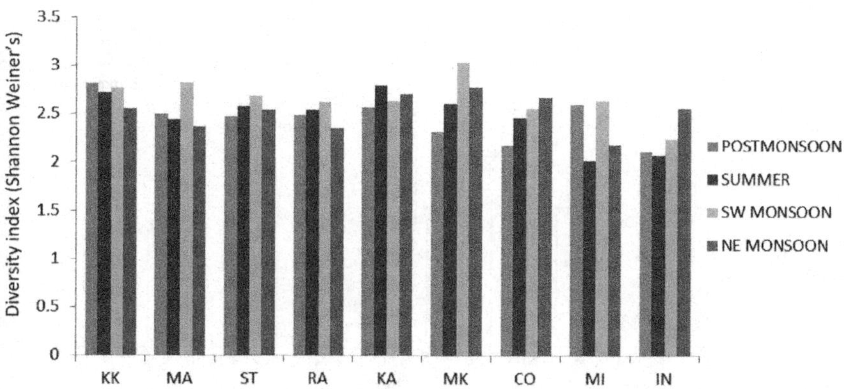

Figure 5.14 Shannon diversity index along the study area (Meiofauna).

5.23.2 Multivariate analysis

5.23.2.1 Cluster analysis (Bray–Curtis similarity)

Figs. 5.15 and 5.16 display the results of hierarchical clustering and MDS ordination on meiofaunal abundance representing all the stations during postmonsoon, summer, SW, and NE monsoons.

The similarity index plot shows that there is a distinct separation from the stations during the postmonsoon season and the stations during summer and NE monsoon seasons. During SW monsoon, the stations show a mixed response, and no grouping of similarity is observed. The MDS plot defends with a significant grouping of NE monsoon season having similar macrofaunal assemblages in the sampling stations followed by postmonsoon and summer season. The highest similarity (92%) is seen between Sanguthurai and Rajakkamangalam Thurai representing the postmonsoon season. The species abundance in these two stations has similar environments during the postmonsoon period. The species similarity is nearly 60% between different stations in all four seasons (Fig. 5.16). A clear grouping is observed in the samples representing postmonsoon, summer, and NE monsoon, with 65% of similarity, but the SW monsoon shows mixed similarity with other seasons as well as stations. This reveals the disturbed condition of the sampling sites with regard to seasons due to mining as external climatic disturbance is present during the SW monsoon. The grouping analysis of postmonsoon and summer seasons has a much higher similarity of

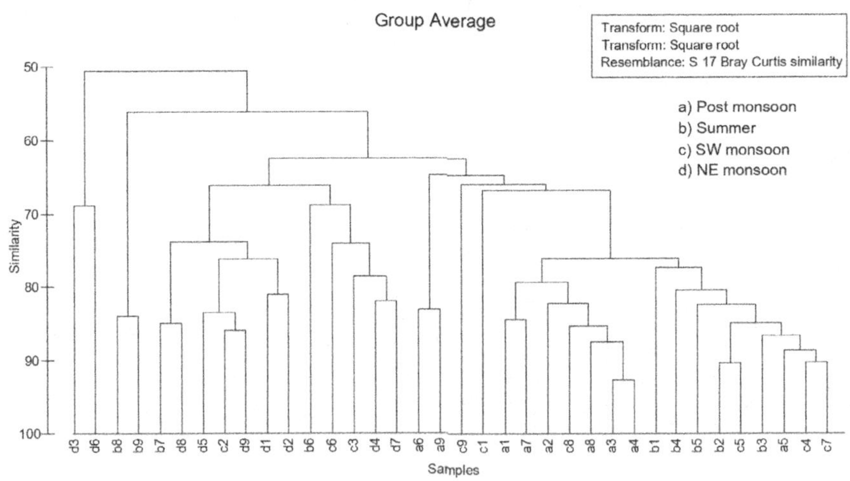

Figure 5.15 Bray curtis similarity in all the stations (Meiofauna).

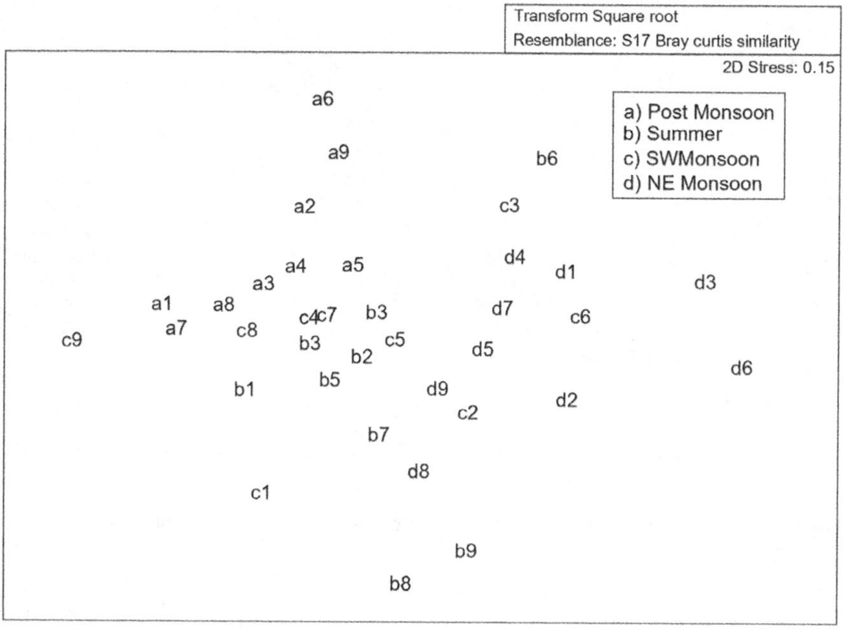

Figure 5.16 MDS plot for meiofauna overall distribution.

75% concerning sampling stations. This shows that the meiofauna in these stations has similar environmental conditions.

5.23.2.2 K-dominance

The outcome of the K-dominance plot shows that Halacaroidea is dominant throughout the study area. The second dominant group is the Copepoda followed by Polychaeta, Ostracoda, Foraminifera, Oligochaeta, Nematoda, Amphipoda, and Cladocera. This shows that Halacaroidea is the most diverse species as they can adapt to climatic and environmental changes in the study area. The sensitive groups include Amphipoda and Cladocera. A slight change in their climatic or habitat conditions can alter their distribution in the study area (Fig. 5.17).

5.23.2.3 BIO-ENV analysis

The BIO-ENV analysis of meiofauna in the postmonsoon season reveals that temperature, salinity, and sand influence the abundance of the meiofaunal community (r = 0.977). They are also influenced by physicochemical parameters like temperature, placer minerals (r = 0.975), pH (r = 0.974), and silt (r = 0.73). The influence of physical

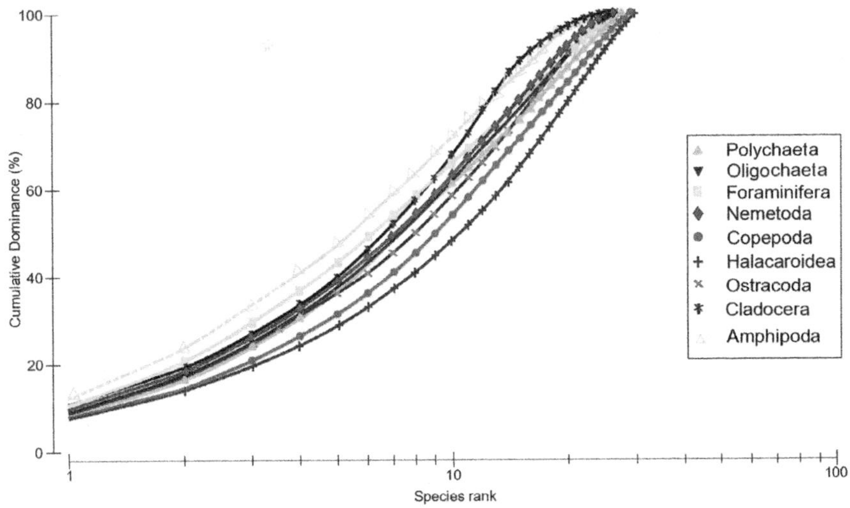

Figure 5.17 K-dominance plot discriminating the meiofaunal assemblages.

factors like temperature, salinity, and pH support the meiofaunal assemblages, and the influence of placer minerals and sediment texture support the habitat. However, mining practices during these seasons harm the meiofaunal abundance (Table 5.9).

During summer, temperature, DO, sand, and silt have the most influence (r = 0.548), followed by temperature, salinity, pH, DO, sand, silt, and clay (r = 0.547), then temperature, salinity, pH, DO, silt, and clay (r = 0.546) and temperature, salinity, pH, DO, sand, silt, and clay (r = 0.545). This shows that the above factors followed by DO have their influence during the postmonsoon season. Sediment texture is the base for their healthy habitat and temperature, pH, and DO are the base for their abundance as oxygen availability is reduced during summer.

During the SW monsoon, the major factors controlling their abundance are salinity, DO, and silt (r = 0.601) followed by salinity and DO (r = 0.586), and then salinity and pH (r = 0.584), and temperature and pH (r = 0.580). The influence of biogenic factors such as DO, salinity, and pH govern species abundance. The sediments from estuaries and streams also contribute to the diversity of faunal assemblages. The salinity is affected or diluted during the monsoons and this in turn affects meiofaunal habitat.

During the NE monsoon, a similar environmental condition is seen in the study area. Hence similar results as that of the SW monsoon are

Table 5.9 Bio-Env analysis of meiofauna along the study area with respective seasons.

No. of variables	Parameters	Spearman correlation
Postmonsoon		
3	Temperature, salinity, and sand	0.977
2	Temperature, placer minerals	0.975
2	Salinity and pH	0.974
2	Salinity and silt	0.973
Summer		
4	Temperature, DO, sand, silt	0.548
7	Temperature, salinity, pH, DO, sand, silt, and clay	0.547
6	Temperature, salinity, pH, DO, silt, and clay	0.546
7	Temperature, salinity, pH, DO, sand, silt, and clay	0.545
SW monsoon		
3	Salinity, DO, and silt	0.601
2	Salinity and DO	0.586
2	Salinity and pH	0.584
2	Temperature and pH	0.580
NE monsoon		
4	Salinity, pH, silt, and clay	0.601
4	Salinity, DO, silt, and clay	0.586
3	Salinity, DO, and sand	0.585
2	Salinity and pH	0.584

seen in this season also. The most influential factors are salinity, pH, silt, and clay ($r = 0.601$) followed by salinity, DO, silt, and clay ($r = 0.586$), then salinity, DO, and sand ($r = 0.585$), and salinity and pH ($r = 0.584$). The BIO-ENV results reveal that biogenic factors like

DO, followed by physical factors like temperature, salinity, pH, and habitat bases like placer minerals, sand, silt, and clay play major roles in the distribution of species. The results reveal that natural or anthropogenic factors affect and have adverse effects on the diversity and abundance of the meiofaunal communities.

5.24 Principal component analysis

5.24.1 Seawater

The point source pollutant identification and principal component analysis were done to find the impact of natural as well as human-induced factors on the marine environment (Jolliffe, 1986; Rubio et al., 2000). Principal component analysis (PCA) is done on the logarithmic form of the metal data, and Varimax rotation (Gotelli & Ellison, 2004) is used to maximize the sum of the variance of the factor coefficients. Positive extracted factor loading shows that the contribution of the variables increases with increasing loading in dimension and negative loading specifies a decrease.

PCA is done separately for all the seasons, on the normalized data sets containing 16 variables. It is done to compare the compositional pattern between water quality parameters and to recognize the factors that influence them. Eigenvalue measures the importance of the PC. PCs with the highest eigenvalues are the most important. Eigenvalues of 1.0 or greater are considered significant. Classification of factor loading is "strong," "moderate," and "weak," corresponding to absolute loading values of > 0.75, 0.75–0.50, and 0.50–0.30, respectively (Liu et al., 2003). Table 5.10 shows that in the postmonsoon season, three factors account for 78.29% of the total variance. PC1 accounts for 40.70% of the total variance with positive factor loadings of salinity, TDS, Cl, Na, Ca, Mg, and SO_4. The rest of the parameters are negatively loaded. In PC2, positive loadings of almost all parameters can be found except temperature, TDS, DO, Mg, Na, and PO_4 that are negatively loaded and contribute 21.27% of the total variability. In PC3, the positive and negative loadings are equal in number accounting for 16.86% total variability. However, PC1 factor (salinity, TDS, Cl, Na, Ca, Mg, and SO_4; anthropogenic factor), PC2 factor (pH, EC, BOD, and NO_2; natural process factor), and PC3 factor (DO and K; intermediate factor) might have come from different sources and the other metals might have

Table 5.10 Principal component analysis for physicochemical properties of seawater with respective seasons.

Factor loadings	Postmonsoon			Summer			SW monsoon			NE monsoon		
	PC1	PC2	PC3	PC1	PC2	PC3	PC1	PC2	PC3	PC1	PC2	PC3
Temperature	0.059	−0.248	**−0.892**	**0.889**	−0.265	0.214	0.004	**−0.879**	0.111	**0.425**	**−0.676**	−0.346
pH	0.084	**0.87**	0.094	−0.315	**−0.83**	0.213	0.325	**0.865**	0.195	−0.006	0.155	**0.924**
EC	−0.077	**0.954**	−0.001	−0.266	**−0.822**	0.162	0.325	**0.865**	0.195	−0.001	−0.01	**0.97**
Salinity	**0.959**	0.034	0.011	**0.805**	**0.536**	0.199	**0.943**	0.058	−0.289	**0.573**	**0.726**	0.025
TDS	**0.936**	−0.059	−0.063	−0.099	**0.884**	0.262	0.103	**−0.683**	**0.492**	**0.5**	0.249	0.126
DO	0.236	−0.378	**0.807**	**0.929**	0.19	0.162	**0.786**	0.075	**−0.429**	−0.15	**0.429**	**0.708**
BOD	0.143	**0.854**	−0.135	**−0.708**	−0.165	**−0.591**	0.145	−0.115	**−0.869**	**0.761**	0.119	0.052
Cl	**0.958**	0.188	0.057	**0.684**	**0.542**	**0.406**	**0.838**	0.395	−0.249	0.175	**0.897**	0.146
Na	**0.856**	−0.038	−0.261	**0.811**	**0.507**	0.106	**0.921**	−0.251	0.092	**0.654**	**0.718**	−0.049
K	−0.313	0.154	**0.848**	0.266	−0.177	**0.533**	−0.23	−0.33	**0.665**	**0.684**	−0.087	**−0.47**
Ca	**0.957**	0.102	0.094	0.155	−0.055	**0.644**	**0.929**	0.26	−0.131	**0.93**	0.161	−0.056
Mg	**0.826**	0.022	−0.273	**0.898**	−0.071	0.168	−0.203	**0.417**	**0.752**	0.098	**0.428**	0.037
SO_4	**0.95**	0.102	−0.105	−0.116	**0.866**	0.174	−0.399	**0.713**	**0.46**	**0.462**	**−0.515**	0.385
PO_4	−0.319	**−0.441**	0.28	−0.062	**−0.479**	**−0.694**	0.192	−0.143	**−0.826**	0.286	−0.311	**0.739**
NO_2	−0.012	**0.636**	0.183	**−0.505**	0.326	**0.595**	**−0.56**	−0.132	**0.636**	**−0.75**	−0.043	−0.078
NO_3	0.521	0.091	**−0.492**	−0.368	−0.067	**−0.652**	−0.307	**−0.614**	**−0.557**	0.12	**0.634**	−0.066
Eigen values	6.51	3.3	2.72	5.42	4.25	2.8	4.86	4.25	4.06	4.03	3.57	3.39
%Variation	40.7	20.6	16.99	33.85	26.55	17.47	30.4	26.58	25.36	25.19	22.32	21.19
Cum.%Variation	40.7	61.3	78.29	33.85	60.4	77.87	30.4	56.98	82.35	25.19	47.51	68.7

Extraction method: principal component analysis. Rotation method: varimax with Kaiser mormalization.

originated from similar sources. pH, salinity, Cl, and Ca parameters are positively loaded in all the PCs.

During summer, three factors make up for 77.87% of the total variance. PC1 contributes 33.85% of the total variability and the first factor has a good relationship with temperature, salinity, DO, Cl, Na, and Mg which are the noticeable parameters with positive loadings and is the most important component. PC2 contributes 26.55% of the total variance wherein all the parameters show positive factor loadings except temperature, pH, EC, BOD, K, Ca, Mg, PO_4, and NO_3 which are negative loadings. PC3 contributes 17.47% of the total variability, the majority of the variables are positively loaded, and BOD, PO_4, and NO_3 are negatively loaded. Salinity, DO, Cl, and Na parameters are found to be positively loaded in all the PCs.

During the SW monsoon, three factors contribute 82.35% of the total variance. PC1 contributes 30.40% of the total variance with positive factor loadings of salinity DO, Cl, Na, and Ca. NO_2 is negatively loaded. In PC2, positive loadings are more than that of negative loadings. pH, EC, Mg, and SO_4 are positively loaded and temperature, TDS, and NO_3 are negatively loaded. PC2 contributes 26.58% of the total variability, whereas in PC3 the parameters like TDS, K, Mg, SO_4, and NO_2 are positively loaded, and DO, BOD, PO_4, and NO_3 are negatively loaded and contribute 25.36% of total variability. pH and EC parameters are positively loaded in all the PCs.

During NE monsoon three factors contribute to 68.70% of the total variance. PC1 contributes 25.19% of the total variance with positive factor loadings of temperature, salinity, TDS, BOD, Na, K, Ca, and SO_4 and NO_2 is negatively loaded. PC2 contributes 22.32% of the total variance with positive factor loadings of salinity, DO, Cl, Na, Mg, and NO_3. Temperature and SO_4 are negatively loaded. In PC3, the positively loaded parameters are pH, EC, DO, and PO_4. K is negatively loaded in PC3 and contributes to 21.19% of the total variability. Salinity, TDS, BOD, Cl, and Mg are positively loaded in all the PCs.

5.24.2 Marine sediments

The PCA of the sediment quality data is tabulated in Table 5.11. During postmonsoon, the eigenvalue explains 75.71% of the total variability. PC1 contributes 32.39% of the total variance. Positive factor loadings are more

Table 5.11 Principal component analysis for physicochemical properties of marine sediments with respective seasons.

Factor loadings	Postmonsoon			Summer			SW monsoon			NE monsoon		
	PC1	PC2	PC3	PC1	PC2	PC3	PC1	PC2	PC3	PC1	PC2	PC3
Temperature	**0.698**	−0.099	**0.488**	−0.171	**0.617**	0.052	**0.474**	**0.434**	0.184	−0.087	0.136	**0.777**
pH	**0.778**	0.278	−0.326	**0.876**	−0.019	−0.312	0.166	**−0.718**	0.074	−0.041	0.004	−0.291
TOC	**0.735**	−0.114	**−0.434**	**0.888**	−0.012	0.035	0.014	**0.495**	**0.63**	0.145	0.387	0.013
Carbohydrate (CHO)	**0.685**	0.334	**−0.447**	0.168	0.101	**0.877**	**0.972**	0.03	0.034	**0.94**	0.191	−0.079
Amino acid	−0.078	−0.195	**0.795**	**0.924**	0.054	0.094	**0.902**	0.11	0.241	**0.864**	0.175	0.173
Protein	**0.896**	0.164	0.301	0.014	0.116	**0.954**	**0.943**	−0.105	−0.08	**0.941**	−0.002	−0.12
Lipid	**0.8**	−0.196	−0.173	−0.359	−0.396	**0.599**	**0.78**	0.007	**0.53**	**0.937**	0.083	−0.107
$CaCO_3$	0	**0.957**	0.07	0.085	**0.832**	0.032	−0.207	**−0.916**	−0.16	0.022	**0.97**	−0.053
Sand	−0.005	**−0.962**	−0.081	−0.155	**−0.879**	0	0.219	**0.918**	0.171	−0.207	**−0.765**	**0.41**
Silt	−0.078	0.147	**0.503**	0.021	**0.591**	−0.121	−0.124	−0.01	**−0.887**	−0.052	**0.718**	0.371
Clay	0.08	**0.897**	−0.178	0.245	0.156	−0.396	**0.429**	0.163	0.017	0.187	**0.414**	**−0.562**
Eigen values	3.56	2.98	1.78	2.69	2.4	2.32	3.8	2.68	1.63	3.5	2.45	1.38
%Variation	32.39	27.12	16.19	24.44	21.85	21.09	34.55	24.37	14.77	31.85	22.31	12.51
Cum.%Variation	32.39	59.51	75.71	24.44	46.3	67.39	34.55	58.92	73.69	31.85	54.16	66.66

Extraction method: principal component analysis. Rotation method: varimax with Kaiser normalization.

dominant than negative factors. Temperature, pH, TOC, CHO, protein, and lipid are positive factors derived from anthropogenic factors. PC2 contributes 27.12% of the total variance and positive factors are more dominant than negative factors. $CaCO_3$ and clay are positively loaded and are formed by natural processes. PC3 contributes 16.19% of the total variability. Here negative factors are more dominant than positive factors. Temperature, amino acid, protein, $CaCO_3$, and silt are positively loaded. TOC and CHOs are negatively loaded. Protein and $CaCO_3$ are positively loaded in all the PCs. During summer, the eigenvalues explain 67.39% of the cumulative percentage variability. PC1 contributes 24.44% of the total variance where positive loadings are dominant than negative loadings. pH, TOC, and amino acid are positively loaded and are derived from anthropogenic processes. PC2 contributes 21.85% of the total variability wherein temperature, $CaCO_3$, and silt are positively loaded and sand is negatively loaded. $CaCO_3$ and silt are mostly derived from natural processes. PC3 contributes 21.09% of the total variability. CHOs, proteins, and lipids are positively loaded. CHO, amino acid, protein, and $CaCO_3$ are positively loaded in all the PCs. During SW monsoon, the eigenvalues explain 73.69% of the cumulative percentage variability. PC1 contributes 34.55% of the total variance wherein positive loadings are dominant than negative loadings. Temperature, CHO, amino acid, protein, lipid, and clay are positively loaded and are derived from anthropogenic factors. PC2 contributes 24.37% of the total variability in which negative factors are smaller in number than the positive factors. Temperature, TOC, and sand are positively loaded. pH and $CaCO_3$ are negatively loaded. This indicates that they are derived from natural processes. PC3 contributes 14.77% of the total variability. TOC and lipid are positively loaded, silt is negatively loaded. Temperature, TOC, CHO, amino acid, lipid, sand, and clay are positively loaded in all the PCs. During NE monsoon, the eigenvalues explain 66.66% of the cumulative percentage variability. PC1 accounts for 31.85% of the total variance where positive loadings are more dominant than negative loadings. CHOs, amino acids, protein, and lipid are positively loaded and are derived from anthropogenic factors. PC2 accounts for 22.31% of the total variability where negative factors are less in number than the positive factors. $CaCO_3$, silt, and clay are positively loaded and sand is negatively loaded. PC3 accounts for 12.51% of the total variability. Temperature and sand are positively loaded, and clay is negatively loaded. TOC and amino acids are positively loaded in all the PCs.

5.25 Conclusion

The SW coast of Tamil Nadu has a delicate ecosystem with many anthropogenic interferences in the majority of the beaches. Anthropogenic activities in the region in the last few decades have greatly influenced the marine ecosystem bringing damage to it. The geochemical properties of seawater and marine sediments are widely influenced in the distribution of faunal species including the coastal and marine environment. The salinity of the ecosystem and the offshore environment is influenced by various environmental factors such as runoff, precipitation, evaporation, upwelling, and surface current patterns. Nutrient content and anthropogenic impact including habitat changes also immensely affect faunal diversity. Three important objectives have been set to reach the goal of the present study (1) the seasonal wise physicochemical parameter monitoring in coastal sediments and water, (2) the macro- and meiofaunal diversity monitoring at mining (Manakudy, Kadiapattinam, Manavalakurichi, and Midalam) and nonmining sites (Kanyakumari, Sanguthurai, Rajakkamangalam Thurai, Colachel, and Inayam) along the study region, and (3) finally intercorrelating the relationship of physicochemical parameters and habitat changes by mining activities and their impact on faunal diversity.

In the study, various multivariate statistical methods have been utilized to evaluate the seasonal variations in surface water quality and sediment quality in the mining and nonmining areas. The results indicate that seasonal variations in the four seasons are noticeably different. The analytical results show that the surface water of the SW coast is drastically affected by weathering, hydrologic, and anthropogenic processes. The chemical composition of surface water and sediment in each season is influenced by external forces and their interactions, dissolution, and dilution, and anthropogenic inputs. The observed data and their PCA analysis indicate that human-induced activities especially coastal heavy mineral mining, strongly play a major role in benthic faunal diversity and reduce species-level distribution. Beach placer mining brings massive changes in the coastal geomorphology, and leads to the loss of physical, chemical, and biological fertility, resulting in changes to the entire ecosystem.

The outcome of the study reveals that postmonsoon and summer seasons have high species diversity due to favorable climatic conditions and the availability of nutrients. Some beaches do not face the same trend due to the effect of mining or other anthropogenic activities. However, the SW and NE monsoons show significant variation in faunal diversity due to the effect of short-term monsoonal fluctuations. These monsoonal fluctuations are observed all along the study region except in placer mining sites. Continuous monitoring and further research may be carried out in the area to determine the long-term effects of anthropogenic impacts and to evaluate the effectiveness of minimizing human activity to upgrade the coastal ecosystem on the west coast of India.

Supplementry Material

Supplementry material.

Abbreviations

%	percentage
°C	degree Celsius
ASTM	American Standard Tested Mesh
BIO-ENV	biological and environmental
BOD	biological oxygen demand
Ca	calcium
$CaCO_3$	calcium carbonate
CCl_4	carbon tetrachloride
CHO	carbohydrate
Cl	chloride
cm	centimeter
CO_2	carbon dioxide
DO	dissolved oxygen
EC	electrical conductivity
H_2SO_4	sulfuric acid
HCO_3	bicarbonate
IREL	Indian Rare Earth Limited
km	kilometers
K	potassium
MDS	multidimensional scaling
Mg	magnesium
m	meter
mm	millimeter
MT	million tons
Na	sodium
NE	northeast
N	normality
NO_2	nitrite
NO_3	nitrate

OC	organic carbon
PCA	principal component analysis
pH	potential of hydrogen
PO_4	phosphate
SO_4	aulfate
SW	southwest
TDS	total dissolved solids
TOC	total organic carbon
TOM	total organic matter
φ	Phi

References

AFNOR. (1997). *Water quality (Volume 1): Terminology, sampling and evaluation methods* (3rd ed., p. 34) Paris: Afnor.

Aller, R. (1988). Benthic fauna and biogeochemical processes in marine sediments: The role of burrow structures. In T. H. Blackburn, & J. Sørensen (Eds.), *Nitrogen cycling in coastal marine environments* (pp. 301–338). Chichester, UK: John Wiley & Sons Ltd.

AMD. (2000). Beach and Inland Placer sand deposits in India, Atomic Minerals Directorate for Exploration and Research, Department of Atomic Energy, Government of India, p. 24.

Angusamy, N., & Rajamanickam, G. V. (2006). Depositional environment of sediments along the southern coast of Tamil Nadu, India. *Oceanologia*, *48*(1), 87–102.

APHA. (1995). *Standard methods for the examination of water and waste water* (19th ed.). Washington, DC: American Public Health Association.

Azam, F. (1998). Microbial control of oceanic carbon flux: The plot thickens. *Science (New York, N.Y.)*, *280*, 694–696.

Barboza, F. R., & Defeo, O. (2015). Global diversity patterns in sandy beach macrofauna: A biogeographic analysis. *Scientific Reports*, *5*, 14515.

Bessa, F., Gonçalves, S., Franco, J. N., André, J. N., Cunha, P. P., & Marques, J. C. (2014). Temporal changes in macrofauna as response indicator to potential human pressures on sandy beaches. *Ecological Indicators*, *41*, 49–57.

Bligh, E. G., & Dyer, W. J. (1959). A lipid method of total lipid extraction and purification. *Canadian Journal of Biochemistry*, *37*, 911–917.

Borja, A., Franco, J., & Pérez, V. (2000). A marine biotic index to establish the ecological quality of soft-bottom benthos within European estuarine and coastal environments. *Marine Pollution Bulletin*, *40*, 1100–1114.

Borreswar, R. C. (1957). Beach erosion and concentration of heavy minerals. *Journal of Sedimentary Petrology*, *27*, 143–147.

Branch, G. M., Thompson, R. C., Crowe, T. P., Castilla, J. C., Langmead, O., & Hawkins, S. J. (2008). *Rocky intertidal shores: Prognosis for the future. Aquatic ecosystems: Trends and global prospects* (pp. 209–225). Cambridge: Cambridge University Press. Available from https://doi.org/10.1017/CBO9780511751790.020.

Bray, J. R., & Curtis, J. T. (1957). An ordination of the upland forest communities of Southern Wisconsin. *Ecological Monographies*, *27*, 325–349.

Brazeiro, A. (2001). Relationships between species richness and morphodynamics in sandy beaches: What are the underlying factors? *Marine Ecology Progress Series*, *224*, 35–44.

Brown, A. C., & McLachlan, A. (1990). *Ecology of sandy shores.* Amsterdam, The Netherlands: Elsevier.

Bruckner, A., & Burrows, F. M. (2005). Chapter 3: Restoration monitoring of coral reefs. Chapter 3. In G. W. Thayer, T. A. McTigue, R. J. Salz, D. H. Merkey, F. M. Burrows, & P. F. Gayaldo (Eds.), *Science-based restoration monitoring of coastal habitats: Tools for monitoring coastal habitats: Vol. 2* (pp. 3.1–3.72). Silver Spring.

Byrne, M. (2011). Impact of ocean warming and ocean acidification on marine invertebrate life history stages: Vulnerabilities and potential for persistence in a changing ocean. *Oceanography and Marine Biology: An Annual Review*, *49*, 1–42.

Carvalho, A. N., Pereira, F., Bosnic, I., Taborda, R., Drago, T., & Gaspar, M. B. (2018). Sedimentary dynamics and benthic macrofauna distribution: Insights from the shoreface in southern Portugal. *Journal of Sea Research*, *137*, 9–25.

Chandrasekar, N., & Rajamanickam, G. V. (2001). Nature of distribution of heavy minerals along the beaches of central Tamil Nadu coast. *Journal of the Indian Association of Sedimnetologists*, *20*(2), 167–180.

Chandrasekaran, S., Viswanathan, G., Krishnan, S., Kumar, V., Ali, M. A., & Parthasarathy, T. N. (1997). Geomorphic control on the development of heavy minerals placers along the south west coast of India. *Indian Journal of Geomorphology*, *2*(1), 99–110.

Clarke, K. R., & Gorley, R. N. (2001). *PRIMER Version 6 user manual/tutorial*. Plymouth, UK: Primer-E Ltd.

Clarke, K. R., & Warwick, R. M. (1994). Relearning the ABC: Taxonomic changes and abundance/biomass relationships in disturbed benthic communities. *Marine Biology*, *118*, 739–744.

Cornwall, C. E., Revill, A. T., Hall-Spencer, J. M., Milazzo, M., Raven, J. A., & Hurd, C. L. (2017). Inorganic carbon physiology underpins macroalgal responses to elevated CO_2. *Scientific Reports*, 7, 46297. Available from https://doi.org/10.1038/srep46297.

Constable, A. J. (1999). Ecology of benthic macro-invertebrates in soft-sediment environments: A review of progress towards quantitative models and predictions. *Austral Ecology*, *24*, 452–476.

Crain, C. M., Kroeker, K., & Halpern, B. S. (2008). Interactive and cumulative effects of multiple human stressors in marine systems. *Ecology Letters*, *11*, 1304–1315. Available from https://doi.org/10.1111/j.1461-0248.2008.01253.x.

Creese, R. G., & Kingsford, M. J. (1998). Organisms of reef and soft substrata intertidal environments. In M. Kingsford, & C. Battershill (Eds.), *Studying temperate marine environments: A handbook for ecologists*. Christchurch: Canterbury University Press.

D'Andrea, A. F., & DeWitt, T. H. (2009). Geochemical ecosystem engineering by the mud shrimp *Upogebia pugettensis* (Crustacea: Thalassinidae) in Yaquina Bay, Oregon: Density-dependent effects on organic matter remineralization and nutrient cycling. *Limnology and Oceanography*, *54*, 1911–1932.

Danovaro, R., Dell'Anno, A., & Pusceddu, A. (2004). Biodiversity response to climate change in a warm deep sea. *Ecology Letters*, 7, 821–828.

Defeo, O., & McLachlan, A. (2005). Patterns, processes and regulatory mechanisms in sandy beach macrofauna: A multi-scale analysis. *Marine Ecology Progress Series*, *295*, 1–20, ISI:000230671800001.

Defeo, O., Mclachlan, A., Shoeman, D. S., Schlacher, A., Dugan, J., Jones, A., Lastra, M., & Scapini, F. (2009). Threats to sandy beach ecosystems: A review. *Estuarine, Coastal and Shelf Science*, *81*, 1–12.

Desprez, M. (2000). Physical and biological impact of marine aggregate extraction along the French coast of the eastern English Channel: Short- and long-term post-dredging restoration. *ICES Journal of Marine Science*, *57*, 1428–1438.

Deutsch, C. A., Tewksbury, J. J., Huey, R. B., Sheldon, K. S., Ghalambor, C. K., Haak, D. C., et al. (2008). Impacts of climate warming on terrestrial ectotherms across latitude. *Proceedings of the National Academy of Sciences of the United States of America*, *105*, 6668–6672. Available from https://doi.org/10.1073/pnas.0709472105.

Doney, S. C., Ruckelshaus, M., Emmett Duffy, J., Barry, J. P., Chan, F., English, C. A., et al. (2012). Climate change impacts on marine ecosystems. *Annual Review of Marine Science, 4*, 11–37. Available from https://doi.org/10.1146/annurev-marine-041911-111611.

Duarte, C. M. (1995). Submerged aquatic vegetation in relation to different nutrient regimes. *Ophelia, 41*, 87–112.

Dubois, M., Gilles, K. A., Hamilton, J. K., Rebers, P. A., & Smith, F. (1956). Colorimetric method for determination of sugars and related substances. *Analytical Chemistry, 28*, 350–356.

Dugan, J. E., Hubbard, D. M., & Quigley, B. J. (2013). Beyond beach width: Steps toward identifying and integrating ecological envelopes with geomorphic features and datums for sandy beach ecosystems. *Geomorphology, 199*, 95–105.

Dulvy, N. K., Rogers, S. I., Jennings, S., Stelzenmüller, V., Dye, S. R., & Skjoldal, H. R. (2008). Climate change and deepening of the North Sea fish assemblage: A biotic indicator of warming seas. *The Journal of Applied Ecology, 45*, 1029–1039.

Folk, R. L. (1974). The natural history of crystalline calcium carbonate: effect of Mg-content and salinity. *Journal of Sedimentary Petrology, 44*, 40–53.

Folk, R. L., & Ward, W. C. (1957). Brazos river bar: A study in the significance of grain size parameters. *Journal of Sedimentary Petrology, 27*, 3–27.

Fraschetti, S., Terlizzi, A., & Boero, F. (2008). How many habitats are there in the sea (and where)? *Journal of Experimental Marine Biology and Ecology, 366*, 109–115.

Friedman, G. M. (1967). Dynamic processes and statistical parameters compared for size frequency distribution of beach and river sands. *Journal of Sedimentary Petrology, 37*, 327–354.

Gaudette, H. E., Flight, W. R., Toner, L., & Folger, D. W. (1974). An inexpensive titration method for the determination of organic carbon in recent sediments. *Journal of Sedimentary Petrology, 44*, 249–253.

Giere, O. (2009). *Meiobenthology. The microscopic motile fauna of aquatic sediments* (2nd ed., p. 527) Heidelberg: Springer Verlag.

Godson, P.S. (2016). Evaluating the Impact of Beach Placer Mining on Sandy Beach Macro and Meiofaunal Community along the Southern Tamil Nadu Coast, Unpublished data.

Gotelli, N. J., & Ellison, A. M. (2004). *A primer of ecological statistics* (p. 510) Sunderland: Sinauer Associates, Inc.

Gray, J. S. (1981). *The ecology of marine sediments. An introduction to the structure and function of benthic communities. Cambridge studies in modern biology* (2, p. 185). Cambridge: Cambridge University Press.

Greig-Smith, P. (1983). *Quantitative plant ecology*. Berkeley, CA: University of California Press.

Halpern, B. S., Walbridge, S., Selkoe, K. A., Kappel, C. V., Micheli, F., D'Agrosa, C., et al. (2008). A global map of human impact on marine ecosystems. *Science, 319*, 948–952.

Harley, C. D. G., Hughes, A. R., Hultgren, K. M., Miner, B. G., Sorte, C. J. B., Thornber, C. S., et al. (2006). The impacts of climate change in coastal marine systems. *Ecology Letters, 9*, 228–241. Available from https://doi.org/10.1111/j.1461-0248.2005.00871.x.

Hawkins, S. J., Moore, P. J., Burrows, M. T., Poloczanska, E., Mieszkowska, N., Herbert, R. J. H., et al. (2008). Complex interactions in a rapidly changing world: Responses of rocky shore communities to recent climate change. *Climate Research, 37*, 123–133. Available from https://doi.org/10.3354/cr00768.

Herbert, D., Phipps, P. J., & Strange, R. E. (1971). Chemical analysis of microbial cells. *Methods in Microbiology, 5*, 209–344. Available from https://doi.org/10.1016/S0580-9517(08)70641-X.

Higgins, R. P., & Thiel, H. (1988). *Introduction to the study of meiofauna* (p. 488) Washington, DC: Smithsonian Institution Press.

Hoegh-Guldberg, O., Jacob, D., Taylor, M., Bindi, S., Brown, I., Camilloni, A., et al. (2018). Impacts of 1.5 C of Global Warming on Natural and Human Systems. https://www.ipcc.ch/sr15.

Hoegh-Guldberg, O., Mumby, P. J., Hooten, A. J. J., Steneck, R. S., Greenfield, P., Omez, E., et al. (2007). Coral reefs under rapid climate change and ocean acidification. *Science, 318*, 1737–1742.

Holmquist, J. G. (1997). Disturbance and gap formation in a marine benthic mosaic: Influence of shifting macroalgal patches on seagrass structure and mobile invertebrates. *Marine Ecology Progress Series, 158*, 121–130.

Hopkins, F. E., & Mudge, S. M. (2004). Detecting anthropogenic stress in an ecosystem: 2. Macrofauna in a sewage gradient. *Environmental Forensics, 5*, 213–223.

Hughes, T. P., Kerry, J. T., Baird, A. H., Connolly, S. R., Chase, T. J., Dietzel, A., et al. (2019). Global warming impairs stock–recruitment dynamics of corals. *Nature, 568*, 387–390. Available from https://doi.org/10.1038/s41586-019-1081-y.

Ingole, B. S., Sivadas, S., Nanajkar, M., Soutya, S., & Nag, A. (2009). A comparative study of macrobenthic community from harbours along the central west coast of India. *Environmental Monitoring and Assessment, 154*, 135–146.

Jaramillo, E., Contreras, H., Duarte, C., & Quijón, P. (2001). Relationships between community structure of the inter-tidal macroinfauna and sandy beach characteristics along the Chilean coast. *Marine Ecology, 22*, 323–342.

Jolliffe, I. (1986). *Principal component analysis*. New York: Springer Verlag.

Joydas, T. V., & Damodaran, R. (2009). In faunal macrobenthos along the shelfwaters of the west coast of India, Arabian Sea. *Indian Journal of Marine Sciences, 38*, 191–204.

Kochert, G. (1978). Carbohydrate determination by the phenol sulfuric acid method. In J. A. Helebust, & J. S. Craig (Eds.), *Handbook of phycologia method* (pp. 56–97). Cambridge: Cambridge University Press.

Kristensen, E., Penha-Lopes, G., Delefosse, M., Valdemarsen, T., Quintana, C. O., & Banta, G. T. (2012). What is bioturbation? The need for a precise definition for fauna in aquatic sciences. *Marine Ecology Progress Series, 446*, 285–302.

Kroeker, K. J., Kordas, R. L., & Harley, C. D. G. (2017). Embracing interactions in ocean acidification research: Confronting multiple stressor scenarios and context dependence. *Biology Letters, 13*, 20160802. Available from https://doi.org/10.1098/rsbl.2016.0802.

Lambshead, P. J. D., & Hodda, M. (1994). The impact of disturbance on measurements of variability in marine nematode populations. *Vie et Milieu, 44*, 21–27.

Littler, M. M., & Littler, D. S. (1985). Nondestructive sampling. In M. M. Littler, & D. S. Littler (Eds.), *Handbook of physiological methods. Ecological field methods: Macroalgae* (pp. 161–175). Cambridge: Cambridge University Press.

Liu, C. W., Lin, K. H., & Kuo, Y. M. (2003). Application of factor analysis in the assessment of groundwater quality in a Blackfoot disease area in Taiwan. *Science of Total Environment, 313*, 77–89.

Loring, D. H., & Rantala, R. T. T. (1992). Manual for the geochemical analyses of marine sediments and suspended particulate matter. *Earth-Science Reviews, 32*, 235–283.

Loveson, V. J., & Rajamanickam, G. V. (1988). Progradation as an evidence around a submerged ancient port, Peripattanam, Tamil Nadu. *Indian Journal Land Systems and Ecological Studies, 12*, 94–98.

Mange, M. A., & Maurer, H. F. W. (1992). *Heavy minerals in colour* (p. 147) London Chapman and Hall.

Manikannan, R., Asokan, S., & Ali, A. H. M. S. (2011). Seasonal variations of physico-chemical properties of the Great Vedaranyam Swamp, Point Calimere Wildlife Sanctuary, South-East coast of India. *African Journal of Environmental Science and Technology, 5*(9), 673–681.

Manoharan, J., Varadharajan, D., Thilagavathi, B., & Priyadharsini, S. (2011). Biodiversity and abundance of benthos along the South East Coast of India. *Advances in Applied Science Research, 2*(6), 554–562.

Margalef, R. (1958). Temporal succession and spatial heterogeneity in phytoplankton. In Buzzati-Traverso (Ed.), *Perspectives in marine biology* (pp. 323–347). Berkeley: University California Press.

McLachlan, A. (1983). Sandy beach ecology. A review. In A. McLachlan, & T. Erasmus (Eds.), *Sandy beaches as ecosystems* (pp. 321–380). Junk: The Hague.

McLachlan, A., & Dorvlo, A. (2005). Global patterns in sandy beach macrobenthic communities. *Journal of Coastal Research, 214*, 674–687.

Menge, B. A. (1976). Organization of the New England Rocky Intertidal Community: Role of predation, competition, and environmental heterogeneity. *Ecological Monographs, 46*, 355–393. Available from https://doi.org/10.2307/1942563.

Mieszkowska, N. (2016). *Climate change: Observed impacts on planet earth* (2nd ed., pp. 213–229). The Netherland: Elsevier. Available from https://doi.org/10.1016/B978-0-444-63524-2.00014-2.

Millero, F., Woosley, R., DiTrolio, B., & Waters, J. (2009). Effect of ocean acidification on the speciation of metals in seawater. *Oceanography, 22*, 72–85. Available from https://doi.org/10.5670/oceanog.2009.98.

Murray, J. W. (1973). *Distribution and ecology of living benthic foraminiferids* (p. 244) London: Heinemann Educational Books.

Muxika, I., Borja, A., & Bald, J. (2007). Using historical data, expert judgement and multivariate analysis in assessing reference conditions and benthic ecological status, according to the European Water Framework Directive. *Marine Pollution Bulletin, 55*, 16–29.

Naeem, S. (2008). Species redundancy and ecosystem reliability. *Conservation Biology, 12*, 39–45. Available from https://doi.org/10.1111/j.1523-1739.1998.96379.x.

Nigam, R. (1984). Living benthonic forminifera in tidal environment; Gulf of Khambhat (India). *Marine Geology, 58*, 415–425.

Nixon, S. W. (1995). Coastal marine eutrophication: A definition, social causes, and future concerns. *Ophelia, 41*, 199–219.

Palacin, C., Martin, D., & Gilli, J. M. (1991). Features of spatial distribution of benthic infauna in Mediterranean shallow water Bay. *Marine Biology, 110*, 315–320.

Parsons, T. R., Maita, Y., & Lalli, C. M. (1984). *A manual of chemical and biological methods for seawater analysis* (p. 173) Oxford: Pergamon Press.

Peterson, C. H., Hickerson, D. H. M., & Johnson, G. G. (2000). Short-term consequences of nourishment and bulldozing on the dominant large invertebrates of a sandy beach. *Journal of Coastal Research, 16*, 368–378.

Pettijohn, F. J., Potter, P. E., & Siever, R. (1972). *Sand and sandstones* (p. 158) New York: Springer-Verlag.

Pielou, E. C. (1975). *Ecological diversity*. New York: Wiley, [A general book on ecological diversity. Entropy-based diversity measures. Models (for the distribution of species)].

Porri, F., Hill, J. M., & McQuaid, C. D. (2011). Associations in ephemeral systems: The lack of trophic relationships between sandhoppers and beach wrack. *Marine Ecology Progress Series, 426*, 253–262.

Przeslawski, R., Byrne, M., & Mellin, C. (2015). A review andmeta-analysis of the effects of multiple abiotic stressors on marine embryos and larvae. *Global Change Biology, 21*, 2122–2140. Available from https://doi.org/10.1111/gcb.12833.

Queirós, A. M., Taylor, P., Cowles, A., Reynolds, A., Widdicombe, S., & Stahl, H. (2015). Optical assessment of impact and recovery of sedimentary pH profiles in ocean acidification and carbon capture and storage research. *International Journal of Greenhouse Gas Control, 38*, 110–120.

Rajamanickam, G. V., Chandrasekar, N., Angusamy, N., & Loveson, V. J. (2004). Status of beach placer mineral exploration in India. In V. J. Loveson, & D. D. Misra (Eds.), *National workshop and seminar on sustainable development of coastal placer minerals* (pp. 10–19). New Delhi: Allied Publishers Pvt. Ltd.

Rao, N. S., & Misra, S. (2009). Sources of monazite sand in southern Orissa beach placer, eastern India. *Journal of the Geological Society of India, 74*, 357e362.

Rasheed, M., Badran, M. I., & Hüttel, M. (2003a). Particulate matter filtration and seasonal nutrient dynamics in permeable carbonate and silicate sands of the Gulf of Aqaba, Red Sea. *Coral Reefs, 22*, 167–177.

Reis, R. D. S., & Rizzo, A. E. (2019). Human trampling effect on benthic fauna of sandy beaches with different intensities of use in Rio de Janeiro, Brazil. *Oecologia Australis, 23*, 28–38.

Reyes-Martínez, M. J., Ruíz-Delgado, M. C., Sanchez-Moyano, J. E., & Garcia-García, F. J. (2015). Response of intertidal sandy-beach macrofauna to human trampling: An urban vs. natural beach system approach. *Marine Environmental Research, 103*, 36–45.

Rubio, B., Nombela, M. A., & Vilas, F. (2000). Geochemistry of major and trace elements in sediments of the Ria de Vigo (NW Spain): An assessment of metal pollution. *Marine Pollution Bulletin, 40*, 968–980.

Sala, E., & Knowlton, N. (2006). Global marine biodiversity trends. *Annual Review of Environment and Resources, 31*, 93–122.

Schlacher, T. A., WM, A., Schoeman, D. S., Olds, A. D., Huijbers, C. M., & Connolly, R. M. (2015). Golden opportunities: A horizon scan to expand sandy beach ecology. *Estuarine, Coastal and Shelf Science, 157*, 1–6.

Shannon, C. E. (1949). The mathematical theory of communication. In C. E. Shannon, & W. Weaver (Eds.), *The mathematical theory of communication* (pp. 3–91). Urbana: University of Illinois Press.

Sivadas, M., Abdussamad, E. M., Jasmine, S., Rohit, P., Koya, K. P. S., Ghosh, S., Joshi, K. K., Dhokia, H. K., Prakasan, D., & Bineesh, K. K. (2012). Assessment of the fishery and stock of striped bonito, *Sarda orientalis* (Temminck and Schlegel, 1844) along Kerala coast with a general description of its fishery from Indian coast. *Indian Journal of Fisheries, 59*(2), 57–61.

Smale, D. A., Wernberg, T., & Vanderklift, M. A. (2017). Regional-scale variability in the response of benthic macroinvertebrate assemblages to a marine heatwave. *Marine Ecology Progress Series, 568*, 17–30. Available from https://doi.org/10.3354/meps12080.

Smeets, E. & Weterings, R. (1999). Environmental Indicators: Typology and overview. Technical report No 25. EEA, p. 19.

Snelgrove, P. V. R. (1998). The biodiversity of macrofaunal organisms in marine sediments. *Biodiversity and Conservation, 7*, 1123–1132.

Soundarapandian, P., Premkumar, T., & Dinakaran, G. K. (2009). Studies on the physico-chemical characteristic and nutrients in the Uppanar Estuary of Cuddalore, South East Coast of India. *Current Research Journal of Biological Science, 1*(3), 102–105.

Thrush, S. F., Hewitt, J. E., Kraan, C., Lohrer, A. M., Pilditch, C. A., & Douglas, E. (2017). Changes in the location of biodiversity–ecosystem function hot spots across the seafloor landscape with increasing sediment nutrient loading. *Proceedings of the Royal Society B: Biological Sciences, 284*, 20162861.

Van Dalfsen, J. A., Essink, K., Toxvig Madsen, H., Birklund, J., Romero, J., & Manzanera, M. (2000). Differential response of macrozoobenthos to marine sand extraction in the North Sea and the western Mediterranean. *ICES Journal of Marine Science, 57*, 1439–1445.

Van der Heide, T., van Nes, E. H., Geerling, G. W., Smolders, A. J. P., Bouma, T. J., & van Katwijk, M. M. (2007). Positive feedbacks in seagrass ecosystems: Implications for success in conservation and restoration. *Ecosystems, 10*, 1311–1322.

Veloso, V. G., Sallorenzo, I. A., Ferreira, B. C. A., & De Souza, G. N. (2010). Atlantorchestoidea brasiliensis (Crustacea: Amphipoda) as an indicator of disturbance caused by urbanization of a beach ecosystem. *Brazilian Journal of Oceanography, 58*, 13–21.

Vijayakumar, S. K., Rajesh, K. M., Mendon, R., & Hariharan, V. (2000). Seasonal distribution and behaviour of nutrients with reference to tidal rhythm in the Mulki estuary, southwest coast of India. *Journal of the Marine Biological Association of India, 42*, 21–31.

Vincx, M. (1996). Meiofauna in marine and fresh water sediments. In G. S. Hall (Ed.), *Methods for the examination of organismal diversity in soils and sediments* (pp. 187–195). New York: CAB International.

Walton, W. R. (1952). Techniques for recognition of living foraminifera. *Contributions of the Cushman Foundation for Foraminiferal Research, 3*(2), 56–60.

Watling, L. (1991). The sedimentary milieu and its consequences for resident organisms. *American Zoologist, 31*, 789–796.

Wernberg, T., Russell, B. D., Moore, P. J., Ling, S. D., Smale, D. A., Campbell, A., et al. (2011). Impacts of climate change in a global hotspot for temperate marine biodiversity and ocean warming. *Journal of Experimental Marine Biology and Ecology, 400*, 7–16. Available from https://doi.org/10.1016/j.jembe.2011.02.021.

Woodin, S. A., & Jackson, J. B. C. (1979). Facilitative and inhibitory interactions among estuarine meiobenthic harpacticoid copepods. *Ecology, 68*, 1906–1919.

Yachi, S., & Loreau, M. (1999). Biodiversity and ecosystem productivity in a fluctuating environment: The insurance hypothesis. *Proceedings of the National Academy of Sciences of the United States of America, 96*, 1463–1468. Available from https://doi.org/10.1073/pnas.96.4.1463.

Zweng, R. C., Koch, M. S., & Bowes, G. (2018). The role of irradiance and C-use strategies in tropicalmacroalgae photosynthetic response to ocean acidification. *Scientific Reports, 8*, 9479. Available from https://doi.org/10.1038/s41598-018-27333-0.

CHAPTER 6

Physico-chemical factors regulating marine benthos structure and function

Seerangan Manokaran[1], Thadickal V. Joydas[1] and Ajmal Khan[1,2]
[1]Center for Environment and Marine Studies, Research Institute, King Fahd University of Petroleum and Minerals (KFUPM), Dhahran, Saudi Arabia
[2]Centre of Advanced Study in Marine Biology, Annamalai University, Parangipettai, Tamil Nadu, India

6.1 Introduction

Physicochemical characteristics such as the nature and type of sediment, its geological origin, topography, depth, light availability, temperature, salinity, dissolved oxygen, pH, redox potential, organic matter, contaminants, microplastics, and exposure to the combined effects of wave and tidal action determine the benthic faunal assemblages. Among the above parameters, sediment grain size has prime importance as it directly establishes the spatial and structural status of benthic faunal habitats and indirectly determines the physical and chemical properties of the sediment (Peterson, 1913; Townsend & Cammen, 1988; Levin et al., 2001; Ganesh & Raman, 2007; Thrush et al., 2013; Manokaran et al., 2015; Joydas et al., 2019). The saturation of water in the sediment and its flow play a vital role in structuring the macrobenthos assemblages. The flow ensures the supply of oxygen and nutrients to the various layers of the substratum. However, in the deeper layers of the sediment, the flow is reduced. The poorly sorted sediments (e.g., sand mixed with gravel and silt) become tightly packed and here the interstitial pore volume is often reduced to only 20% of the total volume. Well-sorted (coarse) sediments contain up to 45% pore volume (Giere, 2009). In the intertidal area, due to reduced pore water content, the occurrence of macro and meiofauna can get restricted. Moreover, the reduced capillary force causes steep gradients of many abiotic factors, such as temperature and salinity often with negative impacts on the benthos.

The importance of such environmental parameters in determining the status of benthic communities is well known, to the extent that the

Ecology and Biodiversity of Benthos
DOI: https://doi.org/10.1016/B978-0-12-821161-8.00005-2

physical and chemical characteristics of the seabed environmental parameters are used to provide a good estimate of the types of the benthic community likely to be found inhabiting the seabed. Due to anthropogenic disturbances, contaminants have also joined the rank and file of major physicochemical parameters besides depth, temperature, salinity, dissolved oxygen, pH, turbidity, nutrients, and organic matter, in regulating the structure and function of marine benthos (Manokaran et al., 2015; Ajmal Khan et al., 2017; Joydas et al., 2019).

6.2 Physical factors

The major physical parameters influencing the community structure and distribution of benthos include depth, temperature, salinity, turbidity, and sediment texture.

6.2.1 Depth

Important abiotic factors structuring the organization in the rocky and sandy subtidal infaunal and epifaunal communities differ predictably with the depth factor. Deeper depth (continental shelf, slope, and abyssal plain, submarine canyon and trenches) environments are extreme and more challenging than the shallower ones (Sanders, 1968; Jackson, 1972; Ballesteros & Zabala, 1993; Ansari et al., 2017a, 2017b). Depth is the major gradient in structuring density, diversity, and size of deep-sea megafaunal assemblages (Merrett & Haedrich, 1997; Rex & Etter, 1998; Colloca et al., 2003; Ganesh & Raman, 2007; Papiol et al., 2012). Depth is a proxy for a combination of several abiotic factors (temperature, salinity, light, water current, nutrients, and sedimentation) that influences benthic organisms (Garrabou et al., 2001; Cartes et al., 2011; Papiol et al., 2012). Therefore, the depth can be considered as one of the physical stress gradients.

Significant changes in zonation pattern of infaunal macrobenthic communities (polychaetes and mollusks) occur with depth from near shore to 2000 or 2500 m and it decreases rapidly over higher depths in all the oceans (Lalli & Parsons, 1997; Heyns et al., 2016; Gopal et al., 2020; Xu et al., 2020). Different assemblages of the benthic communities with distinct species dominance and composition are the main features with change in depth. There is some evidence suggesting that the density, number of species, and diversity of macrobenthos generally decrease with a rise in depth (Joydas & Damodaran, 2009; Ingole et al., 2010; Ajmal

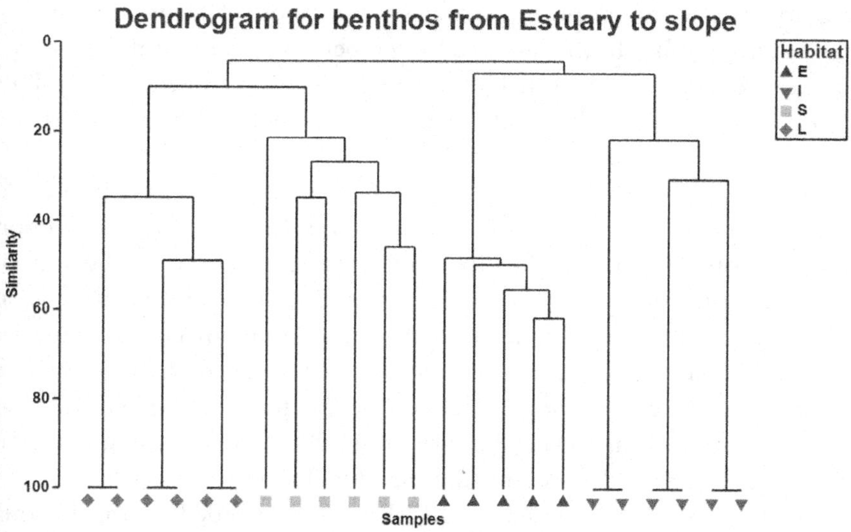

Figure 6.1 Serial changes in species composition with increase in depth in the Bay of Bengal. *E*, Estuary; *I*, inshore; *S*, continental shelf; and *L*, continental slope. *From Khan, A., Manokaran, S., & Lyla, S. (2017). Changes in macrobenthic community structure from estuary to continental slope in the south-east coast of India. Journal of the Marine Biological Association of the United Kingdom, 97(1), 161–180.*

Khan et al., 2017). Joydas and Damodaran (2013) stated that benthic composition varies significantly between shallower and deeper waters. The family and generic levels data also show community shifts in relation to depth variations.

Ajmal Khan et al. (2017) described the serial changes occurring in species composition with an increase in depth from the estuary through, inshore, continental shelf to continental slope regions (Fig. 6.1). In studies related to deep-sea macrobenthic infauna, the depth size relations are a key feature both at community and species levels. Several deep-sea studies from small infauna such as polychaetes and cumaceans (Grassle et al., 1979) through gastropods (Rex, 1973) to megafauna (Day & Pearcy,1968; Haedrich et al., 1980; Stefanescu et al., 1993) exhibited the existence of boundaries of greater faunal renewal at certain depths (Ajmal Khan et al., 2017).

6.2.2 Temperature

Air temperature regulates the aquatic temperature (freshwater, estuarine, and marine). The temperature is a contributing factor to the gradient in salinity. Higher temperature generates evaporation of seawater changing the salinity profile. In this way, water temperature is one of the most significant environmental variables determining the biogeographic shifts (Southward et al., 2004) and phenology of benthos (Beukema et al., 1998; Philippart et al., 2003; Kirby et al., 2007).

Biogeographic shifts show the link between species distribution and sea temperature. The latitudinal shift noticed of late in species distribution in the North Atlantic is due to changes in temperature arising out of global climate conditions (Poloczanska et al., 2013). Such shifts have been observed in the rocky intertidal habitats also. Southern epifaunal species such as *Osilinus lineatus* and *Gibbula umbilicalis* have been observed to have north and northeastern range extensions in Britain's rocky intertidal areas due to climate change (Mieszkowska et al., 2006). While the population size has increased, a decrease in adult size has also been noticed.

Such biogeographical shifts have also been recorded in the benthic community of Britain waters (Southward et al., 2004). Changes in macrobenthic infaunal density and composition are influenced by the rising temperature in the North Sea and Bay of Bengal (Birchenough et al., 2013; Ajmal Khan et al., 2018). Crustacean assemblages have become dominant besides the appearance of undocumented species in the soft substratum. The same pattern has been recorded in the rocky bottom with an increase in warm water epifaunal barnacle due to a rise in seawater temperature (Birchenough et al., 2013). The benthic infaunal density and structures have declined due to a rise in temperature during summer in the Gulf of Salwa and several organisms found during winter could not survive (Joydas et al., 2015).

Phenology deals with the study of species life-cycle events and their relationship with climatic conditions. These include (1) reproductive output, (2) larval transport and settlement besides, and (3) recruitment and postrecruitment development of benthic organisms. These play a vital role in benthic composition, diversity, and functional aspects (Birchenough et al., 2015). Biotic and abiotic (direct—physiological responses and indirect—variations in trophic relationships) factors modulate the life cycle processes, which are influenced by climate change. Change in temperature affects the metabolic activity and contributes to higher biomass followed by the production of more bivalve eggs in spring (Strasser et al., 2003) whereas rising temperature reduces the reproductive activity and delays spawning among intertidal bivalves (Philippart et al., 2003). Hence the recruit density varies highly.

In the deep oceans, the temperature in the abyssal region ranges between 0.01°C and 4°C (Yasuhara & Danovaro, 2016) and in the bathyal from 2°C to 8°C except in some seafloor of the warm seas which have >20°C temperature (Roder et al., 2013; Sweetman et al., 2017). The earth model system has shown that some abyssal ecosystems are warming (0.01°C–1°C decade-1) and the approximate ocean's heat uptake into the deep ocean (>2000 m) is 19% (Southern Ocean; Purkey & Johnson, 2010; Sweetman et al., 2017). In the next 84 years, the temperature changes will be the highest in abyssal plains (Mora et al., 2013). Increasing higher temperatures in bathyal depths are also expected as per the modeled data (3.6°C, 4.4°C, and 3.7°C) in deep oceans by 2100 (Purkey & Johnson, 2010; Mora et al., 2013; Somavilla et al., 2013). In the polar region, the temperature has been predicted to rise in bathyal and abyssal depths at 0.1°C–3.7°C by 2100 (Sweetman et al., 2017). Changes in size distribution and dominance of macro and megafaunal communities due to temporal variations in food inputs as a consequence of climate change are happening in abyssal areas (Ruhl et al., 2014; Sweetman et al., 2017).

The relative importance of archaea in benthic ecosystem functions might have been altered by deep-water temperature variations in the polar region. The warming of polar benthic regions will also create new habitats for invasive species. The invasion of durophagous predators due to the continuous temperature rise has the prospects of moderating the shelf benthos. However, it is the cause of the erosion of the indigenous marine life in Antarctica.

6.2.3 Salinity

Benthic organisms (both macro and meio) are distributed in all salinity regimes from freshwater to brine seep areas as well as estuarine, intertidal, deep-sea slopes, and abyssal plains because of their adaptations to a wide range of salinities. Many benthic studies have indicated the adaptation ranges of benthos in different habitats due to salinity fluctuations (Giere & Pfannkuche, 1978; Moens & Vincx, 2000) especially during the monsoon season when heavy rainfall brings high river discharge into the estuarine habitats which lead to alterations in salinity conditions of the estuary (Feebarani et al., 2016) (Fig. 6.2). The benthic faunal composition has been reported to change in the zones having critical salinity variations (brackish water, estuary, and lagoon) (Boesch et al., 1976; Matthews &

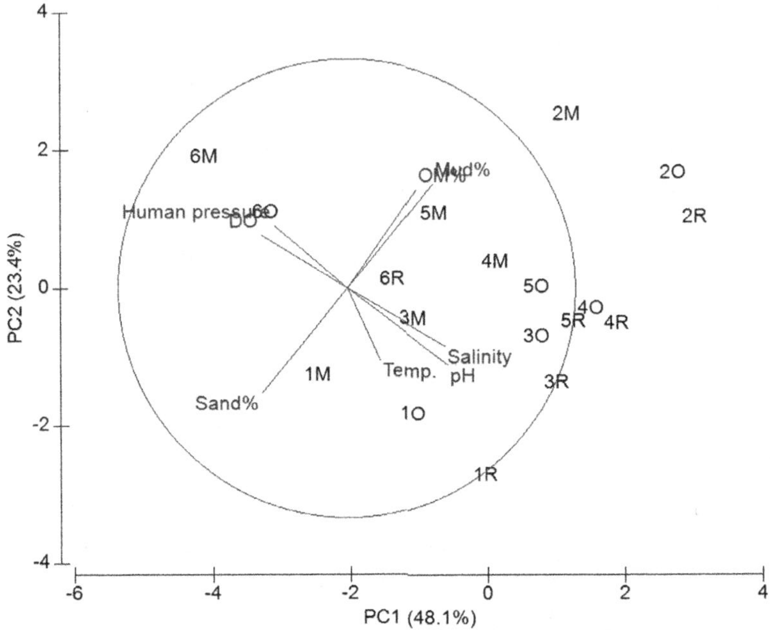

Figure 6.2 Principal component analysis (PCA) showing station clustering based on environmental parameters (salinity and dissolved oxygen), and human pressure value. *R*, premonsoon; *M*, monsoon; and *O*, postmonsoon. *From Feebarani, J., Joydas, T.V, Damodaran, R., & Borja, A. (2016). Benthic quality assessment in a naturally- and human-stressed tropical estuary. Ecological Indicators, 67, 380–390.*

Fairweather, 2004; Ansari et al., 2017a, 2017b) as salinity gradients strongly determine the occurrence and infaunal species composition (Ingole et al., 1998; Richmond et al., 2007; Rolston & Dittmann, 2009). To undertake effective management measures, the assessment of estuarine health is important. For example, the Cochin estuary is a tropical estuary that faces anthropogenic industrial stress as has been proved by the assessment of benthic communities. It showed moderate-disturbance class and moderate poor ecological status of benthic communities. An increase in salinity can cause the death of larvae and juveniles because of the dehydration of cells and the decline in turgor pressure (Einav et al., 2002). Salinity changes in estuaries occur over tidal cycles. Floods also cause salinity variations radically, which are detrimental to the sedentary benthic communities (Ritter et al., 2005).

In the interstitial space of the intertidal area, the salinity sharply decreases vertically and horizontally in relation to temperature. In the muddy shore, the water-saturated fine sediment particles with a reduced capacity of water exchange along the vertical layer make the surface salinity go up to hypersaline conditions by evaporation. When rainfall occurs, the salinity comes down suddenly to zero. These fluctuations are highest in the uppermost layer of the seabed. In the depths of 2–5 cm, the salinity fluctuations with temperature are minimal.

The meiofaunal density declines after rains in the surface sediment layer (Alongi, 1990). In the high-water area of the mudflat (North Sea), salinity levels increase up to 40 PSU during warm summer. However, the salinity at the depth below 5 cm comes down to 30 PSU. Below this depth range, the salinity is similar to brackish water conditions (around 20 PSU). In the well-drained beaches having high permeability due to the presence of coarser sediment, rains influence the salinity down to the depth of 30 cm.

In the sublittoral and nearshore coastal area, depending on the local geological, climatic, and geographical conditions, the salinity can vary in its influence on the benthic community. The Gulf of Salwa in the Saudi Arabian coastline of the Arabian Gulf is a unique habitat characterized by very limited water circulation, long residence time and high evaporation (Purser & Seibold, 1973; Basson et al., 1977; Joydas et al., 2015) leading to high salinity ranges from about 55 PSU in the entrance of the Gulf to 70PSU in the southern region (Basson et al., 1977; Joydas et al., 2015). Here the macrobenthic infauna is characterized by low species composition (1990; Riera et al., 2013; Joydas et al., 2015) (Fig. 6.3). Benthic infaunal communities in the southern region of the Gulf of Salwa are under stress due to higher salinity (60 PSU) and temperature (35°C). Benthos such as *Fabricia* sp., *Heteromastus filiformis*, *Platynereis insolita*, and *Nereis* sp. 1 and nonpolychaetes as chironomid larvae, podocops, and cumaceans are reported to be capable of surviving under these extreme conditions (Joydas et al., 2015).

In the continental margin area, depth is the main factor in structuring megafaunal assemblages. The general pattern observed in salinity is that it decreases in the outer shelf region during summer. It increases rapidly during the fall strengthening the upwelling-favorable easterly winds (Conlan et al., 2008). In relation to the changes in salinity, benthic communities are reported to vary.

Generally, in the deeper waters (beyond 300 m), in view of the homogeneous conditions in salinity with temperature, the benthic community gets adapted to this stable condition having limited food availability (Thiel, 1975; Gage, 1978; Janssen & Taviani, 2015). However, few

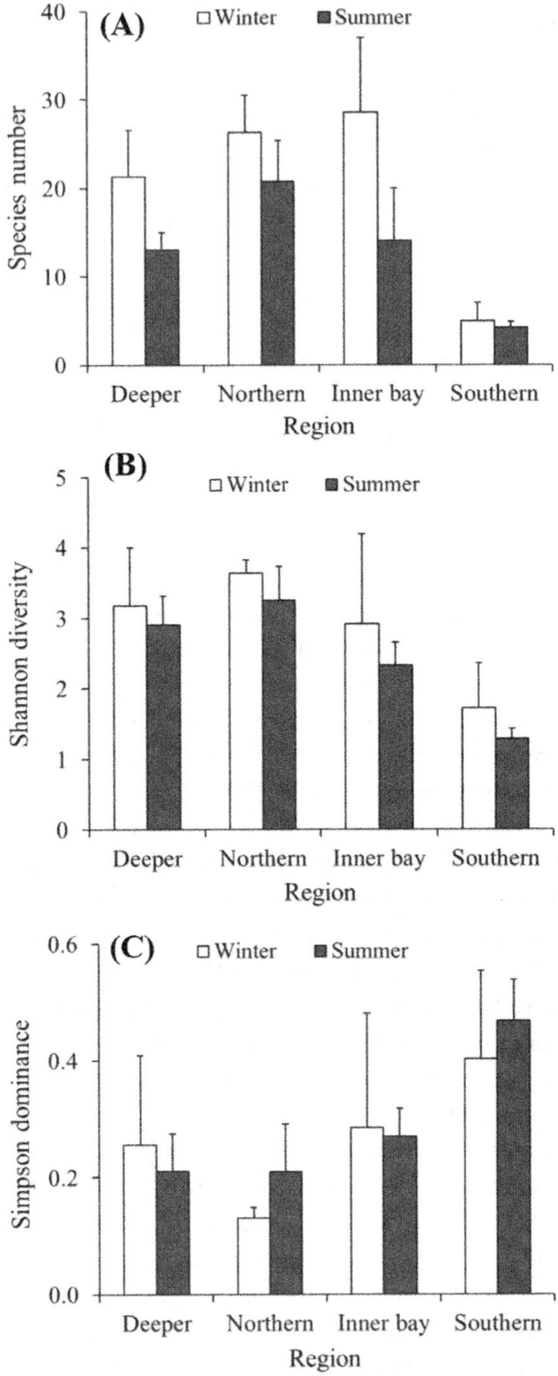

Figure 6.3 Spatial and seasonal variations (mean ± SE) in community structure of macrobenthic polychaetes: (A) species number, (B) species diversity index, and (C) species dominance index. *From Joydas, T. V., Qurban, M. A., Manikandan, K. P. Ashraf, T. T. M., Ali, S. M., Al-Abdulkader, K., Qasem, A., Krishnakumar, P. K. (2015). Status of macrobenthic communities in the hypersaline waters of the Gulf of Salwa Arabian Gulf. Journal of Sea Research, 99, 34–46.*

large deep-water areas (Arctic, Mediterranean, and Red sea) of world oceans having salinity changes show variations in composition, richness, diversity, and standing stock of infaunal benthic community (Cusson et al., 2007; Papiol et al., 2012; Joydas et al., 2018).

6.2.4 Turbidity

Turbidity is the amount of reduction in water transparency due to suspended solids, organic matter content, and algal cells (Maltese et al., 2013). Rivers, creeks, and lagoons have connections with the sea through the estuarine system in the coastal regions and these bring sediments, which vary in relation to seasons including the monsoon floods (Doxaran et al., 2006). The sediments are then dispersed by the currents. These then settle and are re-suspended by wave and current energy. That way, the turbidity varies remarkably over the intertidal and tidal zone (Uncles et al., 2002; Wolanski & Spagnol, 2003; Maltese et al., 2013). Human activities in the coastal area in particular dredging impacts the water body and influence the turbidity regime occasionally and locally (Schoellhamer, 1996; Maltese et al., 2013) In line with the changes in turbidity, alternations in the marine food web also take place (Harvey & White, 2008).

Turbidity can be considered as one of the water quality parameters of the coastal area due to its influence on planktonic and benthic algal productivity besides underwater vegetation (seagrass) and coral reefs (Maltese et al., 2013). The impacts on infaunal and epifaunal benthic biota are caused due to alteration in the physical structure of sea substratum (sedimentation and eutrophication) and variations in the physical properties of water (temperature, dissolved oxygen, and light penetration) (Pennock & Sharp, 1994; Moreno-Madriñan et al., 2010; Maltese et al., 2013; Bessell-Browne et al., 2017). During the period of higher levels of turbidity, mortality can occur due to suspended materials interfering with the feeding of suspension and deposit feeders (bivalves, tubicolous polychaetes, and sea urchins) through clogging of their feeding organs (Nicholls et al., 2003).

Benthic organisms (macro and meio) in the estuarine habitat are more affected by dredging and suspended sediment carried from upstream which causes higher turbidity (Mattos & de Almeida, 2016). Even though river discharges influence the benthic ecosystem, dredging impact causes more turbidity, harming the biota. However, higher turbid conditions are favorable to opportunistic species such as *Heleobia australis* at Itajaí-Açu river estuary in Brazil (Mattos & de Almeida, 2016). High turbidity in the oligohaline zone due to higher inputs of organic content and nutrients has been reported to have high microbial load leading to hypoxic conditions often (Goosen et al., 1999; Ysebaert et al., 2003). Due to highly variable environmental conditions, constant disturbances are there to the benthic community not allowing it to progress beyond the early stage of succession from oligohaline to mesohaline (Ysebaert et al., 2003). Turbidity is also one of the key factors influencing the composition, structure, biomass availability, and growth rate of infaunal benthic communities in lagoons (Meerhoff et al., 2013) and intertidal region (Raghunathan et al., 2003; Bhadja et al., 2014).

Turbidity has been found to show a negative correlation with water depth as a load of suspended material brought by river runoff gets diluted from the nearshore to offshore (Xu et al., 2018). The infaunal community assemblages have also been found to vary depending upon turbidity. The density of *Sternaspis scutata, Odontamblyopus rubicundus*, and *Oratosquilla oratoria* has been found high in the nearshore (lower depth) area with higher turbidity whereas *Plesionika izumiae* and *P. argentatus* have been found abundant in the offshore area (higher depth) having lower turbidity off the coast of Zhejiang in East China Sea (Xu et al., 2018).

The hydrodynamics of mesoscale structures is one of the key factors determining the export of materials to benthos in oceanic waters. Storm and flood events also create press-type disturbances (bring terrigenous material) and mobilization of bottom sediments also makes the pulse type process toward the edge of the continental shelf which is displaced by turbid currents (Ogston et al., 2000; Hughes et al., 2009; Nishijima et al., 2013). Additional allochthonous (turbid sediments, phytodetritus, leaves, flowers, and roots of mangroves and seagrasses) materials are distributed in the continental margin. Active canyons here are the major topographic features with unstable environmental conditions, which are related to tidal currents, episodic slumps, gravity, turbidity, and flushing flow of sediments (Gardner, 1989; Canals et al., 2006; Dde Stigter et al., 2007). These canyons play a key role in transporting mesoscale structures, storms, and floods as well as additional allochthonous materials from shelf to abyssal depths. These materials are being gradually degraded by bacterial associations (Rowe et al., 1982; Palanques et al., 2008; Gunton et al., 2015) (Fig. 6.4). Due to this process, much higher turbidity is flowing over the canyon depositing fine sediments in the adjacent slope, which can enrich organic content in the surrounding area of the canyon

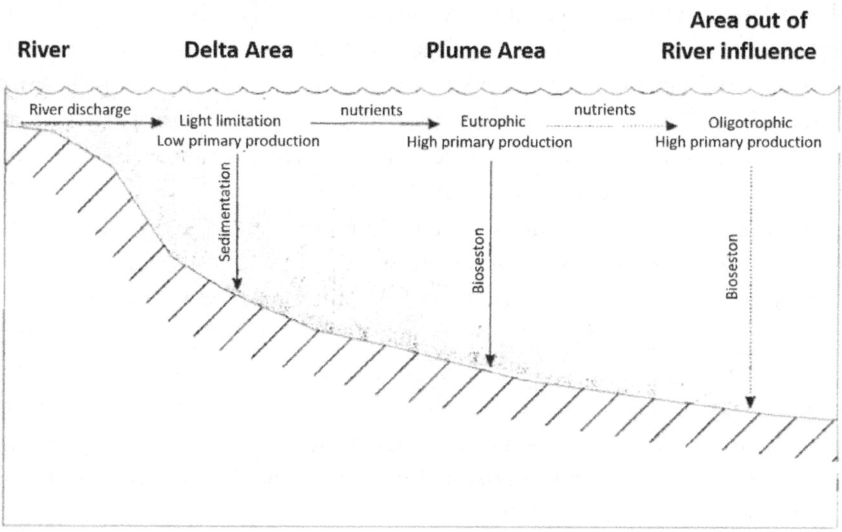

Figure 6.4 A generalized model for the influence of the discharge of a large river on the sedimentation processes on the shelf. Arrows indicate nutrient transport and sedimentation. *From Rhoads, D. C., Boesch, D. K., Zhican, T., Fengshan, X., Liqiang, H., & Nilsen, K. J. (1985). Macrobenthos and sedimentary facies on the Changjiang delta platform and adjacent continental shelf, East China Sea.* Continental Shelf Research, 4, *189–213 and Wijsman, J. W. M., Herman, P. M. J., & Gomoiu, M. T. (1999). Spatial distribution in sediment characteristics and benthic activity on the Northwestern Black Sea shelf.* Marine Ecology Progress Series, 181, *25–39.*

(Cunningham et al., 2005). Bottom-currents associated with turbidity inside the canyons are believed to reduce the infaunal density as found in the canyons and abyssal plains of Nova Scotia Rise (western North Atlantic) (Aller, 1989; Gage et al., 1995; Koho et al., 2007). A physical disturbance that creates more organic content is shown to increase the density of macrofauna four-time than adjacent plain (Cascadia Channel in the northeast Pacific Ocean) as well as canyons (Whittard Canyon, Northeast Atlantic) (Griggs et al.,1969; Thistle et al., 1985,1991).

6.2.5 Sediment texture

The composition of the sediment in the sea substratum varies in relation to the source materials either geological (biogenous, cosmogenous, terrigenous and hydrogenous, carbonate, clay, quartz, etc.) or biological (benthic ecosystem) or both (Eleftheriou & McIntyre, 2005).

Sediment texture crucially depends on the environmental energy (waves, currents, tides, depth, and types of suspended matter) as well as the balance between accumulation and erosion. The process in the balance of temporal and climatic changes can significantly be related to sediment accumulation and erosion processes (Widdows et al., 2000). Physical and biological aggregations of grains change the size and properties of the sediment. Besides, in the silt-rich sediments, due to the process of adsorption of organic molecules over the particle surfaces, besides electrostatic, and interparticle attractions, the sediment behaves cohesively compared to the coarser grain sizes (Eleftheriou & McIntyre, 2005). Fine sediments, which are packed tightly and have less interstitial space restrict water movement for particle transports. Such sediments require high energy for particle movement (Gray & Elliott, 2009). Coarse sediments mixed loosely have more interstitial spaces for high water movement with particle transportation. Sediment composition (grain size) influences the occurrence of macro and meiofaunal organisms' in the benthic substratum (Eleftheriou & McIntyre, 2005).

The proportion of suspension feeders has been found to increase with increasing grain size while that of deposit feeders decreased. Conversely, the percentage of epifauna, generally suspension feeders decreased from gravel to mud whereas the proportion of infauna mainly deposit feeders increased (Sanders, 1968; Decho, 1990; Manokaran et al., 2013).

Van Colen et al. (2014) investigated the changes in macrobenthic biomass and sediment composition during 2003 and 2010 in two contrasting habitats (between the deep tidal inlet channel and the intertidal habitat surrounding the tidal pond) in the Wester Schelde estuary of the Southern Bight in the North Sea at the border between Belgium and The Netherlands. The model constructed by them relates the occurrence and biomass of macrobenthic species with sediment median grain size. The spatial and seasonal distribution of benthic fauna is related to variation in the composition of water and sediment as well as dietary habits (Snelgrove & Butman, 1994; Mannino & Montagna, 1997; Ourives et al., 2011). Deposit feeders and burrow builders alter the physicochemical properties of the sediments and make high volumes of turnover of the sediment by bioturbation, which support faunal colonization (Peterson, 1991; Jayachandran et al., 2019). The relationship has also been found among the combined biotic factors such as microbial activity, organic matter content, competition, predation, bioturbation, amensalism, larval settlement dynamics, and abiotic factors including grain size, hydrodynamics, and chemistry of water-sediment interface layers together with specific processes of sediment accumulation, resuspension, and transport (Hall, 1994; Manokaran et al., 2013).

In the shallow coastal habitats, the benthic infaunal and epifaunal composition, density, and diversity are strongly related to the habitat characteristics including morphology and physical drivers such as wave, tidal, and sediment dynamics (Ganesh & Raman, 2007; Gray & Elliott, 2009). This relation has been found prevailing at different depth contours from 30 to 100 m in the North Sea (Thrush et al., 2005), shelf waters of Bay of Bengal (Ganesh & Raman, 2007; Manokaran et al., 2013), Mediterranean (Coma et al., 2000), and North Pacific ocean (Kaufmann & Smith, 1997). Roditi-Elasar et al. (2019) have stated that the higher macrofaunal abundance and taxon richness may be related to a more heterogeneous sea bed provided by the higher diversity in grain size existing in the Akhziv submarine canyon (eastern Levant Basin, Mediterranean Sea). Durden et al. (2015) have studied the benthic megafaunal biomass and diversity in the deep sea (abyssal hill and plain sites) of the Mid-Atlantic Ridge. In the abyssal area, the sediment movement has been found to be due to local hydrodynamics such as historical turbidity current and modern sediment movement by currents making physical disturbances. These are related to macro and meiobenthic communities (Young et al., 2001; Durden et al., 2015). Megafaunal assemblages and trophic compositions in plains and hills are influenced by sediment particle size distribution (Durden et al., 2015).

6.3 Chemical factors

The important chemical parameters determining the composition and function of benthic communities include factors such as dissolved oxygen, nutrients, organic matter, redox potential index, and contaminants (heavy metal, petroleum hydrocarbons, and microplastics).

6.3.1 Dissolved oxygen

Benthic composition and community structure are also influenced by oxygen levels in the bottom waters (Rosenberg, 2001). In the coastal area, eutrophication has an important say (negative) in the level of oxygen (Diaz & Rosenberg, 1995). The overall volume of oxic microniches grossly exceeds that of the "classical" two-dimensional sediment surface. Considering the complexity and dynamics of the "oxygen realm" and its eco-physiological implications on small infaunal benthos, further microscale investigations are needed especially studies under natural field conditions.

Coastal hypoxic conditions influence feeding, growth, and individual health of benthic infaunal animals (Wu, 2002). The concentration of oxygen varies in the natural environment and is patchy due to the tidal cycle, bioturbation, and other activity of animals. The low oxygen level can force benthic infaunal organisms to move close to the sediment surface (Wu, 2002, Selvaraj et al., 2019) and can make deposit feeders become suspension feeders (Diaz & Rosenberg, 1995; Wu, 2002; Levin et al., 2009). The hypoxic condition can become lethal to many organisms and can eliminate sensitive organisms (Rabalais & Baustian, 2020) (Fig. 6.5). It reduces richness and changes the structure and functional attributes of benthic in and epifaunal communities from intertidal to deep oceans (Diaz & Rosenberg, 1995; Wu, 2002; Levin et al., 2009; Islam et al., 2013).

Persistent low oxygen level is evident in the mid-water oxygen minimum zones (OMZs), defined as regions where oxygen concentrations are less than $0.5\ \mathrm{mL\,L^{-1}}$ (or about 7.5% of saturation; $<22\ \mu\mathrm{M}$). Such regions referred to as oxygen minimum layers or oxygen-deficient zones

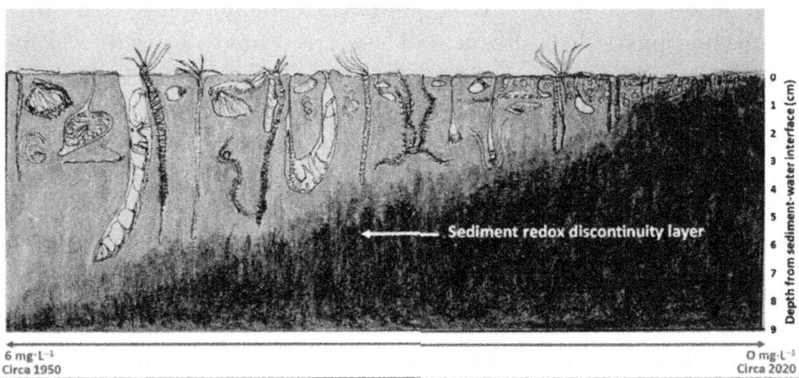

Callianassa (c) *Anadara* (b) *Echinocardium* (e) *Abra* (b) *Scoloplos* (p) *Armandia* (p)	*Mulinia* (b) *Diopatra* (p) *Anadara* (b) *Tellina* (b) *Scoloplos* (p)	*Chaetopterus* (p) *Scoloplos* (p) *Armandia* (p) *Solen* (b) *Diopatra* (p) *Alphaeus* (c)	*Sabella* (p) *Hemipholis* (e) *Mulinia* (b) *Capitella* (p) *Armandia* (p) *Paraprionospio* (p)	*Macoma* (b) *Pectinaria* (p) *Owenia* (p) *Magelona* (p)	*Armandia* (p) *Capitella* (p) *Sabellla* (p) *Paraprionospio* (p) *Magelona* (p)	*Magelona* (p) *Paraprionospio* (p) *Ampharete* (p) Sulphide- oxidizing bacteria

Figure 6.5 Changes in benthic infauna (by genera) and sediment structure including the redox discontinuity layer from well-oxygenated waters on the left, likely representing years circa 1950, to no oxygen on the right, representing years circa 2020. (Letters are (b) bivalve, (c) crustacean, (e) echinoderm, and (p) polychaete.) *Modified from Pearson, T. H., & Rosenberg, R. (1978). Macrobenthic succession in relation to organic enrichment and pollution of the marine environment.* Oceanography and Marine Biology: An Annual Review, 16, *229–311; Rabalais, N. N., & Baustian, M. M. (2020). Historical shifts in benthic infaunal diversity in the Northern Gulf of Mexico since the appearance of seasonally severe hypoxia.* Diversity, 12, *49.*

are present at different water depths ranging from shelf to upper bathyal zones (100–1300 m). Oxygen minima are created through biochemical oxygen consumption, with circulation affecting their distribution and position within the water column (Kamykowski & Zentara, 1990; Levin, 2003). Even though such zones are formed in the water column, their effects are strongly noticed in the benthic substratum (Devol & Hartnett, 2001; Joydas & Damodaran, 2014). Here benthic infaunal organisms are found to have morphological and physiological adaptations in gills and parapodia that help to increase oxygen uptake. In such places, low diversity and high dominance of small polychaetes have been found (Childress & Seibel, 1998; Levin, 2003; Joydas & Damodaran, 2014).

The benthic habitat is heterogeneous in the OMZs due to oxygen level in the bottom water, the gradient in sulfide levels, characteristics of the substratum, bacterial mats, and variations in organic content and level of pH (Gooday et al., 2010). In some places of the OMZs, the substrata are colonized by encrusting organisms on hard grounds, phosphorites, carbonates/biotic substrata, and additional habitats. However, much of the heterogeneity in OMZs is attributed to strong sea-floor oxygen gradients, reinforced by variations in sediment nature and organic matter (Gooday et al., 2010). In the Arabian Sea, a clear pattern in physicochemical and biological parameters between OMZ and non-OMZ regions has been observed (Gooday et al., 2010; Joydas & Damodaran, 2014). Infaunal species assemblage has been constituted by organisms such as *Magelona cincta, Prionospio pinnata, Ancystrosyllis parva, Lumbrineris latrielli,* and *Pista* spp. in the shallower depths. Among these, only *P. pinnata* could proliferate well in the OMZ due to its high adaptability to low level of oxygen with elaborate branchial structure with more number of pinnules which can take more oxygen besides enzymatic adaptations to anaerobic metabolism. *Prionospio pinnata, Prionospio cirrifera, Prionospio cirrobranchiata* (Family Spionidae), *Paraonis gracilis gracilis, Aricidea fauveli* (Family Paraonidae), *Cirriformia* sp. 1, and *Cirratulus dasylophius* (Family Cirratulidae) have been found dominant in the deeper depth stations of OMZ (Quiroga et al., 2007; Levin et al., 2009; Joydas & Damodaran, 2014) (Fig. 6.6). Many studies have been done on the macrobenthic organisms of the OMZs in the Pacific and Atlantic oceans, besides the Arabian Sea and Bay of Bengal. Many other studies did also show oxygen is the key factor that determines the meiofaunal, macrofaunal, and megafaunal regional composition and diversity pattern (Levin, 2002; Gooday et al., 2010; Ingole et al., 2010; Joydas & Damodaran, 2014; Raman et al., 2015).

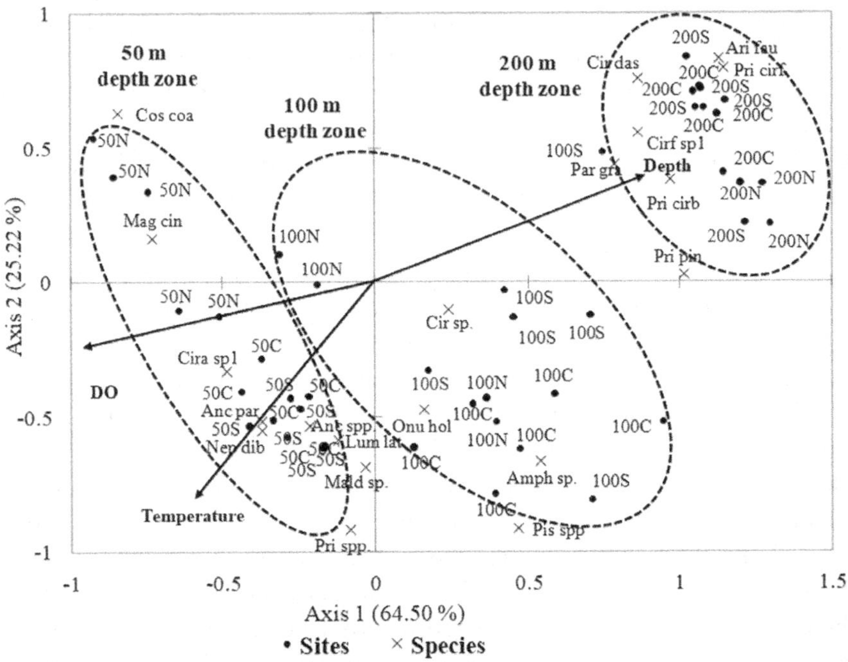

Figure 6.6 Triplot of canonical correspondence analysis showing scores of sites, the 20 most abundant species and explanatory variables. *From Joydas, T. V., & Damodaran, R. (2014). Infaunal macrobenthos of the oxygen minimum zone on the Indian western continental shelf. Marine Ecology, 35, 22–35. doi:10.1111/maec.12052.*

6.3.2 pH

pH is one of the important factors that determine the benthic assemblage. When the level drops under 5 or rises above 9, it is detrimental to the benthic organisms. Variations in pH may compromise connected biological processes including metabolism and growth. However, it also increases the solubility of calcium carbonates and the bioavailability of metals. The increase in the level of CO_2 concentration in the ocean has led to a reduction of around 0.1 unit of pH unit in the surface waters when compared to the preindustrial times (Caldeira & Wickett, 2003). A further decline in the range of 0.3–0.5 is expected by 2100 (Caldeira & Wickett 2005). A set of biogeochemical processes impact the pH value in the marine benthic substratum (sediments) which increase or decrease free proton levels in sediment pore waters. These processes connected mainly to cycles of carbon, oxygen, nitrogen, phosphate, silicate, sulfur, iron, and manganese are associated with activities such as heterotrophic respiration chemoautotrophic process, photosynthesis, precipitation, and dissolution of calcium carbonate. These influence the pH balance in marine sediments (Cai et al., 1995; Stahl et al., 2006; Silburn et al., 2017). Ocean acidification causes major impacts on the skeleton building process of marine in and epibenthic organisms (Hendriks et al., 2010; Kleypas et al., 1999). The coral reefs will be very much affected, as the net accumulation of calcium carbonate will be impacted. Reduction in the level of carbonate mineral aragonite due to acidification will come in the way of calcification (Kleypas et al., 1999).

The benthic environment is a reservoir for biodiversity. However, ocean acidification reduces it through its impact on major biological processes including growth rate, respiration, calcification, and metabolic activity. Moreover, chemical changes linked with ocean acidification impact individuals at a physiological level especially through disruption of extracellular acid-base balance. Among the benthic invertebrates, epifaunal organisms including corals, echinoderms, and mollusks are having heavily calcified structures and low capacities to defend the set points of acid–base regulation. Therefore they suffer the most. However, macrobenthos (Selvaraj et al., 2019) especially crustaceans are comparatively resistant to ocean pH variations (Hendriks et al., 2010). In intertidal regions, crustaceans and gastropods are linked with pH variations and expose differences in metabolic activities, morphological behaviors (Widdicombe & Spicer, 2008). Similarly, deep-sea animals indicate the slow rate of ion exchange among their gills (Goffredi & Childress, 2001; Widdicombe & Spicer, 2008). Ocean warming can relate to ocean acidification in species with magnesium-calcite shells or skeletons as its content in calcite generally increases with seawater temperature (Figuerola et al., 2020).

The epibenthic taxa such as bryozoans, bivalves, sea urchins, asteroids, pteropods, and stony corals are coming under high vulnerability whereas coccolithophores and foraminiferans are under medium vulnerability. The negative impact of ocean acidification on these taxa can play a major role in the ecosystem functioning which can modify benthic structure and biodiversity (Figuerola et al., 2020).

6.3.3 Nutrients

Nutrients arrive from terrestrial inputs, atmospheric deposition, and as well through decomposition of organic substances in the coastal environments. However, these can become intensified due to human interference

(Nixon, 1995) through sewage inputs, run-off of nutrient-rich fertilizers, cleaning products, and animal wastes from the land increasing a load of nitrogen and phosphorus in the coastal environments (Nixon, 1995; Vitousek et al., 1997). For example, the Bay of Bengal is traditionally considered to have relatively poorer biological productivity than its western counterpart—the Arabian Sea in the Indian Ocean (Madhupratap et al., 2003). Although the riverine flux may bring in nutrients, they are thought to be lost to the deep because of their narrow shelf (Sen Gupta et al., 1977; Radhakrishna et al., 1978). Comparatively heavier cloud cover during the summer monsoon and light inhibition due to turbidity could be the other reasons. Being a cyclone-prone region, the episodic events are likely to churn up the area, injecting nutrients to the shallow euphotic zone thereby enhancing production in the upper layers (Madhupratap et al., 2003). These have the potential to stimulate primary productivity in the coastal waters (Vitousek et al., 1997; Cloern, 2001). The high nutrients are being connected to eutrophication resulting in algal blooms, which lead to hypoxic and anoxic conditions. These significantly alter seafloor communities and processes (Gladstone-Gallagher et al., 2020) through both bottom-up and top-down processes, which modify the ecosystem architecture and interaction networks (Thrush et al., 2014; Douglas et al., 2017; Gladstone-Gallagher et al., 2020).

The biological components in the coastal waters including intertidal regions filter the nutrients before they could reach the deeper waters. It happens through feeding and subsequently through degradation, metabolization, and remineralization transforming and retaining nutrients. Bioturbation and bioirrigation are helpful indirectly in retaining the nutrients through P-binding and by facilitating N removal by nitrification-coupled denitrification in sediment (Karlson et al., 2007; Asmala et al., 2017; Carstensen et al., 2020).

In the intertidal shallower coastal ecosystem, the synergistic effect of bioturbating Chironomids and microphytobenthos strongly affects the internal nutrient cycling as well as the use of bottom-up control of pelagic primary production. However, in heterogeneous areas, the role of benthic in and epifaunal species/functional traits for nutrient cycling in sediment is highly complex (Benelli et al., 2018; Carstensen et al., 2020). Gladstone-Gallagher et al. (2020) have reported that nutrient enrichment can change sediment properties and macrofaunal composition, which lead to sediment erosion as well as benthic-pelagic coupling resulting in the release of nutrients and chlorophyll-a.

By studying the biological traits in the coastal area, the impacts of primary and secondary producers on the turnover of nutrients can be evaluated. The turnover rates are based on the association of two biotas having the dominant role as the primary producers with short-lived annuals including cyanobacteria and filamentous algae, which grow, and take up nutrients rapidly as well as vascular plants and macroalgae which grow slow but store and retain nutrients on longer time scales. However, the role of vegetation in filtering coastal nutrients depends on local environmental conditions and the composition of species' biological traits.

In the inner coastal area, benthic communities are more prone to eutrophication, indicating their biological traits, which enhance carbon and nutrient turnover (Villnäs et al., 2019; Carstensen et al., 2020). However, in deep seas, there is a less nutrient-enriched area with long-lived, and deep burrowing organisms which may have a major role in nutrient retention and removal from the coastal filter (Carstensen et al., 2020) (Fig. 6.7).

The deep-sea environmental conditions play a major role in the cycling of various nutrients including nitrogen, silica, phosphorus, hydrogen, and sulfur. In nitrogen cycling, microbially mediated processes such

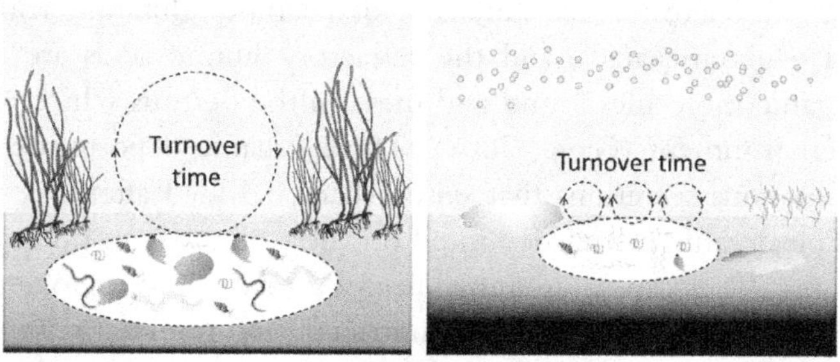

Figure 6.7 The effects of plants and animals on turnover time in the coastal filter, which are largely dependent on what biological traits the organisms express. Large and long-lived species are likely to promote the removal and/or retention of carbon and nutrients (left). In contrast, species characterized as small and short-lived will enhance the turnover rates of carbon and nutrients within the coastal zone (right). Increasing eutrophication can result in such a shift with major implications for the coastal nutrient filter. Images by Tracey Saxby, Dieter Tracey, Jane Thomas, Integration and Application Network, University of Maryland Center for Environmental Science (https://ian.umces.edu/imagelibrary/). *From Carstensen, J., Conley, D. J., Almroth-Rosell, E., Asmala, E., Bonsdorff, E., Fleming-Lehtinen, V., Gustafsson, B. G., Gustafsson, C., Heiskanen, A. S., Janas, U., Norkko, A., Slomp, C., Villnäs, A., Voss, M., & Zilius, M. (2020). Factors regulating the coastal nutrient filter in the Baltic Sea.* Ambio, 49, *1194–1210. https://doi.org/10.1007/s13280-019-01282-y.*

as nitrification, nitrogen fixation, denitrification, and anaerobic ammonium oxidation take place in both benthic and pelagic environments. These provide the sources and sinks as well facilitate the transformation of nitrogen in the nitrogen pools (Gruber & Sarmiento, 1997; Wuchter et al., 2006; Thurber et al., 2014). In the water column and sediments, nitrogen is selectively separated from dissolved organic materials by biological process (Hunter et al., 2012). However, nitrogen cycling in the meio and macrobenthos discharge nutrients including nitrate, nitrite, and ammonium back into the water column (Laverock et al., 2011). These nutrients are returned to the surface water by the upwelling process.

6.3.4 Organic matter

Detritus can be defined as dead organic materials from any trophic level from the ecosystem. Most of the detritus occurring in sediments are derived from dead plankton and decaying plankton blooms. It is trapped by huge filtering systems of seashores (Riedl, 1971; Olafsson et al., 1999; Giere, 2009) and deposited on the seafloor in the intertidal, coastal, sublittoral, and deep-sea areas. The detritus has essential fatty acids derived from plants. Unbalanced nonhumic substances such as sugars (glucose, galactose, and sucrose), free amino acids (alanine, glutamic acid, aspartic acid, and ß-glutaric acid), and the refractory humic acids are the other major organic molecules found and these in the detritus which enrich the benthic environment (Giere, 2009). These organic deposits are accumulated by mucous secretions that ensure bacterial load after a few days or weeks. The organic matter deposited on the seafloor goes through one of three routes: (1) it can get remineralized by macro and meiobenthos into CO_2, dissolved organic carbon and nutrients, and remixed with the water (returned to the hydrosphere); (2) can get stored in the benthic biomass; or (3) can get buried in the deeper sediments and isolated from the atmosphere for hundreds of years (Klages et al., 2004).

Organic carbon significantly supports increased density and diversity of macro and meiobenthic organisms (Lambshead & Gooday, 1990; Witte et al., 2003; Ansari et al., 2014; Ajmal Khan et al., 2018). In the nearshore area, seagrass and benthic macroalgae serve (52% and 46%, respectively) as the source of detritus which promotes the density of benthic communities (Duarte et al., 2005; Giere, 2009; Kowalczuk et al., 2019). There is a positive relationship between these organic deposits and infaunal surface feeders. These deposits also indirectly induce infaunal subsurface feeder during their reproduction periods (Ólafsson & Elmgren, 1997; Giere, 2009; Ansari et al., 2015). However, more density indicates a decrease in richness and biomass due to depletion of oxygen and increase in toxic substances (Pearson & Rosenberg, 1978; Diaz & Rosenberg, 1995; Gray et al., 2002) which co-vary about other physicochemical parameters (higher size of sediment particles which give more surface area for absorption of organic substances) (Landrum & Robbins, 1990; Hyland et al., 2005; Kowalczuk et al., 2019). A graphic model has been derived relating the pattern of benthic communities with organic content (Pearson & Rosenberg, 1978). The relationship among organic content, the response of benthic organisms, and various physicochemical parameters such as depletion of oxygen and other sediment contaminants have been demonstrated with modification of the conceptual model (Hyland et al., 2005). Under high organic content load, benthic assemblages are characterized by the dominance of pollutant tolerant opportunistic species. Riera et al. (2013) tested macrobenthic community structure with organic-enriched treatments in sandy sediments and found a reduction in density and species richness between control and organic-enriched sediments. These changes have occurred due to a decrease of sensitive tanaid species *Apseudes talpa* in organically enriched treatments (Riera et al., 2013).

6.3.5 Redox potential index

Redox potential (RP) is one of the most frequently used physicochemical variables in benthic studies and it can be used for assessing the quality of benthic habitat (Diaz et al., 2004; Birchenough et al., 2012; Gerwing et al., 2018). However, information on RP only vaguely indicates the supply of oxygen level in the porewater and that way measurement of RP can send false signals. Therefore it has to be used in combination with other variables during the assessment of benthic quality.

In a typically yellowish colored sandy sediment, a positive EH value is obtained, whereas in soft sediment (mud) with rich organic matter having gray-to-black layers beneath a thin brighter surface layer, negative EH values are obtained. That way positive values indicate oxidized sediment whereas negative values show reduced sediment (Gerwing et al., 2018). The transition zone between positive and negative values is the redox potential discontinuity (RPD) layer that is frequently linked with disturbances including enrichment of organic matter or hypoxia/anoxia in the shallower area (Pearson & Rosenberg, 1978; Diaz & Rosenberg, 1995;

Gerwing et al., 2018). The visually assessed RPD is phrased as the apparent redox potential discontinuity or aRPD which shows a color transition from brownish/red to gray/green or black sediment (Teal et al., 2009; Birchenough et al., 2012; Sturdivant et al., 2012; Gerwing et al., 2018). The occurrence level of color-inducing ions, sediment microbes decomposition activity, organic carbon, and chlorophyll-a concentration can influence the RPD depth (Kristensen, 2000; Teal et al., 2010; Hunting & Kampfraath, 2013). Activities of macro and meiobenthic organisms such as bioturbation, irrigation, and biodeposition can also affect RPD depth by providing more oxygenated water into the deep sediment which creates a three dimensional mosaic of oxygen conditions (Kristensen, 2000; Solan & Kennedy, 2002; Sturdivant et al., 2012). The measured and apparent RPD can alter the particle bioturbation in depths, which is connected to the benthic community (Solan et al., 2004; Teal et al., 2010).

The successional stages of the infaunal benthic community can also be assessed using the aRPD along with species composition. A shallower depth aRPD with reduced species composition indicates the early successional stage while the deep vertical distribution of the RPD high functional diversity indicates the late-successional stage (Pearson & Rosenberg, 1978; Birchenough et al., 2012). The metrics such as bioturbation potential index (burrows, tubes, and feeding voids) and apparent redox discontinuity layer (aRPD) provide valuable complementary information. They indicate well the status of the sedimentary environment and functions including carbon and nutrient cycling rates as well as benthic faunal bioturbation activity and species identity (Rosenberg, 2001; Birchenough et al., 2012). Besides the above, the aRPD depth has been found to be a good relative indicator of sediment pore water Eh conditions and DO content. Sediment with a deeper aRPD depth has more available DO, and the pore water has higher Eh values (more oxidized or less reduced) than sediment with a shallower aRPD depth. As such, the aRPD depth is an effective measure for inclusion in assessments of benthic habitat quality, especially for aerobic metazoans.

6.3.6 Contaminants

6.3.6.1 Heavy metal

Heavy metal concentrations in benthic sediments are major pollution indicators due to their long-term stability and toxicity in the marine environment (Saeki et al., 1993; Pourang, 1996; Chakraborty & Owens, 2014). Heavy metals are discharged to the environment through anthropogenic inputs which include industrial effluents, domestic discharges, and mining (Chakraborty & Owens, 2014; Ahmed et al., 2015) which get absorbed and accumulated in the sediments in the estuarine, intertidal, and offshore areas (Reddy et al., 2007; Barua et al., 2017). Hence sediments serve as a major physical pool of metals and that way indicates the health of the aquatic environment including sediment characteristics, chemistry of the water, hydrographic and biological factors (Barua et al., 2017). The effect of these metals on marine organisms such as epibenthic animals such as crustaceans, coelenterates, and mollusks shows alarming levels of impact (Barua et al., 2017). Therefore quantification of metal levels in the sediment can be used as a good indicator for chronic contamination especially in the estuarine shallow water environments (Tomlinson et al., 1980; Maltby, 1999).

Millward and Grant (1995) have found higher tolerance among benthic animals in highly contaminated estuaries. Infaunal and epifaunal organisms showed differential responses to metal contaminants due to several factors including the response of individuals to habitats, the intensity of the stress, and adaptation duration (Thrush et al., 2008; Manokaran et al., 2015). Benthic organisms respond to a higher level of metal contamination both at individual and community levels (Selvaraj et al., 2019). Reduced density and body size besides dominance of tolerant taxa are the indicators of heavy metal contamination (Ryu et al., 2011; Zhang et al., 2013; Hu et al., 2019).

Effects of metal contaminants on ecosystem functioning can be assessed by studying the feeding mode of benthic communities (Piló et al., 2015). Infaunal benthic organisms have been found highly affected in the metal-contaminated sites where subsurface deposit feeders, second-order opportunistic species, and tube builders are represented by species like *Cirriformia tentaculata* and *Cirratulus chrysoderma*. However, in less contaminated sites, different species have been found associated with seagrass as well as crawlers, carnivorous, suspension feeders, and herbivorous (Piló et al., 2015; Hu et al., 2019). Rabaoui et al. (2015) has studied the relationship between macrobenthic invertebrates and heavy metal enrichment in the Gulf of Gabes, Tunisia, and recorded a low number of species, density, and diversity in metal-polluted sites.

6.3.6.2 Petroleum hydrocarbons

Hydrocarbon is also one of the major contaminants introduced in the marine environment (Spormann & Widdel, 2000; Head et al., 2006;

Ferrando et al., 2015). These are derived from natural sources including deep forest fires, natural petroleum seeps, and anthropogenic inputs (oil production, sewage discharges, and urban runoff) (Kim et al., 1999). After reaching the marine systems, hydrocarbons are deposited at high concentrations and adsorbed. They have a prolonged presence in the sediment, especially under anoxic conditions. Their presence affects in many ways the benthic communities (Engraff et al., 2011; Lee & Lin, 2013; Ferrando et al., 2015). Chemical parameters including indices, ratios as well as the occurrence of compounds are used for establishing the pollution level and to distinguish between biogenic and manmade inputs and to calculate hydrocarbons effect on macrobenthic (in and epifauna) composition such as diversity, retrogression to indicator species, and decrease of individual sizes (Gomez Gesteira & Dauvin, 2005; Seo et al., 2014; Zabbey & Uyi, 2014; Ferrando et al., 2015). Venturini et al. (2008), Ferrando et al. (2015), and Woke and Benson (2018) have observed that a higher level of oil contamination produced an impact of 40%–55% on density besides reduction in infaunal species richness and sediment reworking activity. Sensitive species disappeared and opportunistic species such as *Oligochaeta* sp. 1, *Paranebalia* sp., and *Ostracoda* sp. 2 became more prominent. Hydrocarbon has been reported to affect polychaete size and reproductive capacity (Forbes et al., 1996). Among diverse benthic taxa, amphipods are important due to their sensitivity to stress, particularly those generated from oil pollution in different regions including intertidal, nearshore, and offshore (Lotufo et al., 2016; Joydas et al., 2017; Yuewen & Adzigbli, 2018; Podlesińska & Dąbrowska, 2019). Joydas et al. (2017) have made a comparison of infaunal macrobenthic community depth-wise (nearshore and offshore) and region-wise (northern, central, and southern) in the western Arabian Gulf affected by the 1991 oil spill. After a decade from the spill incident, a low level of contamination has been found in the surface sediments of the open waters, whereas the sheltered bays showed still alarming levels.

6.3.6.3 Microplastics

Microplastics are one of the recently encountered contaminants in the marine environment. The production of plastics since 1950 has increased from 1.7 to 348 million metric tons (MT) in 2017. It is predicted that the production will reach 33 billion MT by 2050 (Crawford & Quinn, 2017 Fang et al., 2018). Plastic waste can be divided into five-dimensional scale such as nanoplastics (<1 μm), small microplastics (1.00 mm–1 μm), large microplastics (5−1.01 mm), mesoplastics (200−5.01 mm), and macroplastics (>200 mm) (Andrady, 2017). Among these, microplastics are the most abundant compound by density as well as proportion which have reached 90% now (Eriksen et al., 2014; Salvador Cesa et al., 2017; Sfriso et al., 2020).

The microplastics reach the marine environment through the following ways: primary microplastics are discharged through various open use sources including glitter, face waves, synthetic plastics [low and higher density polyethylene (PE), polypropylene (PP), polyvinyl chloride (PVC), polystyrene (PS), and polyethylene terephthalate (PET)] (Andrady & Neal, 2009; Hidalgo-Ruz et al., 2012; Napper & Thompson, 2016) and secondary microplastics which include fragmented pieces of macroplastic through interaction among biogeological agents (Artham et al., 2009; Muthukumar et al., 2011; Sfriso et al., 2020). The high density of primary and secondary polymer particles get abundantly aggregated in marine snow and sink to the benthic substratum and get accumulated in the sediment thus becoming accessible to the benthic invertebrate organisms in the intertidal, nearshore, and offshore regions (Porter et al., 2018). While infaunal benthic animals particularly suspension feeders (*Tellina* sp.) and deposit feeders (*Sternaspis scutata* and *Magelona cinta*) are taking food, they can accidentally or selectively ingest these sedimentary plastic particles (Cauwenberghe et al., 2015; Naidu et al., 2018; Wright et al., 2013). Ingestion of microplastics by lower trophic groups shows the potential activity of bioaccumulation of pollutants in the food chain (Teuten et al., 2009). Hence microplastic bioaccumulation takes place from lower to higher trophic groups (Farrell & Nelson, 2013; Nelms et al., 2018). Sfriso et al. (2020) found the filter feeders and grazers to have three to five times higher polymer values than omnivores and predators and identified polymers such as nylon (86%) and polyethylene (5%) with the remaining being constituted by polytetrafluoroethylene, polyoxymethylene, phenolic resin, polypropylene, polystyrene resin, and XT polymer in the Terra Nova Bay, Ross Sea, Antarctica.

6.4 Synergistic effect of environmental parameters on marine benthos

The distribution of macro and meiobenthic communities is not influenced by a single environmental parameter; but by a combination of various physicochemical parameters. Sediment pattern, bathymetry as well as hydrodynamic conditions have a broad range of influence on the estuarine and coastal ecosystems through the upwelling process (Bolam et al., 2008; Dutertre et al., 2013; Ansari et al., 2017a, 2017b). Many studies are available on the major environmental parameters, which influence both the infaunal and epifaunal community and its assemblages in different habitats. Dutertre et al. (2013) stated that sediment characteristics such as mud content, mean grain size, sorting index, current velocity, wave agitation, temperature, the concentration of SPM, and oxygen combinedly influence macrofaunal distribution in the south Brittany of the Gulf of Biscay. Joydas et al. (2019) who studied the infaunal benthic community assemblages in the deepwater of the Red sea have found the environmental factors including depth, latitude, organic matter, and sand content to influence. Ajmal Khan et al. (2017) found a strong relationship between macrobenthos and environmental factors including total organic carbon, median particle diameter, depth, salinity, dissolved oxygen, temperature besides sand, and silt/clay composition from estuarine to deep slope habitats. The same pattern was found in the macrobenthic community assemblages of Portugal waters. Sediment and other properties (nutrients, oxygen, and salinity) have explained 59%–72% of the variability in structural parameters of benthos including density, number of species, diversity, and assemblages.

6.5 Conclusion and recommendation

This chapter emphasizes how physicochemical parameters influence benthic-invertebrates directly by and large and indirectly. How these changes will affect the other biological ecosystems has also been indicated. This chapter also indicates the challenges in benthic biodiversity management. Physicochemical parameters also have their role on key species in the food chain, which may have cascading impacts propagating through the entire food chain including demersal components. The role of environmental parameters on benthos has been studied in many habitats in particular the role of salinity, dissolved oxygen concentration, temperature, and sediment parameters including contaminants, organic matter, nutrients, etc. These parameters influence the benthic communities on spatial and temporal scales, which make them appropriate for the application at regional and global scales. Benthic larval, propagule supply, proximity to diversity hotspots are also considered. Many studies have highlighted the influence of the combined effect of environmental parameters on macro and meiobenthic communities.

To consolidate information on the role of environmental parameters on marine benthos and to move further, the following gaps in our knowledge have to be probed: (1) functional diversity and molecular approaches on benthos in deeper depths have to be concentrated; (2) Soft bottom communities in submarine canyons are to be covered in view of the fragmentary knowledge available; (3) Macro and meiobenthic species are having high stress in oxygen minimum zones due to their proliferation limits and minimal respiration. Their physiological adaptations and genetic make-up have to be focused on; (4) Impacts due to climate variability at the population, community, and ecosystem levels have to be probed in view of the vast changes happening the world over and (5) Experiments in field and laboratory levels will facilitate deciphering the influence of specific parameter responses and will help to validate the models that have been developed for better management.

Acknowledgement

We thank the Center for Environment & Marine Studies (CEMS) at the Research Institute of King Fahd University of Petroleum & Minerals, Dhahran, Saudi Arabia, for providing research facilities.

References

Ahmed, M. K., Baki, M. A., Islam, M. S., Goutam Kumar Kundu, M. D., Habibullah-Al-Mamun., Sarkar, S. K., & Muzammel Hossain, M. D. (2015). Human health risk assessment of heavy metals in tropical fish and shellfish collected from the river Buriganga, Bangladesh. *Environmental Science and Pollution Research*, *22*, 15880–15890. Available from https://doi.org/10.1007/s11356-015-4813-z.

Ajmal Khan., Manokaran, S., & Lyla, S. (2017). Changes in macrobenthic community structure from estuary to continental slope in the south-east coast of India. *Journal of the Marine Biological Association of the United Kingdom*, *97*(1), 161–180.

Ajmal Khan, S., Ravikumar, B., Lyla, S., & Manokaran, S. (2018). Impact of the 2004 tsunami on the macrofauna of the continental slope of the southeast coast of India. *Marine Ecology*, *8*(39), e12527. Available from https://doi.org/10.1111/maec.12527.

Aller, J. Y. (1989). Quantifying sediment disturbance by bottom currents and its effect on benthic communities in a deep-sea western boundary zone. *Deep Sea Research Part A. Oceanographic Research Papers*, *36*(6), 901–934.

Alongi, D, M (1990). Community dynamics of free-living Nematodes in some tropical mangrove and sandflat habitatS. *Bulletin of Marine Science*, *46*(2), 358–373.

Andrady, A. L. (2017). The plastic in microplastics: A review. *Marine Pollution Bulletin*, *119*(1), 12–22. Available from https://doi.org/10.1016/j.marpolbul.2017.01.082.

Andrady, A. L., & Neal, M. A. (2009). Applications and societal benefits of plastics. *Philosophical Transactions of the Royal Society B: Biological Sciences*, *364*, 1977–1984. Available from https://doi.org/10.1098/rstb.2008.0304.

Ansari, K. G. M. T., Lyla, P. S., Ajmal Khan, S., & Bhadury, P. (2017a). Multiple spatial scale analysis provide an understanding of benthic macro-invertebrate community structure across a lagoonal ecosystem. *Wetlands*, *37*, 277–287. Available from https://doi.org/10.1007/s13157-016-0866-0.

Ansari, K. G. M. T., Lyla, S., Ajmal Khan, S., & Bhadury, P. (2017b). Trophic diversity, size and biomass spectrum of Bay of Bengal nematodes: A study case on depth and latitudinal patterns. *Continental Shelf Research*, *148*, 139–149. Available from https://doi.org/10.1016/j.csr.2017.08.018.

Ansari, K. G. M. T., Manokaran, S., Raja, S., Lyla, P. S., & Ajmal khan. (2014). Interaction of free-living marine nematodes in the artificial mangrove environment (southeast coast of India). *Environmental Monitoring and Assessment*, *186*, 293–305. Available from https://doi.org/10.1007/s10661-013-3374-1.

Ansari, K. G. M. T., Pattnaik, A. K., Rastogi, G., & Bhadury, P. (2015). An inventory of free-living marine nematodes from Asia's largest coastal lagoon, Chilika, India. *Wetlands Ecology and Management*, *23*, 881–890. Available from https://doi.org/10.1007/s11273-015-9426-2.

Artham, T., Sudhakar, M., Venkatesan, R., Nair, C. M., Murty, K. V. G. K., & Doble, M. (2009). Biofouling and stability of synthetic polymers in sea water. *International Biodeterioration and Biodegradation*, *63*, 884–890. Available from https://doi.org/10.1016/j.ibiod.2009.03.003.

Asmala, E. J., Carstensen, D. J., Conley, C. P., Slomp, J., Stadmark., & Voss, M. (2017). Efficiency of the coastal filter: Nitrogen and phosphorus removal in the Baltic Sea. *Limnology and Oceanography*, *62*, S222–S238. Available from https://doi.org/10.1002/lno.10644.

Ballesteros, E., & Zabala, M. (1993). El bentos: el marc físic. In J. A. Alcover, E. Ballesteros, & J. J. Fornós (Eds.), *Història Natural de l'arxipèlag de Cabrera. Monografies de la Societat d'Història Natural de les Balears* (Vol. 2, pp. 663–685). Societat d'Història Natural de les Balears.

Barua, P., Rahman, S. H., & Molla, M. H. (2017). Heavy metals effluence in sediments and its impact on macrobenthos at Shipbreaking Area of Bangladesh. *Asian Profile*, *45*(2), 167–180.

Basson, P. W., Burchard, I. E., Hardy, J. T., & Price, A. R. G. (1977). *Biotopes of the Western Arabian Gulf* (p. 284) ARAMCO, Dhahran.

Benelli, S., Bartoli, M., Zilius, M., Vybernaite-Lubiene, I., Ruginis, T., Petkuviene, J., & Fano, E. A. (2018). Microphytobenthos and chironomid larvae attenuate nutrient recycling in shallow-water sediments. *Freshwater Biology*, *63*, 187–201. Available from https://doi.org/10.1111/fwb.13052.

Bessell-Browne, P., Fisher, R., Duckworth, A., & Jones, R. (2017). Mucus sheet production in Porites: An effective bioindicator of sediment related pressures. *Ecological Indicators*, 77, 276–285.

Beukema, J. J., Honkoop, P. J. C., & Dekker, R. (1998). Recruitment in *Macoma balthica* after mild and cold winters and its possible control by egg production and shrimp predation. *Hydrobiologia*, *376*, 23–34.

Bhadja, P., Poriya, P., & Kundu, R. (2014). Community structure and distribution pattern of intertidal invertebrate macrofauna at some anthropogenically influenced Coasts of Kathiawar Peninsula (India). *Advances in Ecology*, *2014*, 547395. Available from https://doi.org/10.1155/2014/547395.

Birchenough, S. N., Reiss, H., Degraer, S., Mieszkowska, N., Borja, Á., Buhl-Mortensen, L. Braeckman, U., Craeymeersch, J., De Mesel, I., Kerckhof, F., Kröncke, I., Parra, S. Rabaut, M., Schröder, A., Van Colen, C., Van Hoey, G., Vincx, M., & Wätjen, K (2015). Climate change and marine benthos: A review of existing research and future directions in the North Atlantic. *WIREs Climate Change*, *6*, 203–223. Available from https://doi.org/10.1002/wcc.330.

Birchenough, S. N. R., Bremner, J., Henderson, P., Hinz, H., Jenkins, S., Mieszkowska, N., Roberts, J. M., Kamenos, N., & Plenty, S. (2013). Impacts of climate change on shallow and shelf subtidal habitats. *MCCIP Science Review, 2013*, 193–203.

Birchenough, S. N. R., Parker, R. E., McManus, E., & Barry, J. (2012). Combining bioturbation and redox metrics: Potential tools for assessing seabed function. *Ecological Indicators, 12*, 8–16.

Boesch, D. F., Wass, M. C., & Virnstein, R. W. (1976). The dynamics of estuarine benthic communities. In M. L. Wiley (Ed.), *Estuarine processes: Uses, stresses and adaptations to the estuary* (Vol. 1, pp. 177–196). Academic Press.

Bolam, S. G., Eggleton, J., Smith, R., Mason, C., Vanstaen, K., & Rees, H. L. (2008). Spatial distribution of macrofaunal assemblages along the English Channel. *Journal of the Marine Biological Association of the United Kingdom, 88*, 675–687.

Cai, W. J., Reimers, C. E., & Shaw, T. (1995). Microelectrode studies of organic carbon degradation and calcite dissolution at a California continental rise site. *Geochimica et Cosmochimica Acta, 59*, 497–511. Available from https://doi.org/10.1016/0016-7037(95)00316-R.

Caldeira, K., & Wickett, M. E. (2003). Anthropogenic carbon and ocean pH. *Nature, 425*, 365.

Caldeira, K., & Wickett, M. E. (2005). Ocean model predictions of chemistry changes from carbon dioxide emissions to the atmosphere and ocean. *Journal of Geophysical Research, 110*, C09S04. Available from https://doi.org/10.1029/2004JC002671.

Canals, M., Puig, P., de Madron, X. D., Heussner, S., Palanques, A., & Fabres, J. (2006). Flushing submarine canyons. *Nature, 444*, 354–357.

Carstensen, J., Conley, D. J., Almroth-Rosell, E., Asmala, E., Bonsdorff, E., Fleming-Lehtinen, V., Gustafsson, B. G., Gustafsson, C., Heiskanen, A. S., Janas, U., Norkko, A., Slomp, C., Villnäs, A., Voss, M., & Zilius, M. (2020). Factors regulating the coastal nutrient filter in the Baltic Sea. *Ambio, 49*, 1194–1210. Available from https://doi.org/10.1007/s13280-019-01282-y.

Cartes, J. E., Mamouridis, V., & Fanelli, E. (2011). Deep-sea supra benthos assemblages (Crustacea) off the Balearic Islands (western Mediterranean): Mesoscale variability in diversity and production. *Journal of Sea Research, 65*, 340–354.

Cauwenberghe, V., Devriese, L., Galgani, F., Robbens, J., & Janssen, C. R. (2015). Microplastics in sediments: A review of techniques, occurrence and effects. *Marine Environmental Research, 111*, 5–17.

Chakraborty, S., & Owens, G. (2014). Metal distributions in seawater, sediment and marine benthic macroalgae from the South Australian coastline. *International Journal of Environmental Science and Technology, 11*, 1259–1270. Available from https://doi.org/10.1007/s13762-013-0310-4.

Childress, J. J., & Seibel, B. A. (1998). Life at stable low oxygen levels: Adaptations of animals to oceanic oxygen minimum layers. *The Journal of Experimental Biology, 201*, 1223–1232.

Cloern, J. E. (2001). Our evolving conceptual model of the coastal eutrophication problem. *Marine Ecology Progress Series, 210*, 223–253.

Colloca, F., Cardinale, M., Belluscio, A., & Ardizzone, G. D. (2003). Structure and diversity of demersal assemblages in the Central Mediterranean Sea. *Estuarine, Coastal and Shelf Science, 56*, 469–480.

Coma, R., Ribes, M., Gili, J. M., & Zabala, M. (2000). Seasonality in coastal benthic ecosystems. *Trends in Ecology & Evolution, 15*, 448–453.

Conlan, K., Aitken, A., Hendrycks, E., McClelland, C., & Melling, H. (2008). Distribution patterns of Canadian Beaufort Shelf macrobenthos. *Journal of Marine Systems*, 7, 864–886.

Crawford, C. B., & Quinn, B. (2017). *Microplastic pollutants* (1st ed.). Elsevier. Available from https://www.elsevier.com/books/microplasticpollutants/crawford/978-0-12-809406-8.

Cunningham, M. J., Hodgson, S., Masson, D. G., & Parson, L. M. (2005). An evaluation of along- and down-slope sediment transport processes between Goban Spur and Brenot Spur on the Celtic Margin of the Bay of Biscay. *Sedimentary Geology, 179*, 99–116.

Cusson, M., Archambault, P., & Aitken, A. (2007). Biodiversity of Arctic continental shelf benthic assemblages: Historical data from the Canadian Arctic Archipelago. *Marine Ecology Progress Series, 331*, 291–304.

Day, D. S., & Pearcy, W. G. (1968). Species associations of benthic fishes on the continental shelf and slope off Oregon. *Journal of the Fisheries Research Board of Canada., 25* (12), 2665–2675.

Dde Stigter, H., Boera, W., Mendesb, P., Jesusc, C., Thomsenb, L., van den Bergha, G., & van Weeringa, T. (2007). Recent sediment transport and deposition in the Nazaré Canyon, Portuguese continental margin. *Marine Geology, 246*(2–4), 144–164.

Decho, A. W. (1990). Microbial exopolymer secretions in ocean environments: Their role (s) in food webs and marine processes. *Oceanography and Marine Biology: An Annual Review, 28*, 73–153.

Devol, A. H., & Hartnett, H. E. (2001). Role of the oxygen-deficient zone in the transfer of organic carbon to the deep ocean. *Limnology and Oceanography, 46*, 1684–1690.

Diaz, R. J., & Rosenberg, R. (1995). Marine benthic hypoxia: A review of its ecological effects and the behavioral responses of benthic macrofauna. *Oceanography and Marine Biology: An Annual Review, 33*, 245–303.

Diaz, R. J., Solan, M., & Valente, R. M. (2004). A review of approaches for classifying benthic habitats and evaluating habitat quality. *Journal of Environmental Management, 73*, 165–181.

Douglas, E. J., Pilditch, C. A., Kraan, C., Schipper, L. A., Lohrer, A. M., & Thrush, S. F. (2017). Macrofaunal functional diversity provides resilience to nutrient enrichment in coastal sediments. *Ecosystems, 20*, 1324–1336. Available from https://doi.org/10.1007/s10021/017/0113/4.

Doxaran, D., Castaing, P., & Lavender, S. J. (2006). Monitoring the maximum turbidity zone and detecting fine-scale turbidity features in the Gironde estuary using high spatial resolution satellite sensor (SPOT HRV, Landsat ETMþ) data. *International Journal of Remote Sensing, 27*, 2303–2321.

Duarte, C. M., Middelburg, J. J., & Caraco, N. (2005). Major role of marine vegetation on the oceanic carbon cycle. *Biogeosciences, 2*, 1–8. Available from https://doi.org/10.5194/bg-2-1-2005.

Durden, J. M., Bett, B. J., Jones, D. O. B., Huvenne, V. A. I., & Ruhl, H. A. (2015). Abyssal hills—Hidden source of increased habitat heterogeneity, benthic megafaunal biomass and diversity in the deep sea. *Progress in Oceanography, 137*, 209–218. Available from https://doi.org/10.1016/j.pocean.2015.06.006.

Dutertre, M., Hamon, D., Chevalier, C., & Ehrhold, A. (2013). The use of the relationships between environmental factors and benthic macrofauna distribution in the establishment of a baseline for coastal management. *ICES Journal of Marine Science, 70*, 294–308.

Einav, R., Harussi, K., & Perry, D. (2002). The footprint of the desalination processes on the environment. *Desalination., 152*, 141–154.

Eleftheriou, A., & McIntyre, A. D. (Eds.), (2005). *Methods for the study of marine benthos* (3rd ed., p. 418). Blackwell.

Engraff, M., Solere, C., Smith, K., Mayer, P., & Dahllöf, I. (2011). Aquatic toxicity of PAHs and PAH mixtures at saturation to benthic amphipods: Linking toxic effects to chemical activity. *Aquatic Toxicology, 102*(3–4), 142–149.

Eriksen, M., Lebreton, L. C. M., Carson, H. S., Thiel, M., Moore, C. J., Borerro, J. C., Galgani, F., Ryan, P. G., & Reisser, J. (2014). Plastic pollution in the world's oceans: More than 5 trillion plastic pieces weighing over 250,000 tons afloat at sea. *PLoS One*, *9*, 0111913. Available from https://doi.org/10.1371/journal.pone.0111913.

Fang, C., Zheng, R., Zhang, Y., Hong, F., Mu, J., Chen, M., Song, P., Lin, L., Lin, H., Le, F., & Bo, J. (2018). Microplastic contamination in benthic organisms from the Arctic and sub-Arctic regions. *Chemosphere*, *209*, 298–306.

Farrell, P., & Nelson, K. (2013). Trophic level transfer of microplastic: *Mytilus edulis* (L.) to *Carcinus maenas* (L.). *Environmental Pollution (Barking, Essex: 1987)*, *177*, 1–3.

Feebarani, J., Joydas, T., Damodaran, R., & Borja, A. (2016). Benthic quality assessment in a naturally- and human-stressed tropical estuary. *Ecological Indicators*, *67*, 380–390.

Ferrando, A., Gonzalez, E., Franco, M., Commendatore, M., Nievas, M., Militon, C., Stora, G., Gilbert, F., Esteves, J., & Cuny, P. (2015). Oil spill effects on macrofaunal communities and bioturbation of pristine marine sediments (Caleta Valdés, Patagonia, Argentina): Experimental evidence of low resistance capacities of benthic systems without history of pollution. *Environmental Science and Pollution Research International*, *22*(20), 15294–15306. Available from https://doi.org/10.1007/s11356-015-4167-6.

Figuerola, B., Hancock A. M., Bax, N., Cummings, V. J., Downey, R., Griffiths, H. J., Smith, J., & Stark, J. S., (2020). Predicting potential impacts of ocean acidification on marine calcifiers from Southern Ocean. SCAR online.

Forbes, V. E., Forbes, T. L., & Holmer, M. (1996). Inducible metabolism of fluoranthene by the opportunistic polychaete *Capitella* sp.I. *Marine Ecology Progress Series*, *132*, 63–70.

Gage, J. D. (1978). Animals in deep sea sediments. *Proceedings of the Royal Society of Edinburgh, Section B: Biological Sciences*, *76*, 77–93.

Gage, J. D., Lamont, P. A., & Tyler, P. A. (1995). Deep-sea macrobenthic communities at contrasting sites off Portugal, preliminary results: I. Introduction and diversity comparisons. *Internationale Revue der gesamten Hydrobiologie und Hydrographie*, *80*(2), 235–250.

Ganesh, T., & Raman, A. V. (2007). Macrobenthic community structure of the northwest Indian shelf, Bay of Bengal. *Marine Ecology Progress Series*, *341*, 59–73.

Gardner, W. D. (1989). Baltimore Canyon as a modern conduit of sediment to the deep sea. *Deep-Sea Research*, *36*, 323–358.

Garrabou, J., Perez, T., Sartoretto, S., & Harmelin, J. G. (2001). Mass mortality event in red coral Corallium rubrum populations in Provence region (France, NW Mediterranean). *Marine Ecology Progress Series*, *217*, 263–272.

Gerwing, T. G., Allen Gerwing, A. M., Allen, S., Campbell, L., Rogers, M., Gray, O., Drewes, M., & Juanes, F. (2018). Potential impacts of logging on intertidal infaunal communities within the Kitimat River estuary. *Journal of Natural History*, *52*, 2833–2855.

Giere, O. (2009). *Meiobenthology. The microscopic motile fauna of aquatic sediments* (2nd ed.). Springer.

Giere, O., & Pfannkuche, O. (1978). An ecophysiological approach to the micro distribution of meiobenthic Oligochaeta IL Phallodrilus mono spermathecus (Tubificidae) from boreal shores in comparison to populations from subtropical beaches. *Kieler Meeresforsch. Sdh.*, *4*, 79–86.

Gladstone-Gallagher, R. V., Mangan, S., Thrush, S. F., & Pilditch, C. A. (2020). Porewater nutrient enrichment alters benthic-pelagic coupling on intertidal sandflats. *Journal of Sea Research*, *159*, 101876. Available from https://doi.org/10.1016/j.seares.2020.101876, ISSN 1385-1101.

Goffredi, S. K., & Childress, J. J. (2001). Activity and inhibitor sensitivity of ATPases in the hydrothermal vent tubeworm Riftia pachyptila: A comparative approach. *Marine Biology*, *138*, 259–265. Available from https://doi.org/10.1007/s002270000462.

Gomez Gesteira, J. L., & Dauvin, J. C. (2005). Impact of the Aegean Sea oil spill on the Rîa of Ares and Betanzos fine sand community (Northwest Spain). *Marine Environmental Research*, *60*, 289–316.

Gooday, A. J., Bett, B. J., Escobar, E., Ingole, B., Levin, L. A., Neira, C., Raman, A. V., & Sellanes, J. (2010). Habitat heterogeneity and its relationship to biodiversity in oxygen minimum zones. *Marine Ecology*, *31*, 125–147.

Goosen, N. K., Kromkamp, J., Peene, J., Van Rijswijk, P., & Van Breugel, P. (1999). Bacterial and phytoplankton production in the maximum turbidity zone of three European estuaries: The Elbe, Westerschelde and Gironde. *Journal of Marine Systems*, *22*, 151–171.

Gopal, A., Abdul Jaleel, K. U., Parameswaran, U. V., Sanjeevan, V. N., Saramma, A. V., Vijayan, A., Saravanan, N., Gupta, G. V. M., & Sudhakar, M. (2020). Distinctive community patterns with exceptional diversity of Polychaetes around a tectonically active Archipelago in the tropical Indian Ocean. *Frontiers in Marine Science*, 7, 710. Available from https://doi.org/10.3389/fmars.2020.00710.

Grassle, J. F., Sanders, H. L., & Smith, W. K. (1979). Faunal changes with depth in the deep-sea benthos. *Ambio Special Report No. 6*, 47–50.

Gray, J. S., & Elliott, M. (2009). *Ecology of marine sediments. From science to management* (2nd ed., p. 225)Oxford University Press.

Gray, J. S., Wu, R. S. S., & Or, Y. Y. (2002). Effects of hypoxia and organic enrichment on the coastal marine environment. *Marine Ecology Progress Series*, *238*, 249–279.

Griggs, G. B., Carey, A. G., Jr., & Kulm, L. D. (1969). Deep-sea sedimentation and sediment-fauna interaction in Cascadia Channel and on Cascadia Abyssal Plain. *Deep-Sea Research*, *16*, 157–170.

Gruber, N., & Sarmiento, J. L. (1997). Global patterns of marine nitrogen fixation and denitrification. *Global Biogeochemical Cycles*, *11*(2), 235–266.

Gunton, L. M., Gooday, A. J., Glover, A. G., & Bett, B. J. (2015). Macrofaunal abundance and community composition at lower bathyal depths in different branches of the Whittard Canyon and on the adjacent slope (3500 m; NE Atlantic). *Deep-Sea Research Part I: Oceanographic Research Papers*, *97*, 29–39.

Sen Gupta, R., De Sousa, S. N., & Joseph, T. (1977). On nitrogen and phosphorus in the western Bay of Bengal. *Indian Journal of Geo-Marine Sciences*, *6*, 107–110.

Haedrich, R. L., Rowe, G. T., & Polloni, P. T. (1980). The megabenthic fauna in the deep sea south of New England, USA. *Marine Biology*, *57*, 165–179.

Hall, S. J. (1994). Physical disturbance and marine benthic communities: Life in unconsolidated sediments. *Oceanography and Marine Biology: An Annual Review*, *32*, 179–239.

Harvey, B. C., & White, J. L. (2008). Use of benthic prey by salmonids under turbid conditions in a laboratory stream. *Transactions of the American Fisheries Society*, *137*, 1756–1763.

Head, I. M., Jones, D. M., & Röling, W. F. M. (2006). Marine microorganisms make a meal of oil. *Nature Reviews Microbiology*, *4*(3), 173–182.

Hendriks, I. E., Duarte, C. M., & Álvarez, M. (2010). Vulnerability of marine biodiversity to ocean acidification: A meta-analysis. *Estuarine, Coastal and Shelf Science*, *86*, 157–164.

Heyns, E. R., Bernard, A. T. F., Richoux, N. B., & Götz, A. (2016). Depth-related distribution patterns of subtidal macrobenthos in a well-established marine protected area *Marine Biology*, *163*, 39. Available from https://doi.org/10.1007/s00227-016-2816-z.

Hidalgo-Ruz, V., Gutow, L., Thompson, R. C., & Thiel, M. (2012). Microplastics in the marine environment: A review of the methods used for identification and quantification. *Environmental Science & Technology*, *46*(6), 3060–3075.

Hu, C., Yang, X., Gao, L., Zhang, P., Li, W., Dong, J., Li, C., & Zhang, X. (2019) Comparative analysis of heavy metal accumulation and bioindication in three seagrasses: Which species is more suitable as a bioindicator? *The Science of the Total Environment*, *669*, 41–48. Available from https://doi.org/10.1016/j.scitotenv.2019.02.425, ISSN 0048–9697.

Hughes, D. A., Lamont, P. A., Levin, L. A., Packer, M., & Gage, J. D. (2009). Macrofaunal communities and sediment structure across the Pakistan Margin oxygen minimum zone, northeast Arabian Sea. *Deep-Sea Research Part II: Topical Studies in Oceanography, 56*, 434–448.

Hunter, W. R., Veuger, B., & Witte, U. (2012). Macrofauna regulate heterotrophic bacterial carbon and nitrogen incorporation in low oxygen sediments. *ISME Journal, 6*, 2140–2151.

Hunting, E. R., & Kampfraath, A. A. (2013). Contribution of bacteria to redox potential (E h) measurements in sediments. *International Journal of Environmental Science and Technology, 10*, 55–62. Available from https://doi.org/10.1007/s13762-012-0080-4.

Hyland, J., Balthis, L., Karakassis, I., Magni, P., Petrov, A., Shine, J., Vestergaard, O., & Warwick, R. (2005). Organic carbon content of sediments as an indicator of stress in the marine benthos. *Marine Ecology Progress Series, 295*, 91–103.

Ingole, B. S., Ansari, Z. A., & Parulekar, A. H. (1998). Spatial variation in meiofaunal abundance of some coralline beaches of Mauritius. *Tropical Ecology, 39*(1), 103–108.

Ingole, B. S., Sautya, S., Sanitha, S., Singh, R., & Nanajkar, M. (2010). Macrofaunal community structure in the western Indian continental margin including the oxygen minimum zone. *Marine Ecology, 31*(1), 148–166. Available from https://doi.org/10.1111/j.1439-0485.2009.00356.x.

Islam, M. S., Sikder, M. N. A., Al-Imran, M., Hossain, M. B., Mallick, D., & Morshed, M. M. (2013). Intertidal macrobenthic fauna of the Kamafuli Estuary: Relations with environmental variables. *World Applied Sciences Journal, 21*(9), 1366–1373.

Jackson, J. B. C. (1972). The ecology of the molluscs of Thalassia communities, Jamaica, West Indies. II. Molluscan population variability along an environmental stress gradient. *Marine Biology, 14*, 304–337.

Janssen, R., & Taviani, M. (2015). Taxonomic, ecological and historical considerations on the deep-water benthic mollusc fauna of the Red Sea. In N. M. A. Rasul, & I. C. F. Stewart (Eds.), *The Red Sea* (pp. 511–529). Springer-Verlag.

Jayachandran, P. R., Nandan, S. B., Jima, M., Joseph, P., Xavier, N. D., Sreedevi, O. K., & Joseph, K. J. (2019). Macrobenthic functional feeding groups in a microtidal monsoonal estuary (Kodungallur–Azhikode Estuary, India). *Regional Studies in Marine Science, 25*, 100444.

Joydas, T., & Damodaran, R. (2013). Testing depth-related multivariate patterns of macrofauna on the Indian continental shelf using reduced taxonomic resolution and data transformation. *Journal of the Marine Biological Association of the United Kingdom, 93*(1), 37–45. Available from https://doi.org/10.1017/S0025315411001858.

Joydas, T. V., & Damodaran, R. (2009). Infaunal macrobenthos along the shelf waters of the west coast of India, Arabian Sea. *Indian Journal of Geo-Marine Sciences, 38*, 191–204.

Joydas, T. V., & Damodaran, R. (2014). Infaunal macrobenthos of the oxygen minimum zone on the Indian western continental shelf. *Marine Ecology, 35*, 22–35. Available from https://doi.org/10.1111/maec.12052.

Joydas, T. V., Qurban, M. A., Ali, S. M., Albarau, J. F., Rabaoui, L., Manikandan, K. P., Ashraf, M., Papadopoulos, V. P., Giacobbe, S., & Krishnakumar, P. K. (2018). Macrobenthic community structure in the deep waters of the Red Sea. *Deep-Sea Research Part I: Oceanographic Research Papers, 137*, 38–56. Available from https://doi.org/10.1016/j.dsr.2018.05.004, ISSN 0967-0637.

Joydas, T. V., Qurban, M. A., Basali, A., Nazeer, Z., Lopez, M., Manokaran, S., & Cali, N. A. (2019). Benthic communities in the deep-sea environments of the Red Sea. In *International Conference on Benthos*, held on 6–8 March 2019. Kerala, Cochin, India: Cochin University of Science and Technology.

Joydas, T. V., Qurban, M. A., Borja, A., Krishnakumar, P. K., & Al-Suwailem, A. (2017). Macrobenthic community structure in the Northwestern Arabian Gulf, twelve years after the 1991 oil spill. *Frontiers in Marine Science*, *4*, 248. Available from https://doi.org/10.3389/fmars.2017.00248.

Joydas, T. V., Qurban, M. A., Manikandan, K. P., Ashraf, T. T. M., Ali, S. M., Al-Abdulkader, K., Qasem, A., & Krishnakumar, P. K. (2015). Status of macrobenthic communities in the hypersaline waters of the Gulf of Salwa, Arabian Gulf. *Journal of Sea Research*, *99*, 34–46.

Kamykowski, D., & Zentara, S. J. (1990). Hypoxia in the world ocean as recorded in the historical data set. *Deep Sea Research Part A. Oceanographic Research Papers*, *37*, 1861–1874. Available from https://doi.org/10.1016/0198-0149(90)90082-7.

Karlson, K., Bonsdorff, E., & Rosenberg, R. (2007). The impact of benthic macrofauna for nutrient fluxes from Baltic Sea sediments. *Ambio*, *36*, 161–167.

Kaufmann, R. S., & Smith, K. L. J. R. (1997). Activity patterns of mobile epibenthic megafauna at an abyssal site in the eastern North Pacific: Results from a 17-month time-lapse photographic study. *Deep-Sea Research Part I: Oceanographic Research Papers*, *44*, 559–579.

Kim, G. B., Maruya, K. A., Lee, R. F., Lee, J. H., Koh, C. H., & Tanabe, S. (1999). Distribution and sources of polycyclic aromatic hydrocarbons in sediments from Kyeonggi Bay, Korea. *Marine Pollution Bulletin*, *38*, 7–15.

Kirby, R. R., Beaugrand, G., Lindley, J. A., Richardson, A. J., Edwards, M., & Reid, P. C. (2007). Climate effects and benthic-pelagic coupling in the North Sea. *Marine Ecology Progress Series*, *330*, 31–38.

Klages, M., Boetius, A., Christensen, J. P., Deubel, H., Piepenburg, D., Schewe, I., & Soltwedel, T. (2004). The benthos of Arctic seas and its role for the organic carbon cycle at the seafloor. In R. Stein, & R. W. Macdonald (Eds.), *The organic carbon cycle in the Arctic Ocean* (pp. 139–167). Springer.

Kleypas, J. A., Buddemeier, R. W., Archer, D., Gattuso, J. P., Langdon, C., & Opdyke, B. N. (1999). Geochemical consequences of increased atmospheric carbon dioxide on coral reefs. *Science (New York, N.Y.)*, *284*, 118–120.

Koho, K. A., Kouwenhoven, T. J., de Stigter, H. C., & van der Zwaan, G. J. (2007). Benthic foraminifera in the Nazaré canyon, Portuguese continental margin: Sedimentary environments and disturbance. *Marine Micropaleontology*, *66*, 27–51.

Kowalczuk, M. W., Mazurkiewicz, M., Górska, B., Michel, L. N., Jankowska, E., & Zaborska, A. (2019). Organic carbon origin, benthic faunal consumption, and burial in sediments of northern Atlantic and Arctic fjords (60–81°N). *Journal of Geophysical Research: Biogeosciences*, *124*, 3737–3751. Available from https://doi.org/10.1029/2019JG005140.

Kristensen, E. (2000). Organic matter diagenesis at the oxic/anoxic interface in coastal marine sediments, with emphasis on the role of burrowing animals. In G. Liebezeit, S. Dittmann, & I. Kröncke (Eds.), *Life at interfaces and under extreme conditions. Developments in hydrobiology* (Vol. 151). Springer. Available from https://doi.org/10.1007/978-94-011-4148-2_1.

Lalli, C. M., & Parsons, T. R. (1997). *Biological oceanography, an introduction* (2nd ed., p. 306)Butterworth-Heinemann.

Lambshead, P. J. D., & Gooday, A. J. (1990). The impact of seasonally deposited phytodetritus on epifaunal and shallow infaunal benthic foraminiferal populations in the bathyal northeast Atlantic: The assemblage response. *Deep Sea Research Part A. Oceanographic Research Papers*, *37*, 1263–1283.

Landrum, P. F., & Robbins, J. A. (1990). Bioavailability of sediment associated contaminants to benthic invertebrates. In R. Baudo, J. P. Giesy, & H. Muntau (Eds.), *Sediments: Chemistry and toxicity of in-place pollutants* (pp. 237–263). Lewis Publishers.

Laverock, B., Gilbert, J., Tait, K., Osborn, A. M., & Widdicombe, S. (2011). Bioturbation: Impact on the marine nitrogen cycle. *Biochemical Society Transactions, 39*, 315–320. Available from https://doi.org/10.1042/BST0390315.

Lee, L. H., & Lin, H. J. (2013). Effects of an oil spill on benthic community production and respiration on subtropical intertidal sand flats. *Marine Pollution Bulletin, 73*(1), 291–299. Available from https://doi.org/10.1016/j.marpolbul.2013.05.006.

Levin, L. A. (2002). Deep-ocean life where oxygen is scarce. *American Scientist, 90*, 436–444.

Levin, L. A. (2003). Oxygen minimum zone benthos: Adaptation and community response to hypoxia. *Oceanography and Marine Biology: An Annual Review, 41*, 1–45.

Levin, L. A., Etter, R. J., Rex, M. A., Gooday, A. J., Smith, C. R., Pineda, J., Stuart, C. T., Hessler, R. R., & Pawson, D. (2001). Environmental influences on regional deep-sea species diversity. *Annual Review of Ecology, Evolution, and Systematics, 32*, 51–93.

Levin, L. A., Whitcraft, C. R., Mendoza, G. F., Gonzalez, J. P., & Cowie, G. L. (2009). Oxygen and organic matter thresholds for benthic faunal activity on the Pakistan Margin oxygen minimum zone (700–1100 m). *Deep Sea Research Part II: Topical Studies in Oceanography, 56*, 449–471.

Lotufo, G. R., Farrar, J. D., Biedenbach, J. M., Laird, J. G., Krasnec, M. O., Lay, C., Morris, J. M., & Gielazyn, M. L. (2016). Effects of sediment amended with Deepwater Horizon incident slick oil on the infaunal amphipod *Leptocheirus plumulosus*. *Marine Pollution Bulletin, 109*, 253–258.

Madhupratap, M., Gauns, M., Ramaiah, N., Kumar, S. P., Muraleedharan, P. M., de Sousaa, S. N., Sardessaia, S., & Muraleedharan, U. (2003). Biogeochemistry of the Bay of Bengal: Physical, chemical and primary productivity characteristics of the central and western Bay of Bengal during summer monsoon 2001. *Deep Sea Research Part II: Topical Studies in Oceanography, 50*, 881–896.

Maltby, J. (1999). The internal structure of a derived, revised, and amended measure of the Religious Orientation Scale: The 'Age-Universal' I-E Scale-12. *Social Behavior and Personality.*, *27*(4), 407–412.

Maltese, A., Capodici, F., Ciraolo, G., & Loggia, G. L. (2013). Coastal zone water quality: Calibration of a water-turbidity equation for MODIS data. *European Journal of Remote Sensing, 46*(1), 333–347. Available from https://doi.org/10.5721/EuJRS20134619.

Mannino, A., & Montagna, P. A. (1997). Small-scale spatial variation of macrobenthic community structure. *Estuaries, 20*(1), 159–173.

Manokaran, S., Khan, S. A., & Lyla, P. S. (2015). Macrobenthic composition of the southeast continental shelf of India. *Marine Ecology, 36*, 1–15.

Manokaran, S., Khan, S. A., Lyla, S., Raja, S., & Ansari, K. G. M. T. (2013). Feeding guild composition of shelf macrobenthic polychaetes of southeast coast of India. *Tropical Zoology, 26*, 120–139.

Matthews, T. G., & Fairweather, P. G. (2004). Effect of lowered salinity on the survival, condition and reburial of *Soletellina alba* (Lamarck, 1818) (Bivalvia: Psammobiidae). *Austral Ecology, 29*, 250–257.

Mattos, P. R., & de Almeida, T. C. M. (2016). Spatiotemporal distribution of the benthic macrofauna in an urbanized subtropical estuary: Environmental variations and anthropogenic impacts. *Brazilian Journal of Oceanography, 64*(3), 227–238. Available from https://doi.org/10.1590/S1679-87592016111306403.

Meerhoff, E., Rodriguez-Gallego, L., Giménez, L., Muniz, P., & Conde, D. (2013). Spatial patterns of macrofaunal community structure in coastal lagoons of Uruguay. *Marine Ecology Progress Series, 492*, 97–110.

Merrett, N. R., & Haedrich, R. L. (1997). *Deep-sea demersal fish and fisheries* (p. 282) *Chapman and Hall.*

Mieszkowska, N., Kendall, M. A., Hawkins, S. J., Leaper, R., Williamson, P., Hardman-Mountford, N. J., & Southward, A. J. (2006). Changes in the range of some common rocky shore species in Britain—A response to climate change? *Hydrobiologia*, *555*, 241–251.

Millward, R. N., & Grant, A. (1995). Assessing the impact of copper on nematode communities from a chronically metal enriched estuary using pollution-induced community tolerance. *Marine Pollution Bulletin*, *30*, 701–706.

Moens, T., & Vincx, M. (2000). Temperature, salinity and food thresholds in two brackish-water bacterivorous nematode species: Assessing niches from food absorption and respiration experiments. *Journal of Experimental Marine Biology and Ecology*, *243*, 137–154.

Mora, C., Wei, C., Rollo, A., Amaro, T., Baco, A. R., Billett, D., Bopp, L., Chen, Q., Collier, M., Danovaro, R., Gooday, A. J., Grupe, B. M., Halloran, P. R., Ingels, J., Jones, D. O. B., Levin, L. A., Nakano, H., Norling, K., Ramirez-Llodra, E., ... Yasuhara, M. (2013). Biotic and human vulnerability to projected changes in ocean biogeochemistry over the 21st century. *PLoS Biology*, *11*(10), e1001682. Available from https://doi.org/10.1371/journal.pbio.1001682.

Moreno-Madriñan, M. J., Al-Hamdan, M. Z., Rickman, D. L., & Muller-Karger, F. E. (2010). Using the surface reflectance MODIS Terra product to estimate turbidity in Tampa Bay, Florida. *Remote Sensing*, *2*, 2713–2728. Available from https://doi.org/10.3390/rs2122713.

Muthukumar, T., Aravinthan, A., Lakshmi, K., Venkatesan, R., Vedaprakash, L., & Doble, M. (2011). Fouling and stability of polymers and composites in marine environment. *International Biodeterioration & Biodegradation*, *65*(2), 276–284.

Naidu, S. A., Ranga Rao, V., & Ramu, K. (2018). Microplastics in the benthic invertebrates fromthe coastal waters of Kochi, Southeastern Arabian Sea. *Environmental Geochemistry and Health*, *40*, 1377–1383.

Napper, I. E., & Thompson, R. C. (2016). Release of synthetic microplastic plastic fibres from domestic washing machines: Effects of fabric type and washing conditions. *Marine Pollution Bulletin*, *112*(1–2), 39–45. Available from https://doi.org/10.1016/j.marpolbul.2016.09.025.

Nelms, S. E., Galloway, T. S., Godley, B. J., Jarvis, D. S., & Lindeque, P. K. (2018). Investigating microplastic trophic transfer in marine top predators. *Environmental Pollution (Barking, Essex: 1987)*, *238*, 999–1007. Available from https://doi.org/10.1016/j.envpol.2018.02.016.

Nicholls, P., Hewitt, J., & Halliday, J. (2003). Effects of suspended sediment concentrations on suspension and deposit feeding marine macrofauna. ARC Technical Publication Auckland Regional Council. NIWA Client Report: ARC03267. National Institute of Water and Atmospheric Research Ltd, Hamilton.

Nishijima, W., Nakano, Y., Nakai, S., Okuda, T., Imai, T., & Okada, M. (2013). Impact of flood events on macrobenthic community structure on an intertidal flat developing in the Ohta River Estuary. *Marine Pollution Bulletin*, *74*, 364–373.

Nixon, S. W. (1995). Coastal marine eutrophication: A definition, social causes, and future concerns. *Ophelia.*, *41*, 199–219.

Ogston, A. S., Cacchione, D. A., Sternberg, R. W., & Kineke, G. C. (2000). Observations of storm and river flood driven sediment transport on the northern California continental shelf. *Continental Shelf Research*, *20*, 2141–2162. Available from https://doi.org/10.1016/S0278-4343(00)00065-0.

Ólafsson, E., & Elmgren, R. (1997). Seasonal dynamics of sublittoral meiobenthos in relation to phytoplankton sedimentation in the Baltic Sea. *Estuarine, Coastal and Shelf Science*, *45*, 149–164.

Olafsson, E., Modig, H., & van de Bund, W. J. (1999). Species specific uptake of radiolabeled phytodetritus by benthic meiofauna from the Baltic Sea. *Marine Ecology Progress Series*, *177*, 63–72.

Ourives, T. M., Rizzo, A. E., & Boehs, G. (2011). Composition and spatial distribution of the benthic macrofauna in the Cachoeira River estuary, Ilhéus, Bahia, Brazil. *Revista de Biología Marina y Oceanografía*, *46*(1), 17–25.

Palanques, A., Guillen, J., Puig, P., & Durrieu de Madron, X. (2008). Storm-driven shelf-to-canyon suspended sediment transport at the southwestern end of the Gulf of Lions. *Continental Shelf Research*, *28*, 1947–1956.

Papiol, V., Cartes, J. E., Fanelli, E., & Maynou, F. (2012). Influence of environmental variables on the spatio-temporal dynamics of bentho-pelagic assemblages in the middle slope of the Balearic Basin (NW Mediterranean). *Deep Sea Research Part I: Oceanographic Research Papers*, *61*, 84–99.

Pearson, T. H., & Rosenberg, R. (1978). Macrobenthic succession in relation to organic enrichment and pollution of the marine environment. *Oceanography and Marine Biology: An Annual Review*, *16*, 229–311.

Pennock, J. R., & Sharp, J. H. (1994). Temporal alteration between light-and-nutrient-limitation of phytoplankton production in a coastal plain estuary. *Marine Ecology Progress Series*, *111*, 275–288.

Peterson, C. G. J. (1913). Valuation of the sea. II. The animal communities of the sea bottom and their importance for marine zoogeography. *Report of the Danish Biological Station*, *21*, 1–44.

Peterson, C. H. (1991). Intertidal zonation of marine invertebrates in sand and mud. *American Scientist*, *79*, 236–249.

Philippart, C. J. M., van Aken, H. M., Beukema, J. J., Bos, O. G., Cadee, G. C., & Dekker, R. (2003). Climate-related changes in recruitment of the bivalve *Macoma balthica*. *Limnology and Oceanography*, *48*, 2171–2185.

Piló, D., Pereira, F., Carriço, A., Cúrdia, J., Pereira, P., Gaspar, M. B., & Carvalho, S. (2015). Temporal variability of biodiversity patterns and trophic structure of estuarine macrobenthic assemblages along a gradient of metal contamination. *Estuarine Coastal and Shelf Science*, *167*, 286–299. Available from https://doi.org/10.1016/j.ecss.2015.06.018.

Podlesińska, W., & Dąbrowska, H. (2019). Amphipods in estuarine and marine quality assessment—A review. *Oceanologia*, *61*, 179–196.

Poloczanska, E. S., Brown, C. J., Sydeman, W. J., Kiessling, W., Schoeman, D. S., Moore, P. J., et al. (2013). Global imprint of climate change on marine life. *Nature Climate Change*, *3*, 919–925. Available from https://doi.org/10.1038/nclimate1958.

Porter, A., Lyons, B. P., Galloway, T. S., & Lewis, C. (2018). Role of marine snows in microplastic fate and bioavailability. *Environmental Science & Technology*, *52*, 7111–7119.

Pourang, N. (1996). Heavy metal concentrations in surficial sediments and benthic macroinvertebrates from Anzali wetland, Iran. *Hydrobiologia*, *331*, 53–61. Available from https://doi.org/10.1007/BF00025407.

Purkey, S. G., & Johnson, G. C. (2010). Warming of global abyssal and deep Southern Ocean waters between the 1990s and 2000s: Contributions to global heat and sea level rise budgets. *Journal of Climate*, *23*, 6336–6351. Available from https://doi.org/10.1175/2010JCLI3682.1.

Purser, B. H., & Seibold, E. (1973). The principle environmental factors influencing Holocene sedimentation and diagenesis in the Persian Gulf. In B. H. Purser (Ed.), *The Persian Gulf: Holocene carbonate sedimentation and diagenesis in a shallow epicontinental sea* (pp. 1–10). Springer.

Quiroga, E., Quinones, R. A., Gonzalez, R. R., Gallardo, V. A., & Jessen, G. (2007). Aerobic and anaerobic metabolism of *Paraprionospio pinnata* (Polychaeta: Spionidae) in central Chile. *Journal of the Marine Biological Association of the United Kingdom, 87*, 459–463.

Rabalais, N. N., & Baustian, M. M. (2020). Historical shifts in benthic infaunal diversity in the Northern Gulf of Mexico since the appearance of seasonally severe hypoxia. *Diversity, 12*, 49.

Rabaoui, R., Arculeo, M., Mansour, L., & Tlig-Zouari, S. (2015). Occurrence of the lessepsian species *Portunus segnis* (Crustacea: Decapoda) in the Gulf of Gabes (Tunisia): First record and new information on its biology and ecology. *Cahiers de Biologie Marine, 56*, 169–175.

Radhakrishna, K., Devassay, V. P., Bhargava, R. M. S., & Bhattathiri, P. M. A. (1978). Primary production in the northern Arabian Sea. *Indian Journal of Geo-Marine Sciences, 7*, 271–275.

Raghunathan, C., Tewari, A., Sravan Kumar, V. G., Trivedi, R. H., & Khambhaty, Y. (2003). Impact of turbidity on intertidal macrofauna at Gopnath, Mahuva and Veraval coasts (west coast of India). *Indian Journal of Geo-Marine Sciences, 32*, 214–221.

Raman, A. V., Damodaran, R., Levin, L. A., Ganesh, T., Rao, Y. K. V., Nanduri, S., & Madhusoodhanan, R. (2015). Macrobenthos relative to the oxygen minimum zone on the East Indian margin, Bay of Bengal. *Marine Ecology, 36*, 679–700. Available from https://doi.org/10.1111/maec.12176.

Reddy, M. S., Mehta, B., Dave, S., Joshi, M., Karthikeyan, L., Sarma, V. K. S., Basha, S., Ramachandraiah, G., & Bhatt, O. (2007). Bioaccumulation of heavy metals in some commercial fishes and crabs of the Gulf of Cambay, India. *Current Science, 92*, 1489–1491.

Rex, M. A. (1973). Deep-sea species diversity: Decreased gastropod diversity at abyssal depths. *Science (New York, N.Y.), 181*, 1051–1053.

Rex, M. A., & Etter, R. J. (1998). Bathymetric patterns of body size: Implications for deep-sea biodiversity. *Deep Sea Research Part II: Topical Studies in Oceanography, 45*, 103–127.

Rhoads, D. C., Boesch, D. K., Zhican, T., Fengshan, X., Liqiang, H., & Nilsen, K. J. (1985). Macrobenthos and sedimentary facies on the Changjiang delta platform and adjacent continental shelf, East China Sea. *Continental Shelf Research, 4*, 189–213.

Richmond, E. C., Wethey, D. S., & Woodin, S. A. (2007). Climate change and increased environmental variability: Demographic responses in an estuarine harpacticoid copepod. *Ecological Modelling, 209*, 189–202.

Riedl, R. (1971). How much seawater passes through sandy beaches? *Internationale Revue der gesamten Hydrobiologie und Hydrographie, 56*, 923–946.

Riera, R., Sacramento, A., Perez, Ó., Monterroso, Ó., Ramos, E., Rodríguez, M., & Almansa, E. (2013). Effects of organic enrichment on macrofauna community structure: An experimental approach. *Brazilian Journal of Oceanography, 61*(4), 223–229. Available from https://doi.org/10.1590/S1679-87592013000400003.

Ritter, C., Montagna, P. A., & Applebaum, S. (2005). Short-term succession dynamics of macrobenthos in a salinity-stressed estuary. *Journal of Experimental Marine Biology and Ecology, 323*, 57–69.

Roder, C., Berumen, M. L., Bouwmeester, J., Papathanassiou, E., Al-Suwailem, A., & Voolstra, C. R. (2013). First biological measurements of deep-sea corals from the Red Sea. *Scientific Reports, 3*, 2802. Available from https://doi.org/10.1038/srep02802.

Roditi-Elasar, M., Kerem, D., Lazar, M., Barneah, O., Almogi-Labin, A., & Angel, D. L. (2019). Benthic macro-faunal abundance and diversity and sediment distribution in Akhziv submarine canyon and the adjacent slope (eastern Levant Basin, Mediterranean Sea). *Mediterranean Marine Science, 20*, 521. Available from https://doi.org/10.12681/mms.19430.

Rolston, A., & Dittmann, S. (2009). The distribution and abundance of macrobenthic invertebrates in the Murray Mouth and Coorong Lagoons 2006 to 2008 (CSIRO: Water for a Healthy Country National Research Flagship, Melbourne).

Rosenberg, R. (2001). Marine benthic faunal successional stages and related sedimentary activity. *Scientia Marina, 65*, 107–119.

Rowe, G. T., Polloni, P. T., & Haedrich, R. L. (1982). The deep-sea macrobenthos on the continental margin of the Northwest Atlantic Ocean. *Deep Sea Research Part A. Oceanographic Research Papers, 29*, 257–278. Available from https://doi.org/10.1016/0198-0149(82)90113-3.

Ruhl, H. A., Bett, B. J., Hughes, S. J. M., Alt, C. H. S., Ross, E. J., Lampitt, R. S., Pebody, C. A., Smith, K. L., & Billett, D. S. (2014). Links between deep-sea respiration and community dynamics. *Ecology, 95*, 1651–1662. Available from https://doi.org/10.1890/13-0675.1.

Ryu, B. G., Park, G., Yang, J. W., & Baek, K. (2011). Electrolyte conditioning for electrokinetic remediation of As, Cu, and Pb contaminated soil. *Separation and Purification Technology, 79*, 170–176.

Saeki, K., Okazaki, M., & Kubota, M. (1993). Heavy metal accumulations in a semi-enclosed hypereutrophic system: Lake Teganuma, Japan. Part 2. Heavy metal accumulations in sediments. *Water, Air, and Soil Pollution, 69*(1–2), 79–91.

Salvador Cesa, F., Turra, A., & Baruque-Ramos, J. (2017). Synthetic fibers as microplastics in the marine environment: A review from textile perspective with a focus on domestic washings. *The Science of the Total Environment, 598*, 1116–1129.

Sanders, H. L. (1968). Marine benthic diversity: A comparative study. *The American Naturalist, 102*, 243–282.

Schoellhamer, D. H. (1996). Anthropogenic sediment resuspension mechanisms in a shallow microtidal estuary. *Estuarine, Coastal and Shelf Science, 43*(5), 533–548. Available from https://doi.org/10.1006/ecss.1996.0086.

Selvaraj, P., Murugesan, P., Punniyamoorthy, R., Parthasarathy, P., & Marigoudar, S. R. (2019). Assessment of the ecological health of Vellar and Ennore estuarine ecosystems using health indices. *Indian Journal of Geo-Marine Sciences, 48*(10), 1580–1592.

Seo, J. Y., Kim, M., Lim, H. S., & Choi, J. W. (2014). The macrofaunal communities in the shallow subtidal areas for the first 3 years after the Hebei Spirit oil spill. *Marine Pollution Bulletin, 82*(1–2), 208–220.

Sfriso, A. A., Tomio, Y., Rosso, B., Gambaro, A., Sfriso, A., Corami, F., Rastelli, E., Corinaldesi, C., Mistri, M., & Munari, C. (2020). Microplastic accumulation in benthic invertebrates in Terra Nova Bay (Ross Sea, Antarctica). *Environment International, 137*, 105587.

Silburn, B., Kröger, S., Parker, E. R., Sivyer, D. B., Hicks, N., Powell, C. F., Johnson, M., & Greenwood, N. (2017). Benthic pH gradients across a range of shelf sea sediment types linked to sediment characteristics and seasonal variability. *Biogeochemistry, 135*, 69–88. Available from https://doi.org/10.1007/s10533-017-0323-z.

Snelgrove, P. V. R., & Butman, C. A. (1994). Animal-sediment relationships revisited: Cause vs effect. *Oceanography and Marine Biology Annual Review, 32*, 111–177.

Solan, M., Cardinale, B. J., Downing, A. L., Engelhardt, K. A. M., Ruesink, J. L., & Srivastava, D. S. (2004). Extinction and eco-system function in the marine benthos. *Science (New York, N.Y.), 306*, 1177–1180.

Solan, M., & Kennedy, R. (2002). Observation and quantification of in situ animal-sediment relations using time-lapse sediment profile imagery (t-SPI). *Marine Ecology Progress Series, 228*, 179–191.

Somavilla, R., Schauer, U., & Budéus, G. (2013). Increasing amount of Arctic Ocean deep waters in the Greenland Sea. *Geophysical Research Letters, 40*, 4361–4366. Available from https://doi.org/10.1002/grl.50775.

Southward, A. J., Hiscock, K., Kerckhof, F., Moyse, J., & Elfimov, A. S. (2004). Habitat and distribution of the warmwater barnacle *Solidobalanus fallax* (Crustacea: Cirripedia). *Journal of the Marine Biological Association of the United Kingdom, 84*, 1169–1177.

Spormann, A. M., & Widdel, F. (2000). Metabolism of alkylbenzenes, alkanes, and other hydrocarbons in anaerobic bacteria. *Biodegradation, 11*, 85–105.

Stahl, H., Glud, A., Schroder, C. R., Klimant, I., Tengberg, A., & Glud, R. N. (2006). Time-resolved pH imaging in marine sediments with a liminescent planar optode. *Limnology and Oceanography: Methods, 4*, 336–345.

Stefanescu, C., Lloris, D., & Rucabado, J. (1993). Deep-sea assemblages in the Catalan Sea (western Mediterranean) below a depth of 1000 m. *Deep Sea Research Part I: Oceanographic Research Papers, 40*, 695–707.

Strasser, M., Dekker, R., Essink, K., Günther, C. P., Jaklin, S., Kröncke, I., Madsen, P. B., Michaelis, H., & Vedel, G. (2003). How predictable is high bivalve recruitment in the Wadden Sea after severe winter? *Journal of Sea Research, 49*, 47–57.

Sturdivant, S. K., Díaz, R. J., & Cutter, G. R. (2012). Bioturbation in a declining oxygen environment, in situ observations from wormcam. *PLoS One*, 7, e34539.

Sweetman, A. K., Thurber, A. R., Smith, C. R., Levin, L. A., Mora, C., Wei, C.-L., Gooday, A. J., Jones, D. O. B., Rex, M., Yasuhara, M., Ingels, J., Ruhl, H. A., Frieder, C. A., Danovaro, R., Würzberg, L., Baco, A., Grupe, B. M., Pasulka, A., Meyer, K. S., ... Roberts, J. M. (2017). Major impacts of climate change on deep sea benthic ecosystems. *Elementa: Science of the Anthropocene, 5*, 4. Available from https://doi.org/10.1525/elementa.203.

Teal, L., Parker, R., Fones, G., & Solan, M. (2009). Simultaneous determination of in situ vertical transitions of color, pore-water metals, and visualization of infaunal activity in marine sediments. *Limnology and Oceanography, 54*(5), 1801–1810.

Teal, L. R., Parker, E. R., & Solan, M. (2010). Sediment mixed layer as a proxy for benthic ecosystem process and function. *Marine Ecology Progress Series, 414*, 27–40. Available from https://doi.org/10.3354/meps08736.

Teuten, E. L., Saquing, J. M., Knappe, D. R. U., Barlaz, M. A., Jonsson, S., BjÃrn, A., Rowland, S. J., Thompson, R. C., Galloway, T. S., Yamashita, R., Ochi, D., Watanuki, Y., Moore, C., Viet, P. H., Tana, T. S., Prudente, M., Boonyatumanond, R., Zakaria, M. P., Akkhavong, K., ... Takada, H. (2009). Transport and release of chemicals from plastics to the environment and to wildlife. *Philosophical Transactions of the Royal Society of London. Series B, Biological Sciences, 364*, 2027–2045.

Thiel, H. (1975). The size structure of the deep-sea benthos. *Internationale Revue der Gesamten Hydrobiologie, 60*, 575–606.

Thistle, D., Ertman, S. C., & Fauchald, K. (1991). The fauna of the HEBBLE site: Patterns in standing stock and sediment-dynamic effects. *Marine Geology, 99*, 413–422.

Thistle, D., Yingst, J. Y., & Fauchald, K. (1985). A deep-sea benthic community exposed to strong near-bottom currents on the Scotian Rise (western Atlantic). *Marine Geology, 66*, 91–112.

Thrush, S. F., Halliday, J., Hewitt, J. E., & Lohrer, A. M. (2008). The effects of habitat loss, fragmentation, and community homogenization on resilience. *Ecological Applications, 18*, 12–21.

Thrush, S. F., Hewitt, J. E., Parkes, S., Lohrer, A. M., Pilditch, C., Woodin, S. A., Wethey, D. S., Chiantore, M., Asnaghi, V., Juan, S. D., Kraan, C., Rodil, I., Savage, C., & Colen, C. V. (2014). Experimenting with ecosystem interaction networks in search of threshold potentials in real world marine ecosystems. *Ecology, 95*, 1451–1457.

Thrush, S. F., Lundqust, C. J., & Hewitt, J. E. (2005). *Spatial and temporal scales of disturbance to the seafloor: A generalized framework for active habitat management. Benthic habitats and the effects of fishing* (pp. 639–649). American Fisheries Society.

Thrush, S. F., Townsend, M., Hewitt, J. E., Davies, K., Lohrer, A. M., Lundquist, C., & Cartner, K. (2013). The many uses and values of estuarine ecosystems. In J. R. Dymond (Ed.), *Ecosystem services in New Zealand—Conditions and trends*. Manaaki Whenua Press.

Thurber, A. R., Sweetman, A. K., Narayanaswamy, B. E., Jones, D. O. B., Ingels, J., & Hansman, R. L. (2014). Ecosystem function and services provided by the deep sea. *Biogeosciences Discussions*, *11*, 3941–3963. Available from https://doi.org/10.5194/bgd-11-3941-2014.

Tomlinson, D. L., Wilson, J. G., Harris, C. R., & Jeffrey, D. W. (1980). Problems in the assessment of heavy-metal levels in estuaries and the formation of a pollution index. *Helgoländer Meeresuntersuchungen*, *33*, 566–575. Available from https://doi.org/10.1007/BF02414780.

Townsend, D. W., & Cammen, L. M. (1988). Potential importance of the timing of spring phytoplankton blooms to benthic pelagic coupling and recruitment of demersal fishes. *Biological Oceanography*, *5*, 215–229.

Uncles, R. J., Stephens, J. A., & Smith, R. E. (2002). The dependence of estuarine turbidity on tidal intrusion length, tidal range and residence time. *Continental Shelf Research*, *22*, 1835–1856.

Van Colen, C., Verbelen, D., Devos, K., Agten, L., Tomme, J. V., Vincx, M., & Degraer, S. (2014). Sediment-benthos relationships as a tool to assist in conservation practices in a coastal lagoon subjected to sediment change. *Biodiversity and Conservation*, *23*, 877–889. Available from https://doi.org/10.1007/s10531-014-0638-1.

Venturini, N., Muniz, P., Bicego, M., Martins, C., & Tommasi, L. (2008). Petroleum contamination impact on macrobenthic communities under the influence of an oil refinery: Integrating chemical and biological multivariate data. *Estuarine, Coastal and Shelf Science*, *78*, 457–467.

Villnäs, A., Janas, U., Josefson, A. B., Kendzierska, H., Nygård, H., Norkko, J., & Norkko, A. (2019). Changes in macrofaunal biological traits estuarine gradients: Implications for the coastal nutrient filter. *Marine Ecology Progress Series*, *622*, 31–48. Available from https://doi.org/10.3354/meps13008.

Vitousek, P. M., Aber, J. D., Howarth, R. W., Likens, G. E., Matson, P. A., Schindler, D. W., Schlesinger, W. H., & Tilman, D. G. (1997). Human alteration of the global nitrogen cycle: Sources and consequences. *Ecological Applications*, *7*, 737–750.

Widdicombe, S., & Spicer, J. I. (2008). Predicting the impact of ocean acidification on benthic biodiversity: What can animal physiology tell us? *Journal of Experimental Marine Biology and Ecology*, *366*, 187–197.

Widdows, J., Brinsley, M. D., Salkeld, P. N., & Lucas, C. H. (2000). Influence of biota on spatial and temporal variation in sediment erodability and material flux on a tidal flat (Westerschelde estuary). *Marine Ecology Progress Series*, *194*, 23–37.

Wijsman, J. W. M., Herman, P. M. J., & Gomoiu, M. T. (1999). Spatial distribution in sediment characteristics and benthic activity on the Northwestern Black Sea shelf. *Marine Ecology Progress Series*, *181*, 25–39.

Witte, U., Aberle, N., Sand, M., & Wenzhöfer, F. (2003). Rapid response of a deep sea benthic community to POM enrichment: An in situ experimental study. *Marine Ecology Progress Series*, *251*, 27–36.

Woke, G. N., & Benson, T. A. (2018). Impacts of petroleum hydrocarbon pollution on macrobentic fauna of Oluasiri River of Niger Delta. *International Journal of Agriculture and Forestry*, *5*(7), 27–32.

Wolanski, E., & Spagnol, S. (2003). Dynamics of the turbidity maximum in King Sound, tropical Western Australia. *Estuarine, Coastal and Shelf Science*, *56*, 877–890.

Wright, S. L., Thompson, R. C., & Galloway, T. S. (2013). The physical impacts of microplastics on marine organisms: A review. *Environmental Pollution (Barking, Essex: 1987)*, *178*, 483–492.

Wu, R. S. (2002). Hypoxia: From molecular responses to ecosystem responses. *Marine Pollution Bulletin*, *45*, 35–45.

Wuchter, C., Abbas, B., Coolen, M. J. L., Herfort, L., van Bleijswijk, J., Timmers, P., Strous, M., Teira, E., Herndl, G. J., Middelburg, J. J., Schouten, S., & Damsté, J. S. S. (2006). Archaeal nitrification in the ocean. *Proceedings of the National Academy of Sciences of the United States of America*, *103*, 12317–12322.

Xu, Y., Sui, J., Ma, L., Li, X., Wang, H., & Zhang, B. (2020). Temporal variation of macrobenthic community zonation over nearly 60 years and the effects of latitude and depth in the southern Yellow Sea and East China Sea. *The Science of the Total Environment*, *739*, 139760.

Xu, Y., Yu, F., Li, X., Ma, L., Dong, D., Kou, Q., Jixing Sui, J., Gan, Z., Gong, L., Yang, M., Wang, Y., Sun, Y., Wang, J., & Wang, H. (2018). Spatiotemporal patterns of the macrofaunal community structure in the East China Sea, off the coast of Zhejiang, China, and the impact of the Kuroshio Branch Current. *PLoS One*, *13*(1), e0192023. Available from https://doi.org/10.1371/journal.pone.0192023.

Yasuhara, M., & Danovaro, R. (2016). Temperature impacts on deep-sea biodiversity. *Biological Reviews of the Cambridge Philosophical Society*, *91*, 275–287. Available from https://doi.org/10.1111/brv.12169.

Young, T. P., Chase, J. M., & Huddleston, R. T. (2001). Succession and assembly as conceptual bases in community ecology and ecological restoration. *Ecological Research*, *19*, 5–19.

Ysebaert, T., Herman, P. M. J., Meire, P., Craeymeersch, J., & Verbeek, H. (2003). Large scale spatial patterns in estuaries: Estuarine macrobenthic communities in the Schelde estuary. *Estuarine, Coastal and Shelf Science*, *57*, 335–356.

Yuewen, D., & Adzigbli, L. (2018). Assessing the impact of oil spills on marine organisms. *Journal of Oceanography and Marine Research*, *06*(1). Available from https://doi.org/10.4172/2572-3103.1000179.

Zabbey, N., & Uyi, H. (2014). Community responses of intertidal soft bottom macrozoobenthos to oil pollution in a tropical mangrove ecosystem, Niger Delta, Nigeria. *Marine Pollution Bulletin*, *82*, 167–174. Available from https://doi.org/10.1016/j.marpolbul.2014.03.002.

Zhang, X., Wang, H., He, L., Lu, K., Sarmah, A., Li, J., Bolan, N. S., Pei, J., & Huang, H. (2013). Using biochar for remediation of soils contaminated with heavy metals and organic pollutants. *Environmental Science and Pollution Research*, *20*, 8472–8483. Available from https://doi.org/10.1007/s11356-013-1659-0.

CHAPTER 7

Ecology, distribution, and biogeography of benthos

S. Satheesh[1] and M.M. El-Sherbiny[1,2]
[1]Department of Marine Biology, Faculty of Marine Sciences, King Abdulaziz University, Jeddah, Saudi Arabia
[2]Department of Marine Sciences, Faculty of Science, Suez Canal University, Ismailia, Egypt

7.1 Introduction

A marine ecosystem consists of three major units; physical factors, chemical constituents, and biotic factors. The functioning of the marine ecosystem is linked with the interactions between abiotic (physical and chemical factors) and biotic factors (Reiss et al., 2009; Gammal et al., 2019). The benthos is one of the largest components of the marine ecosystems that play key functions (Snelgrove, 1998). The benthic organisms that live in or on the sediment or other solid substrates are important in marine ecosystems because they act as prey for commercially significant fish and also participate in nutrient recycling (Levin et al., 2001). In biogeochemical cycling, benthic organisms contribute to the nutrient fluxes between water and sediment through metabolic activity and physical activity in the sediment (Ehrnsten et al., 2020). During the metabolic activity, a portion of the organic matter eaten by benthic organisms could be converted to inorganic compounds (Ehrnsten et al., 2020).

Among the benthic ecosystems, the coastal habitats are significant due to the abundance of species, inputs of nutrients, and other pollutants from the terrestrial sources (Gattuso et al., 1998). Anthropogenic stressors are one of the significant causes of the degradation of coastal ecosystems throughout the world (Worm et al., 2006; Wu et al., 2017). In addition to anthropogenic pressures, marine ecosystems are under constant threats from climate change-induced effects such as warming and ocean acidification (Wernberg et al., 2014; Wabnitz et al., 2018). Consequently marine ecosystems are experiencing a considerable loss of biodiversity that affects ecosystem function (Worm et al., 2006; Levin & Poe, 2017). The response of benthic as well as plankton food web to multiple environmental stressors is multifaceted and nonlinear (Hjorth et al., 2007;

Ecology and Biodiversity of Benthos
DOI: https://doi.org/10.1016/B978-0-12-821161-8.00006-4

Jedruch et al., 2019). Moreover, changes in the community structure of benthos will affect the trophic relationships (Sokołowski et al., 2012) that modify the interactions between the predator–prey relationships. Hence it is essential to understand the structure and ecology of benthic communities for predicting the distribution patterns about changing environmental conditions (Kang et al., 2019).

Benthic macrofauna is the dominant group at species levels in coastal ecosystems (Snelgrove, 1999). The distribution and abundance of benthic macrofauna communities are greatly influenced by the physical, chemical, and biotic factors (Van Wesenbeeck et al., 2007; Choi et al., 2014; Varfolomeeva, 2013; Schuckel et al., 2015). Most of the benthic fauna showed a biphasic life cycle (pelagic larva and benthic adult stages) and the larval supply is one of the major factors which determine the distribution of benthic organisms (Todd, 1998). Hence the changes in macrobenthic communities over the spatial and temporal scales are due to the factors like recruitment and environmental disturbances (Thrush & Dayton, 2002). Therefore benthic organisms are used as good indicators to assess the impacts of environmental conditions on marine ecosystems (Warwick, 1986; Post et al., 2011). Furthermore, the ecology and diversity of benthic organisms are influenced largely by some key organisms like mangroves (Alongi et al., 1998), seagrasses (Hemminga & Duarte, 2000; Leopardas et al., 2014), seaweeds (Ba-Akdah et al., 2016; Satheesh et al., 2017), coral reefs (Pearman et al., 2018), and sponges (Maldonado et al., 2015) in the coastal habitats (Fig. 7.1). Hence understanding the factors responsible for the variability of community structure over spatial and temporal scales is the major theme in marine benthic ecology (Blanchet et al., 2014). This chapter presents an overview of diversity, biogeography, distribution, and important factors that influence the distribution of benthic organisms in the intertidal and subtidal marine ecosystems.

7.2 Ecology of benthos

Macrofauna and meiofauna are the two common categories of benthic organisms living in marine ecosystems. The macrobenthos is largely dominating the benthic communities in marine ecosystems throughout the world (Snelgrove, 1998) which are divided into mobile and sessile organisms. The mobile organisms include polychaetes, crabs, gastropods, echinoderms, and demersal fishes. The mobile organisms can move for a small distance as in the case of gastropods or move freely like crabs. The sessile

Figure 7.1 Some important benthic organisms in marine ecosystems. (A) Seagrass, (B) macroalgae, (C) sponge, and (D) coral reefs. *Images from Central Red Sea, Saudi Arabia.*

organisms cannot move after the settlement which includes encrusting sponges, corals, anemones, bryozoans, hydrozoans, barnacles, tubeworms, mussels, ascidians, seaweeds, and seagrasses.

Mangroves, seagrass meadows, macroalgal beds, coral reefs, rocky shores, sandy shores, soft sediments, and deep-sea benthic habitats are the common benthic ecosystems in the marine environment. The dominant benthic taxa in these ecosystems also varied based on the habitat type, oceanographic conditions, and biotic interactions. For instance, among the macrofauna, polychaetes constitute an important part of total benthic community compositions in soft-bottom environments (Ward & Hutchings, 1996; Hutchings, 1998; Labrune et al., 2008). However, the abundant benthic taxon on soft-bottom communities varied between the geographical regions. Similarly, the dominant benthic fauna associated with habitats such as seagrass, mangroves, and seaweeds also varied between the geographic regions. While a study conducted on the associated fauna of seagrass beds from the Indonesian waters revealed the dominance of polychaetes (Lin et al., 2018), another study from the seagrass and mangrove habitats of Lakshadweep Islands of India reported mollusks as dominant benthic organisms (Susan et al., 2014).

Also, a study conducted on the Brittany coast along the French Atlantic continental shelf indicated the dominance of polychaetes (244 species) in the sediment communities (Dutertre et al., 2013). The macrobenthic community along the continental slope of the Gulf of Guinea (West Africa) was studied by Pabis et al. (2020) which showed the abundance of polychaetes, bivalves, and amphipods. The macrobenthic community of Visakhapatnam harbor on the east coast of India studied by Musale et al. (2015) revealed the dominance of polychaetes (60 species) and other invertebrates taxa such as crustaceans, mollusks, and oligochaetes contributed 24 species. Polychaetes were the dominant fauna (60.54% among the total species) in the benthic community collected from different depths along the southeast coast of India (Khan et al., 2017). The variations in the abundance and diversity of taxa between benthic ecosystems indicated the need to study the ecology of these habitats.

7.2.1 Ecology of macrobenthic community associated with seagrass meadows and kelp forests

Corals, sponges, seaweeds, and seagrasses are considered important organisms in benthic ecosystems because they provide biogenic habitat to many other organisms. Seagrasses are the flowering plants found in the coastal ecosystems, providing several ecosystems services including acting as a habitat for many invertebrates (Lee et al., 2001; Bostrom et al., 2006; Lundquist et al., 2018). Seagrasses are abundantly distributed in the tropical Indo-Pacific region, which is the native for many species (Short et al., 2007; Lin et al., 2018). Macroalgae are an important component in coastal benthic ecosystems throughout the world (Ba-Akdah et al., 2016; Satheesh et al., 2017). Many studies also confirmed that macroalgal beds increase the coastal biodiversity by providing shelter, food, and protection to a variety of marine organisms (Ba-Akdah et al., 2016; Macreadie et al., 2017). The abundance of benthic invertebrates is higher in the seagrass meadows than the unvegetated habitats in the coastal ecosystems (Bloomfield & Gillanders, 2005).

The seagrass and macroalgal beds alter the predator–prey relationships and act as one of the causative factors for the variability of macrofaunal diversity and abundance (Orth et al., 1984; Schultz et al., 2009). The canopy structure of the macroalgae and seagrass may also influence the distribution of associated benthic organisms (Lee et al., 2001; Cacabelos et al., 2010). Also, the chemical defense of the macroalgae is reported to affect the macrofaunal abundance in algal beds (Steinberg et al., 1998). Seasonal variations in the abundance of epifaunal communities are largely due to the production of secondary metabolites by some macroalgal species (Rickert et al., 2015).

In seagrass, the availability of microhabitats such as rhizome, leaf, and shoots will also influence the abundance of many invertebrates (Włodarska-Kowalczuk et al., 2014; McCloskey & Unsworth, 2015). Previous studies indicated that the abundance of organisms was high at the edges of the seagrass ecosystem than the interior parts (Bologna & Heck, 2002; Bostrom et al., 2010). On the other hand, herbivores have the potential to change the structure of biogenic habitats such as seaweeds and seagrasses (Heck & Valentine, 2006; Bellgrove et al., 2014). Hence the increase in the abundance of herbivores may decrease the biomass of these biogenic habitats (Steneck, 2013) due to grazing. This will further affect the distribution of organisms associated with the macroalgal and seagrass beds (McCloskey & Unsworth, 2015).

7.2.2 Ecology of rocky shore communities

Rocky shores are one of the common benthic habitats in the intertidal regions around the coastal areas throughout the world (Branch et al., 2008; Coutinho et al., 2016). The rocks provide a stable substrate for the attachment of benthic communities such as barnacles, mussels, gastropods, and macroalgae. Among the macrobenthic organisms, crustaceans and mollusks are the most abundant groups on rocky shores (Londono-Cruz et al., 2014; Satyam & Thiruchitrambalam, 2018). The organisms living in the rocky intertidal habitats are possessing physiological adaptations to overcome the changing environmental conditions (Harley & Helmuth, 2003; Denny, 2006; Harley et al., 2009). The rocky shores also provide plenty of niches to the benthic organisms that increase the overall biodiversity of the coastal habitats (Smith, 2013).

The zonation or vertical distribution of rocky shores is mainly determined by the factors such as temperature and desiccation and biotic interactions (Little & Kiching, 1996; Tomanek & Helmuth, 2002). The vertical distribution pattern of organisms on rocky shores has been studied extensively throughout the world (Underwood, 1981; Esqueda et al., 2000; Chappuis et al., 2014; Dolenc-Orbanic & Battelli, 2017). The results showed a spatial distribution of benthic communities depending on their physiological tolerance to abiotic and biotic factors (Benedetti-Cecchi, 2001; Londono-Cruz et al., 2014). Some previous studies stressed the role of interspecific competition for the vertical distribution of organisms on rocky shores. For instance, the competition between two barnacle species (*Semibalanus balanoides* and *Chthalamus stellatus*) was responsible for setting the lower distributional range of the *C. stellatus* in the vertical distribution pattern on rocky shores (Connell, 1961). This coexistence pattern through resource partitioning due to competition was observed in many other barnacles species on rocky shores (Shinen & Navarrete, 2014). Furthermore, Paine (1974) indicated that the lower limit of the mussel beds could be extended if the predator species like sea stars were removed from the rocky shores. In general, the physiological adaptations of the benthic organisms to wave exposure, temperature, and salinity along with the biotic interactions may influence the structure of the macrobenthic community on the rocky shores.

7.2.3 Ecology of coral reefs and associated benthic community

The coral reefs which are found in shallow waters are one of the important habitats in the coastal ecosystems due to their role in ecosystem functioning and economical significance. Coral reefs are distributed mainly in the Indian Ocean, Arabian Gulf, Red Sea, Caribbean/western Atlantic, western Pacific, Central Pacific, and eastern Pacific regions in the world (Glynn, 1993). The corals act as a habitat for many benthic organisms and commercially important fish species. The topographical complexity of the coral reefs provides extra space for other organisms and maintains overall biodiversity in the reef ecosystems (Emslie et al., 2014). The benthic fauna inhabiting the reef habitats include different types of sessile and mobile organisms (Enochs et al., 2011; Milne & Griffiths, 2014; Pearman et al., 2018). Most of these associated benthic organisms are considered cryptic fauna due to their occurrence in the hidden spaces of the coral reef (Glynn & Enochs, 2011).

The diversity and distribution of benthic organisms associated with coral reefs depend on the abundance of coral species. The reduction in coral growth may enhance the overgrowth of reef benthos like macroalgae, coralline algae, bryozoans, sponges, and ascidians (Glynn, 1993). These benthic organisms possess superior colonizing ability on the substrates due to their fast growth. Specifically, many incidents of macroalga overgrowth have been reported in the literature (Lirman, 2001; Liu et al., 2009; Brown et al., 2018). Hence macroalga-coral interaction may influence the coral reef benthic community structure. Generally, herbivory is the key biotic interaction that regulates the macroalgal abundance in the reef habitats. Though many species are involved in the herbivory process of the reefs, fishes are the dominant organisms responsible for grazing activity (Bellwood et al., 2004; Richardson et al., 2020). The reduction in macroalgal community abundance will increase the dominance of the coral species (Richardson et al., 2020).

Furthermore, anthropogenic impacts such as pollution, overexploitation of fishery resources, and coastal development activities are directly or indirectly responsible for the decline of the coral reef communities throughout the world marine ecosystems (Jackson et al., 2001; Hughes et al., 2003; Pandolfi, 2011). Another major factor that induces coral species mortality is the bleaching events that indicate the loss of symbiotic zooxanthellae associated with corals (Glynn, 1993). The mass mortality of coral species due to bleaching was reported from various geographical regions (Glynn, 1993; Magel et al., 2019). Many environmental stressors are reported as a causative factor for the bleaching of the corals at local levels but the warming of the ocean is largely responsible for regional level bleaching events (Heron et al., 2016; Magel et al., 2019). In brief, the decline of coral species paves the way for the shifting of the dominance of corals to macroalgae in many reef ecosystems (Pandolfi et al., 2003; Hughes et al., 2007) which may have profound impacts on the coastal ecosystem functioning.

7.2.4 Ecology of mangrove ecosystems

Mangroves are the dominant coastal ecosystems in most tropical and subtropical coastal regions (Alongi et al., 1998). They grow in the coastal belts and act as a barrier to protect the coastal environments from natural events. The mangroves have adaptations to withstand the harsh environmental conditions of the coastal zones. Specifically, they exclude the excess salt through the leaves and the exposed root system allows the oxygen intake. Mangrove ecosystems are considered highly productive and support a large number of benthic as well as planktonic organisms (Christensen, 1978; Alongi et al., 1998; Kristensen, 2008). Mangroves enhance the benthic biodiversity of coastal ecosystems by providing shelter to several marine invertebrates and fish species. The structural complexity of mangroves such as aerial roots and prop roots act as refugia to juvenile fishes and small invertebrates (Lee et al., 2001). A previous study by Kon et al. (2010) showed that the habitat complexity of the mangroves increased the abundance of the epifaunal community. In addition, organic matter is considered a food source for many marine invertebrates (Beck et al., 2001). Hence the high diversity of benthic organisms in the mangrove ecosystems may be due to the occurrence of shelter (to avoid predation) and abundance of food sources.

The ecological role of the benthic community associated with the mangroves is well understood (Smith, 2013; Kristensen, 2008). Generally, crabs constitute about 80% of total macrofaunal community biomass in the mangrove ecosystems (Dharmapalan et al., 2011). Mollusks and polychaetes were also reported as abundant macrobenthic groups associated with the mangrove forests (Nagelkerken et al., 2008; Murugesan et al., 2018). Grapsid crabs (Grabsidae) and fiddler crabs (Ocypodidae) are the dominant benthic fauna in the mangrove forests (Kristensen, 2008). The spatial distribution of crabs in different zones of the mangrove forests depends on environmental factors, competition, and substratum characteristics (Cannicci et al., 2018).

Grapsid crabs play a significant role in the ecology of the mangrove ecosystems. For instance, Lee (1998) indicated that grapsid crabs affect the mangrove community structure through recycling the organic matter, altering the substrate topography by bioturbation, and selective consumption of mangrove propagules. Another study by Schories et al. (2003) reported that the leaf litter removing capacity of the crab Ucides is an important process in the organic matter turnover in the mangrove ecosystems of North Brazil. The benthic fauna associated with the mangroves also showed some ecologically important complex interactions. The mutualistic association between the sponges and the mangrove *Rhizophora mangle* for sharing the nitrogen and carbon resources resulted in increased growth in both organisms (Ellison et al., 1996).

7.3 Biogeography of marine benthos

7.3.1 Distribution

Most of the classical biogeographical studies on the benthic community were focused on species distributions, the evolution of fauna, and continental shifts (Briggs, 1974; Franz & Merrill, 1980). Historically, biogeography originated from descriptive studies which reported the presence or absence of species (Dana, 1853). Biogeography patterns are useful to identify the biodiversity hotspots which host novel fauna and flora (Briggs & Bowen, 2012). For the last two decades, the biogeographic studies are used to predict the species distributional range shifts due to climate change, the impact of pollution on the distribution of organisms, the effects of invasive alien species, and overexploitation of fishery resources on species distribution (Llanso et al., 2002; Buchsbaum & Powell, 2008; Kirby et al., 2008; Hale, 2010). Latitudinal and depth gradient species diversity in the oceans is the two common biogeographic trends used to analyze the species distribution (Rex et al., 2005). The latitudinal gradient of diversity is used to explain the large-scale patterns in global biodiversity. Many authors reported that the species richness is high in the equator and decreases toward higher latitudes (Pianka, 1996; Rex et al., 2005; Tittensor et al., 2010; Brown, 2014).

Large-scale biogeographic studies were conducted in rocky shore benthic communities to understand the distribution patterns (Bustamante & Branch, 1996; Benedetti-Cecchi et al., 2003; Sink et al., 2005; Fica et al., 2017). Studies indicated that geography and oceanographic conditions are the major factors determining the spatial distribution of organisms on rocky shore ecosystems in southern African rocky shores (Bustamante & Branch, 1996). Another study by Fica et al. (2017) revealed the role of biotic and abiotic factors in regulating the spatial distribution of benthic assemblages on the rocky shores of the Pacific shore of Southern Patagonia, South America, and the West Antarctic Peninsula. The coastal biogeographic patterns of the benthic fauna of the Indian coast revealed distinct variations between eastern and western basins due to physical processes such as coastal length, habitat heterogeneity, and temperature (Sivadas & Ingole, 2016).

The benthic species diversity is generally high in tropical marine ecosystems, though it depends on species and geographic regions. The highest species diversity may be due to the fact that in the shallow tropical waters, the latitudinal gradients of benthic diversity are influenced by the coral reefs (Rex et al., 2005). As stated above (section 7.2.3), coral reefs enhance the overall biodiversity and that may be the major controlling factor for the higher benthic biodiversity in the tropical and subtropical regions. Among the tropical marine ecosystems, the Indo-Pacific is the largest tropical biogeographic region with high species diversity (Roberts et al., 2002).

7.3.2 Marine biogeographic provinces

Marine biogeographic provinces are the regions that represent distinct fauna and flora (endemism) that are separated from other geographical regions by boundaries (Briggs & Bowen, 2012; Costello et al., 2017). Though the criteria for the classification of provinces or regions are constantly updated by many researchers, the minimum value of 10% endemism outlined by Briggs (1974) is used as a benchmark for defining the biogeographic provinces. Earlier, Ekman (1953) classified the marine realm into different regions and subregions based on animal distribution data. The biogeographic provinces were later realigned based on individual species distribution data. For instance, Briggs and Bowen (2012) realigned the biogeographic provinces using the fish distributions data. Recently, Costello et al. (2017) conducted a robust analysis of the biogeographic realms based on the distribution of 65000 species which resulted in a new biogeography map for the species distributions. The new biogeography map reported by Costello et al. (2017) revealed the relationship between environmental factors and the evolution of biota. Also, these maps showed the biogeography of different ecosystems such as coastal, benthic, pelagic, and deep sea.

7.4 Factors affecting the distribution of macrobenthic community

The distributional limits of the benthic species are determined by the environmental factors of the water and sediment. Many factors influence the diversity and distribution of benthic communities in marine environments (Fig. 7.2). Specifically, the regional benthic diversity patterns in latitudinal gradients (tropical, subtropical, and high latitudes) are controlled by various factors at larger spatial and temporal scales (Witman et al., 2004). Furthermore, the physicochemical parameters, hydrodynamic conditions, and biotic interactions are mainly responsible for the abundance and distribution of benthic organisms at spatial or temporal scales (Bolam et al., 2008; Dutertre et al., 2013; Gartner et al., 2013; Kroeker et al., 2016). The physical and chemical parameters of the marine environment showed seasonal variations largely in temperate regions (Dutertre et al., 2013). Hence the season plays a crucial role in the distribution of benthic organisms mainly in temperate regions. In tropical and subtropical seas also, seasonal variations in the distribution of benthic

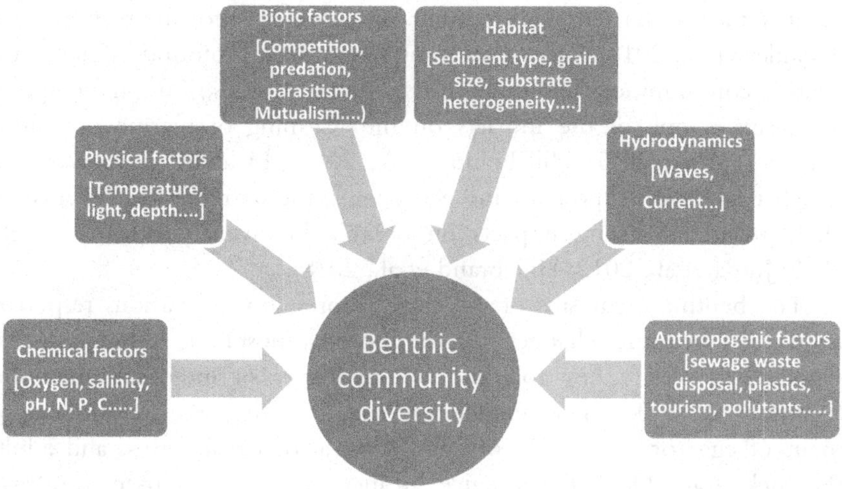

Figure 7.2 Factors affecting the distribution of benthic organisms in coastal ecosystems.

fauna and flora have been widely reported (Shou et al., 2009; Kumar & Khan, 2013).

The distribution of benthic fauna along the ocean gradients is largely affected by the factors such as depth, water currents, dissolved oxygen, the organic carbon content of the sediment, sediment characteristics, and pollution levels (Rousi et al., 2011; Ellis et al., 2012; Solan et al., 2012; Teso et al., 2019; Pabis et al., 2020). The correlation between environmental conditions and the benthic community abundance differs between spatial scales because of the variability of the factors. Also, many mechanisms are involved in the abundance and distribution of benthic organisms. Understanding the processes that control the species diversity and distribution over spatial and temporal scales is one of the major themes in ecology (Witman et al., 2004). Hence the key factors that are reported to affect the distribution of macrobenthic communities are highlighted further.

7.4.1 Temperature

The environmental conditions such as temperature, wave exposure, and emersion time and the biotic factors such as predation and competition are the major factors that determine the structure of benthic organisms in the rocky shores (Coutinho et al., 2016). The effects of temperature on marine organisms are mainly studied to assess the impacts of ocean warming. Warming is one of the impacts of climate change with the increase of ocean surface water temperature by 0.11°C decade^{-1} for the past 50 years (Chaidez et al., 2017). Ocean warming may have a profound influence on benthic communities. The poleward shifting of the distributional range of organisms is one of the impacts on the warming of marine organisms (Chaidez et al., 2017; Hillebrand et al., 2018). In addition to warming which is gradual in most marine ecosystems, the extreme heat events are likely to create strong impacts on benthic biodiversity (Duarte et al., 2012; Jorda et al., 2012; Hillebrand et al., 2018).

The benthic organisms of the rocky shores showed various responses to the temperature changes. Furthermore, atmospheric temperature is reported to play a key role in the physiology of intertidal organisms (Schiel et al., 2004). The temperature variations may affect the development of eggs or the release of eggs, survival of larval forms, and adults (Hiscock et al., 2004). For instance, an increase in water temperature may favor spawning and larval release in many marine benthic organisms (Hiscock et al., 2004). In general, the physiological responses (e.g., growth and reproduction) of the organisms are maximized in optimum temperature conditions. A change in temperature below or above the optimum range may affect the physiological activity of the organisms, which eventually influences the distribution and diversity. The physiological effects of temperature on organisms varied from protein damages to impacts on organ function (Hochachka & Somero, 2002). In many marine ecosystems, the organisms are already living near their thermal limits, so an increase in temperature above the optimum level negatively affects the survival of the organisms (Harley et al., 2006; Chaidez et al., 2017).

The changes in marine benthic community structure in temperate waters are mainly linked to seasonal variations in temperature (Schroeder, 2003). Several previous studies documented a decrease in species abundance and biomass during cold winters in temperate regions (Wieking & Kroncke, 2001; Strasser et al., 2003; Reiss et al., 2006). An increase in temperature may reduce the biodiversity of intertidal habitats (Wethey et al., 2011; Starko et al., 2019). The reduction in biodiversity is largely visible in coral reef communities due to high temperatures (Anthony & Kerswell, 2007). Many previous studies confirmed the variations in the abundance and distribution of benthic organisms due to changes in seawater temperature (Hinz et al., 2011; Hiddink et al., 2015; Hale et al., 2017). However, the effects of temperature on the distribution of benthic communities varied among the marine ecosystems. For instance, tropical and subtropical marine ecosystems may experience an overall biodiversity increase due to the increasing seawater temperature (Mazzuco et al., 2020). While higher productivity in tropical and subtropical ecosystems is one of the factors for higher species diversity, other factors such as temperature-dependent ecological and evolutionary processes may also maintain higher biodiversity (Brown, 2014).

7.4.2 Salinity

The coastal ecosystems experience salinity fluctuations due to the freshwater runoff during rainy seasons. The salinity changes are largely visible in areas around estuarine gradients (Mariano & Barros, 2015; Wolanksi & Elliott, 2015). Hence salinity is described as one of the important factors influencing the distribution and physiology of organisms in the estuarine environments (Matthews & Fairweather, 2004). Salinity affects the diversity and distribution of benthic organisms mainly in intertidal habitats (Smyth & Elliott, 2016). The areas around the estuaries recorded lower diversity and abundance of organisms due to large fluctuations in salinity (Khedhri et al., 2015). The organisms which cannot tolerate a wide range of salinity fluctuations may migrate to offshore regions to avoid the effects of salinity. This is because the salinity changes can create osmotic stress on marine organisms (Danko et al., 2020). Furthermore, marine organisms also need an optimum salinity range for reproduction and low salinity reduced the reproduction rate in many species (Gelin et al., 2001; Steen, 2004). The allocation of more energy for osmotic stress could be one of the reasons for the inhibition or reduction in reproduction rates in low salinity conditions (Danko et al., 2020). Also, large variations in salinity may affect the growth and survival of benthic organisms (Montague & Ley, 1993).

The changes in the salinity may be a causative factor for the shift in the distributional range of some species. For example, in the Baltic Sea, the foundation species and their associated organisms may experience a shift in distributional areas due to the projected decrease in salinity (Vuorinen et al., 2015). Coral reefs and seagrasses are the major benthic habitats that are frequently exposed to salinity variations. Changes in salinity may affect the metabolism and survival capacities of the coral reefs (Brown, 1997; Ferrier-Pages et al., 1999). Coral reefs generally occur in stable salinity environments where freshwater inflow is low (Ferrier-Pages et al., 1999). In seagrasses, the decrease in salinity due to freshwater inputs in the coastal habitats affects the growth and distribution (Lirman & Cropper, 2003).

7.4.3 Nutrients and oxygen

Nutrient concentration of the water and sediment is one of the major limiting factors influencing the abundance and species of benthic organisms. Nitrogen, phosphorus, and organic carbon along with elements are essential for the normal physiological functions of marine organisms. The marine ecosystems around the world are experiencing higher nutrient inputs from various anthropogenic sources. A small increase in the level of nutrients is normally beneficial to benthic organisms (Ellis et al., 2017), but overloading will affect the organisms. Generally, nutrients such as nitrogen, phosphorus, sulfur, and silica are essential for the normal growth of microalgae and macrophytes in benthic ecosystems. Hence primary production in most of the benthic habitats is influenced by nutrient level; usually, the increase in nutrient load is associated with higher production (Andersen et al., 1991). An increase in the nutrient concentration, particularly nitrogen and phosphorus levels in coastal waters is associated with the abundance of microalgal species (Cahoon, 1999; Hagberg & Tunberg, 2000). This increase in microalgal biomass will enhance the food availability of macrobenthic organisms. Simultaneously, the nutrient enrichment increases the turbidity and decreases the light penetration level in the seawater (Drylie et al., 2018). The limited light availability may affect the photosynthesis rate in the primary producers especially seagrass meadows in the coastal ecosystems.

Most marine benthic organisms require oxygen for their metabolism and hence they are abundant on the top layers of the sediment (Middelburg & Levin, 2009). The oxygen depletion level in marine waters and sediments is usually termed anoxia and hypoxia. Anoxia is a condition of the marine ecosystems where complete lack of oxygen and hypoxia is the decrease from the normal oxygen level, normally depending on the sensitivity of the organisms (Rousi et al., 2019). The higher inputs of nutrients to the water column and sediments will lead to hypoxic conditions which are detrimental to the benthos. While higher inputs of nutrient load lead to eutrophication of the ecosystems and affect the benthic community distribution due to hypoxia conditions (Hale et al. 2016), a moderate increase in nutrient content may also affect the benthic communities. For instance, a recent study by Hope et al. (2020) reported that anthropogenic nutrient input and the associated increase in turbidity may change the benthic food web structure. The alterations in benthic food webs will have further consequences in the biotic interactions in the benthic habitat which affects the overall community composition (Hope et al., 2020). Furthermore, the high loading of organic carbon in the sediment will also increase the hypoxia and affect the benthic community structure (Hale et al., 2016). A previous study conducted by Rousi et al. (2019) clearly showed that oxygen level is among one of the critical abiotic factors determining species diversity and distribution of benthic organisms. This study (Rousi et al., 2019) reported the increase in species number with the increased oxygen level in the open coastal Gulf of Finland. In summary, the high nutrient loading and decrease in oxygen level will reduce the benthic biodiversity in marine ecosystems.

7.4.4 Light

Light availability is one of the primary factors influencing the distribution of benthic organisms in marine ecosystems (Miller & Etter, 2008). Specifically, light is an important factor determining the biotic interactions in benthic communities (Irving & Connell, 2002). Many previous studies showed that light is important for the physiology and survival of coral reefs (Osinga et al., 2011; Main & Goodbody-Gringley, 2010; Kuanui et al., 2020). Furthermore, marine algae require light for normal photosynthesis and growth. So, algal species are abundant on the substrates exposed to high light intensity (Gattuso et al., 2006). Light intensity determines the vertical distribution of benthic organisms in coastal habitats. This was proved by many experiments which showed a higher abundance of algae on top sides of the substrates and sessile organisms on shaded surfaces (Baynes, 1999; Irving & Connell, 2002; Pacheco et al., 2010; Khalaman et al., 2018). Furthermore, previous studies largely reported a higher number of invertebrates on shaded areas (Svane & Dolmer, 1995; Glasby, 1999; Carlson et al., 2006).

The larvae of many benthic invertebrates are photopositive during early stages and become photonegative during development (Wendt & Woollacott, 1999; Maldonado et al., 2003). Hence the larval responses to light intensity influence the spatial distribution of benthic organisms mainly in the coastal ecosystems (Svane & Dolmer, 1995; Davies et al., 2014). The availability of space for settlement is limited in high light-exposed areas due to the higher abundance of algae so the sessile organisms are moving to the shaded places where the competition from algal species is low (Miller & Etter, 2008). However, there are some organisms such as ascidians that may compete with algae and are abundant on light-exposed substrates (Terlizzi et al., 2000). Besides the competitive exclusion of animals by the algae, the higher light intensity may also negatively affect the distribution of some benthic marine organisms. For example, high light exposure is detrimental to sponges due to overexposure to UV radiation (Saunders & Connell, 2001).

7.4.5 Sediment properties

Sediment characteristics have been cited as important factors for the diversity and distribution of macrofaunal communities in benthic ecosystems at smaller scales (Van Hoey et al., 2004; Hily et al., 2008; Gray & Elliott, 2009). There is a multifaceted relationship existing between the sediment characteristics and the distribution of benthic organisms (Snelgrove, 1999). Generally, deposit feeder polychaetes and other opportunists are found in higher numbers in the sediment with high organic matter content (Pearson & Rosenberg, 1978). Furthermore, suspension feeders are abundant in high energy-environment in which the grain size of the sediment is large with low organic matter content (Snelgrove, 1999). Neves et al. (2013) reported that the sediment type strongly influences the abundance of mollusks in Guanabara Bay, Brazil. The sediment composition is explained by the variability of bottom benthic communities in intertidal and subtidal regions of north-western Spain (Cacabelos et al., 2009). However, van der Zee et al. (2015) indicated that sediment stabilization rather than the composition of sediment enhanced the diversity of the benthic community.

In the benthic zone, sediment particles move continuously due to the action of waves, tidal currents, and weather conditions (Miller et al., 2002). Benthic organisms living in or on the sediment possess various adaptations to survive the sediment deposition and erosion (Bock & Miller, 1995). For instance, those organisms living in between the sediment particles (infauna) may burrow inside the sediment to avoid the impact of sediment erosion (Miller et al., 2002). Many previous studies reported that the sediment stability in intertidal habitat is mainly dependent on the interaction between microbial mats and grazers (Willows et al., 1998; Paterson & Black, 1999; De Deckere et al., 2000). The extracellular polymeric substances (EPS) produced by the microorganisms in the biofilms play important role in sediment stabilization (De Brouwer et al., 2005). As the EPS production is directly linked with the bacterial community composition and activity (Yallop et al., 2000), the functions of biofilms and microbial mats on the distribution of benthic organisms need to be studied in detail.

The benthic organisms living in the sediments showed vertical distribution patterns and higher density and diversity were observed on the surface layers (Touhami et al., 2018). Several previous studies indicated that the vertical distribution pattern of benthic organisms in the sediment depends on the properties of the sediment and behavior of the organisms (Mucha et al., 2004; Acuna et al., 2012; Muniz et al., 2013). Among the behavior of the organisms, the feeding types play a significant role in the vertical distribution of benthic organisms. Blake and Grassle (1994) reported that the vertical distribution of benthic fauna in the continental slope sediments off Cape Lookout in North Carolina showed a relationship with feeding types. The other biological traits of the organisms which influence the vertical distribution include the size of the species, mobility, competition, and predator avoidance capacities (Rosa & Bemvenuti, 2006; Gogina & Zettler, 2010; Touhami et al., 2018).

7.4.6 Hydrodynamics

Hydrodynamics conditions of marine water influence the spatial and temporal distribution of both benthic macrophytes and macrofauna (Kautsky et al., 1999). The effects of hydrodynamics on the abundance and distribution of benthic fauna and flora are documented widely for different taxonomic groups (Turner et al., 1999; Pfaff et al., 2011; Chollett & Mumby, 2012; Thomson et al., 2012). Current velocity is an important factor that regulates the distribution of benthic macrofauna in intertidal habitats (Van der Wal et al., 2017). The wave exposure is the limiting factor for the abundance of benthic organisms in the rocky shores and intertidal habitats. A previous study by Herkul et al. (2016) reported that wave height has a significant correlation with the benthic variables in the north-eastern Baltic Sea. Furthermore, the physical disturbance from the wave action may increase the turbidity, affect the light penetration and change the nutrient level of the seawater (Herkul et al., 2011). These processes due to the wave energy will affect the benthic habitat availability to the organisms (Denny, 2006). The wave energy varies over temporal and spatial scales because it depends on the topography of the sediment and coastal geometry (Rattray et al., 2015). Since the food availability and larval dispersal of sessile organisms depend on the hydrodynamics of the marine waters (Rattray et al., 2015), changes in these conditions will affect the abundance of the benthic sessile organisms on solid substrates.

7.4.7 Substrate topography

The topographic heterogeneity (fine-scale elevations) in the substrate provides many microhabitats for the benthic organisms (Fig. 7.3). These elevations or irregularities on the substrates change the physical variables at a small scale (Guichard et al., 2001). The topographic heterogeneity affects the distribution of organisms in benthic habitats (Bourget et al., 1994). An increase in topographic heterogeneity may have a positive impact on the diversity and distribution of benthic organisms in intertidal habitats (Price et al., 2019). In high complexity substrates, the benthic assemblages show spatial variability due to the physical characteristics of the substrate (Benedetti-Cecchi & Cinelli, 1997). Furthermore, sessile organisms like macroalgae increase the substrate heterogeneity at fine-scale in intertidal rocky shores which enhance the diversity of associated fauna (Christie et al., 2009). Generally, the effects of topographic heterogeneity on benthic community structure in rocky shores depend on other factors such as hydrodynamics or shading (Guichard et al., 2001). For instance, the topographic heterogeneity may affect the abiotic conditions near substrates. Specifically, the elevations on the substrate could affect the water movement in that area. Furthermore, substrate heterogeneity may reduce the direct sunlight which enhances the shading (Skinner & Coutinho, 2002).

Experiments conducted using artificial substrates to test the effects of substrate topography indicated conflicting results specifying the role of interacting effects of environmental conditions with substrate properties. Many laboratory and field experiments showed that the larvae of sessile organisms preferred to settle on rough substrates than smooth surfaces

Figure 7.3 Substrate heterogeneity in the soft-bottom benthic habitat of the central Red Sea.

(Bologna & Heck, 2000; Skinner & Coutinho, 2002; Petersen et al., 2020). The rough substrates may provide more topographic heterogeneity for the organisms. However, previous studies also reported higher recruitment rates of marine benthic organisms on smooth surfaces than the rough ones (Osman & Whitlatch, 1995; Berntsson et al., 2004). This further confirmed the role of hydrodynamic conditions along with the topographic heterogeneity on the recruitment of benthic communities.

7.5 Role of species interactions in the ecology of benthic community

The species interactions are the major biotic factors affecting the ecology of benthic organisms. Herbivory and predation pressures may exert a strong influence on the community composition of organisms (Menge & Sutherland, 1987). In the rocky shores, competition between the organisms determines the spatial distributions at a small scale (Worm & Karez, 2002). For the sessile organisms, the space for the settlement is one of the limiting sources responsible for intraspecific and interspecific competition. Hence if interspecific competition occurs between two species, the species with a high growth rate will succeed (Levinton, 2009). This is evidenced by the colonization of the encrusting organisms on the available substrate that prevent the settlement of other organisms (Jackson, 1977). Fig. 7.4 explains the colonization of artificial substrate submerged in the seawater

Figure 7.4 Recruitment of encrusting benthic organisms (ascidians and sponges) on aquaculture cage nets submerged in the central Red Sea. Due to the superior space-occupying capacity, these organisms prevented the settlement of other organisms on the cage nets.

by colonial ascidians and sponges. Due to the fast growth rate, these encrusting or colonial organisms are structuring the diversity of the benthic organisms on the hard substrates (Lindeyer & Gittenberger, 2011). This will affect the growth and distribution of inferior competitor species at small scales (Arribas et al., 2016).

In addition, the grazers (herbivores) which plays important role in the transfer of primary production to the higher trophic levels influence the structure and abundance of the macrophytes in the benthic habitats (Poore et al., 2012). A study from the New England rocky shores indicated that herbivory is affecting the abundance of the macroalga *Fucus* sp. (Lubchenco, 1980). Several previous studies also showed that competition and herbivory are the important biotic factors that determine the upper distribution limits of algal species in benthic habitats (Hawkins & Hartnoll, 1985; Underwood, 1985).

Among the biotic interactions, predation is considered to be a significant factor influencing the distribution of prey populations. The control of prey population by predators and primary producers by herbivores are variable across spatial and temporal scales (Moles et al., 2011). These variations in biotic interactions along with the abiotic factors largely explain the distribution patterns of benthic organisms. The tropical marine ecosystems showed higher biodiversity due to the strong biotic interactions which are associated with high speciation rates (Mittelbach et al., 2007). Furthermore, symbiotic associations allow the interacting species to attain new phenotypic traits, metabolic capabilities, and ecological functions (Gilbert et al., 2010). This will help the organisms to occupy new ecological niches (Cavanaugh et al., 2013) which enable their abundance and distribution in the benthic habitats.

7.6 Conclusion

Benthic community structure and distribution vary across spatial scales due to the influence of multiple environmental factors. Our knowledge of the role of environmental as well as biotic factors on benthic community distribution is relatively poor in most of the geographic regions. As marine ecosystems are experiencing multiple stressors both from anthropogenic and natural sources, the effects of these stressors on the benthic community need to be studied. The benthic macrofauna and macrophytes are undoubtedly under the stress from the activities such as pollution, coastal development, dredging, and overexploitation of resources. The shifting of species distribution range due to these stressors as well as climate change-induced effects may have larger consequences to the overall biodiversity and ecosystem functioning. To implement the management strategies for the mitigation of anthropogenic stressors on marine ecosystems, there is a need for understanding the ecology of marine organisms. Long-term monitoring studies of benthic variables to assess the effects of physical, chemical, and biological factors will enhance our understandings of the ecology of benthic communities in marine ecosystems.

References

Acuna, K., Duarte, C., Jaramillo, E., Contreras, H., Manzano, M., & Navarro, J. M. (2012). Vertical distribution of the macroinfauna associated to bivalves in a sedimentary intertidal flat of southern Chile. *Revista de Biología Marina y Oceanografía, 47*(3), 383–393. Available from https://doi.org/10.4067/S0718-19572012000300002.

Alongi, D. M., Sasekumar, A., Tirendi, F., & Dixon, P. (1998). The influence of stand age on benthic decomposition and recycling of organic matter in managed mangrove forests of Malaysia. *Journal of Experimental Marine Biology and Ecology, 225*(2), 197–218.

Andersen, T., Schartau, A. K. L., & Paasche, E. (1991). Quantifying external and internal nitrogen and phosphorus pools, as well as nitrogen and phosphorus supplied through remineralization, in coastal marine plankton by means of a diluation technique. *Marine Ecology Progress Series, 69*, 67–80.

Anthony, K. R. N., & Kerswell, A. P. (2007). Coral mortality following extreme low tides and high solar radiation. *Marine Bioliology, 151*, 1623–1631. Available from https://doi.org/10.1007/s00227-006-0573-0.

Arribas, L. P., Bagur, M., Soria, S. A., Gutierrez, J. L., & Palomo, M. G. (2016). Competition between mussels at the rocky intertidal zone of La Loberia, Rio Negro, Argentina. *Revista del Museo Argentino de CienciasNaturalesnuevaserie, 18*(1), 1–7.

Ba-Akdah, M. A., Satheesh, S., & Al-Sofyani, A. A. (2016). Habitat preference and seasonal variability of epifaunal assemblages associated with macroalgal beds on the Central Red Sea coast, Saudi Arabia. *Journal of the Marine Biological Association of United Kingdom, 96*, 1457–1467.

Baynes, T. W. (1999). Factor structuring a subtidal encrusting community in the southern Gulf of California. *Bulletine of Marine Science, 64*, 419–450.

Beck, M. W., Heck, K. L., Able, K. W., Childers, D. L., Eggleston, D. B., Gillanders, B. M., Halpern, B., Hays, C. G., Hoshino, K., Minello, T. J., Orth, R. J., Sheridan, P. F., & Weinstein, M. P. (2001). The identification, conservation and management of estuarine and marine nurseries for fish and invertebrates. *Bioscience, 51*, 633–641.

Bellgrove, A., Monk, J., & Quinn, G. P. (2014). Effects of micro- and mesograzers on intertidal macroalgal recruitment. *Marine Biology, 161*(5), 1207–1216.

Bellwood, D. R., Hughes, T. P., Folke, C., & Nystrom, M. (2004). Confronting the coral reef crisis. *Nature, 429*, 827–833. Available from https://doi.org/10.1038/nature02691.

Benedetti-Cecchi, L. (2001). Variability in abundance of algae and invertebrates at different spatial scales on rocky sea shores. *Marine Ecology Progress Series, 217*, 79–92.

Benedetti-Cecchi, L., & Cinelli, F. (1997). Spatial distribution of algae and invertebrates in the rocky intertidal zone of the Strait of Magellan: Are patterns general? *Polar Biology, 18*, 337–343.

Benedetti-Cecchi, L., Maggi, E., Bertocci, I., Vaselli, S., Micheli, F., Osio, G. C., & Cinelli, F. (2003). Variation in rocky shore assemblages in the northwestern Mediterranean: Contrasts between islands and the mainland. *Journal of Experimental Marine Biology and Ecology*, *293*(2), 193–215. Available from https://doi.org/10.1016/S0022-0981(03)00220-X.

Berntsson, K. M., Jonsson, P. R., Larsson, A. I., & Holdt, S. (2004). Rejection of unsuitable substrata as a potential driver of aggregated settlement in the barnacle *Balanus improvisus*. *Marine Ecology Progress Series*, *275*, 199–210.

Blake, J., & Grassle, J. (1994). Benthic community structure on the U.S. South Atlantic slope off the Carolinas: Spatial heterogeneity in a current-dominated system. *Deep Sea Research Part II: Topical Studies in Oceanography*, *41*, 835–874. Available from https://doi.org/10.1016/0967-0645(94)90051-5.

Blanchet, H., Gouillieux, B., Alizier, S., Amouroux, J., et al. (2014). Multiscale patterns in the diversity and organization of benthic intertidal fauna among French Atlantic estuaries. *Journal of Sea Research*, *90*, 95–110.

Bloomfield, A. L., & Gillanders, B. M. (2005). Fish and invertebrate assemblages in seagrass, mangrove, saltmarsh and non-vegetated habitats. *Estuaries*, *28*, 63–77.

Bock, M. J., & Miller, D. C. (1995). Storm effects on particulate food resources on an intertidal sandflat. *Journal of Experimental Marine Biology and Ecology*, *187*, 81–101.

Bolam, S. G., Eggleton, J., Smith, R., Mason, C., Vanstaen, K., & Rees, H. (2008). Spatial distribution of macrofaunal assemblages along the English Channel. *Journal of the Marine Biological Association of the United Kingdom*, *88*(4), 675.

Bologna, P. A. X., & Heck, K. L., Jr. (2000). Impacts of seagrass habitat architecture on bivalve settlement. *Estuaries*, *23*, 449–457.

Bologna, P. A. X., & Heck, K. L., Jr (2002). Impact of habitat edges on density and secondary production of seagrass-associated fauna. *Estuaries*, *25*, 1033–1044.

Bostrom, C., Jackson, E. L., & Simenstad, C. A. (2006). Seagrass landscapes and their effects on associated fauna: A review. *Estuarine, Coastal and Shelf Science*, *68*, 383–403. Available from https://doi.org/10.1016/j.ecss.2006.01.026.

Bostrom, C., Tornroos, A., & Bonsdorff, E. (2010). Invertebrate dispersal and habitat heterogeneity: expression of biological traits in a seagrass landscape. *Journal of Experimental Marine Biology and Ecology*, *390*(2), 106–117.

Bourget, E., De Guise, J., & Daigle, G. (1994). Scales of substratum heterogeneity, structural complexity, and the early establishment of a marine epibenthic community. *Journal of Experimental Marine Biology and Ecology*, *181*, 31–35.

Branch, G. M., Thompson, R. C., Crowe, T. P., Castilla, J. C., Langmead, O., & Hawkins, S. J. (2008). Rocky intertidal shores: Prognosis for the future. In N. V. C. Polunin (Ed.), *Aquatic ecosystems* (pp. 209–225). Cambridge University Press.

Briggs, J. C. (1974). *Marine zoogeography*. McGraw-Hill.

Briggs, J. C., & Bowen, B. W. (2012). A realignment of marine biogeographic provinces with particular reference to fish distributions. *Journal of Biogeography*, *39*(1), 12–30. Available from https://doi.org/10.1111/j.1365-2699.2011.02613.x.

De Brouwer, J., Wolfstein, K., Ruddy, G., Jones, T. E. R., & Stal, L. J. (2005). Biogenic stabilization of intertidal sediments: The importance of extracellular polymeric substances produced by benthic diatoms. *Microbial Ecology*, *49*, 501–512. Available from https://doi.org/10.1007/s00248-004-0020-z.

Brown, B. E. (1997). Adaptations of reef corals to environmental stress. *Advances in Marine Biology*, *131*, 221–299.

Brown, J. H. (2014). Why are there so many species in the tropics? *Journal of Biogeography*, *41*, 8–22.

Brown, K. T., Bender-Champ, D., Kubicek, A., Van der Zande, R., Achlatis, M., Hoegh-Guldberg, O., & Dove, S. G. (2018). The dynamics of coral-algal interaction in space and time on the southern Great Barrier Reef. *Frontiers in Marine Science*, *5*, 181. Available from https://doi.org/10.3389/fmars.2018.00181.

Buchsbaum, R., & Powell, J. C. (2008). Symposium review: Long-term shifts in faunal assemblages in eastern North American estuaries. *Reviews in Fish Biology and Fisheries*, *18*, 447–450. Available from https://doi.org/10.1007/s11160-008-9086-x.

Bustamante, R., & Branch, G. (1996). Large scale patterns and trophic structure of southern African rocky shores: The roles of geographic variation and wave exposure. *Journal of Biogeography*, *23*(3), 339–351.

Cacabelos, E., Dominguez, M., & Troncoso, J. S. (2009). Trophic structure of soft-bottom macrobenthos in an inlet in north-western Spain. *Journal of the Marine Biological Association of the United Kingdom*, *89*(3), 439–447. Available from https://doi.org/10.1017/S0025315409003105.

Cacabelos, E., Olabarria, C., Incera, M., & Troncoso, J. S. (2010). Effects of habitat structure and tidal height on epifaunal assemblages associated with macroalgae. *Estuarine, Coastal and Shelf Science*, *89*(1), 43–52.

Cahoon, L. (1999). The role of benthic microalgae in neritic ecosystems. *Oceanography and Marine Biology*, *37*, 47–86.

Cannicci, S., Fusi, M., Cimó, F., Dahdouh-Guebas, F., & Fratini, S. (2018). Interference competition as a key determinant for spatial distribution of mangrove crabs. *BMC Ecology*, *18*, 8. Available from https://doi.org/10.1186/s12898-018-0164-1.

Carlson, R. L., Shulman, M. J., & Ellis, J. C. (2006). Factors contributing to spatial heterogeneity in the abundance of the common periwinkle *Littorina littorea* (L.). *Journal of Molluscan Studies*, *72*, 149–156.

Cavanaugh, C. M., McKiness, Z., Newton, I., & Stewart, F. (2013). Marine chemosynthetic symbioses. In E. Rosenberg, E. DeLong, S. Lory, E. Stackebrand, & F. Thompson (Eds.), *The prokaryotes* (pp. 579–607). Springer.

Chaidez, V., Dreano, D., Agusti, S., Duarte, C. M., & Hoteit, I. (2017). Decadal trends in Red Sea maximum surface temperature. *Scientific Reports*, 7(1), 1–8. Available from https://doi.org/10.1038/s41598-017-08146-z.

Chappuis, E., Terradas, M., Cefali, M. E., Mariani, S., & Ballesteros, E. (2014). Vertical zonation is the main distribution pattern of littoral assemblages on rocky shores at a regional scale. *Estuarine, Coastal and Shelf Science*, *147*, 113–122.

Choi, C., Battley, P. F., Potter, M. A., Murray, A., et al. (2014). Factors affecting the distribution patterns of benthic invertebrates at a major shorebird staging site in the Yellow Sea, China. *Wetlands*, *34*, 1085–1096. Available from https://doi.org/10.1007/s13157-014-0568-4.

Chollett, I., & Mumby, P. J. (2012). Predicting the distribution of Montastraea reefs using wave exposure. *Coral Reefs*, *31*, 493–503. Available from https://doi.org/10.1007/s00338-011-0867-7.

Christensen, B. (1978). Biomass and primary production of *Rhizophora apiculata* Bl. in a mangrove in southern Thailand. *Aquatic Botany*, *4*, 43–52.

Christie, H., Norderhaug, K., & Fredriksen, S. (2009). Macrophytes as habitat for fauna. *Marine Ecology Progress Series*, *396*, 221–233.

Connell, J. H. (1961). The influence of intra-specific competition and other factors on the distribution of the barnacle. *Chthamalusstellatus Ecology*, *42*, 710–723.

Costello, M. J., Tsai, P., Wong, P. S., Cheung, A. K. L., Basher, Z., & Chaudhary, C. (2017). Marine biogeographic realms and species endemicity. *Nature Communications*, *8*(1), 1–10.

Coutinho, R., Yaginuma, L. E., Siviero, F., DosSantos, J. C. Q., Lopez, M. S., Christofoletti, R. A., et al. (2016). Studies on benthic communities of rocky shores on the Brazilian coast and climate change monitoring: Satus of knowledge and challenges. *Brazilian Journal of Oceanography*, *64*, 27–36.

Dana, J. D. (1853). On an isothermal oceanic chart, illustrating the geographical distribution of marine animals. *American Journal of Science and Arts*, *16*, 153–167.

Danko, A., Schaible, R., & Danko, M. J. (2020). Salinity effects on survival and reproduction of hydrozoan *Eleutheria dichotoma*. *Estuaries and Coasts*, *43*, 360–374. Available from https://doi.org/10.1007/s12237-01900675-2.

Davies, T. W., Duffy, J. P., Bennie, J., & Gaston, K. J. (2014). The nature, extent, and ecological implications of marine light pollution. *Frontiers in Ecology and the Environment*, *12*, 347–355.

De Deckere, E. M. G. T., Van de Koppel, J., & Heip, C. H. R. (2000). The influence of corophiumvolutator abundance on resuspension. *Hydrobiologia*, *426*, 37–42.

Denny, M. (2006). Ocean waves, nearshore ecology, and natural selection. *Aquatic Ecology*, *40*, 439–461. Available from https://doi.org/10.1007/s10452-004-5409-8.

Dharmapalan, B., Jyothish, M. S., & Harinandanan, P. V. (2011). Unsung Heroes of Mangrove Ecosystem—Mangrove Crabs. *Science Reporter*, 53–56.

Dolenc-Orbanic, N., & Battelli, C. (2017). The vertical distribution and abundance of *Chthamalusstellatus* Poli and *Chthamalusmontagui* Southward (Crustacea, Cirripedia) at two localities of the Istrian peninsula coast (North Adriatic). *Acta Adriatica*, *58*, 77–88. Available from https://doi.org/10.32582/aa.58.1.6.

Drylie, T. P., Lohrer, A. M., Needham, H. R., Bulmer, R. H., & Pilditch, C. A. (2018). Benthic primary production in emerged intertidal habitats provides resilience to high water column turbidity. *Journal of Sea Research*, *142*, 101–112. Available from https://doi.org/10.1016/j.seares.2018.09.015.

Duarte, C. M., Lenton, T. M., Wadhams, P., & Wassmann, P. (2012). Abrupt climate change in the Arctic. *Nature Climate Change*, *2*, 60–62.

Dutertre, M., Hamon, D., Chevalier, C., & Ehrhold, A. (2013). The use of the relationships between environmental factors and benthic macrofaunal distribution in the establishment of a baseline for coastal management. *ICES Journal of Marine Science*, *70*(2), 294–308. Available from https://doi.org/10.1093/icesjms/fss170.

Ehrnsten, E., Sun, X., Humborg, C., Norkko, A., Savchuk, O. P., Slomp, C. P., Timmermann, K., & Gustafsson, B. G. (2020). Understanding environmental changes in temperate coastal seas: Linking models of benthic fauna to carbon and nutrient fluxes. *Frontiers in Marine Science*, 7, 450. Available from https://doi.org/10.3389/fmars.2020.00450.

Ekman, S. (1953). *Zoogeography of the sea*. Sidgwick and Jackson.

Ellis, J. I., Clark, D., Atalah, J., Jiang, W., Taiapa, C., Patterson, M., Sinner, J., & Hewitt, J. (2017). Multiple stressor effects on marine infauna: Responses of estuarine taxa and functional traits to sedimentation, nutrient and metal loading. *Scientific Reports*, 7(1), 1–16. Available from https://doi.org/10.1038/s41598-017-12323-5.

Ellis, J. I., Fraser, G., & Russell, J. (2012). Discharged drilling waste from oil and gas platforms and its effects on benthic communities. *Marine Ecology Progress Series*, *456*, 285–302.

Ellison, A., Farnsworth, E. J., & Twilley, R. R. (1996). Facultative mutualism between red mangroves and root-fouling sponges in Belizean mangal. *Ecology*, 77, 2431–2444.

Emslie, M. J., Cheal, A. J., & Johns, K. A. (2014). Retention of habitat complexity minimizes disassembly of reef fish communities following disturbance: A large-scale natural experiment. *PLoS One*, *9*(8), e105384.

Enochs, I. C., Toth, L. T., Brandtneris, V. W., Afflerbach, J. C., & Manzello, D. P. (2011). Environmental determinants of motile cryptofauna on an eastern Pacific coral reef. *Marine Ecology Progress Series*, *438*, 105–118.

Esqueda, M. C., Ríos-Jara, E., Michel-Morfin, J. E., & Landa-Jaime, M. V. (2000). The vertical distribution and abundance of gastropods and bivalves from rocky beaches of cuastecomate Bay, Jalisco. *Revista de Biología Tropical*, *48*, 765–775.

Ferrier-Pages, C., Gattuso, J. P., & Jaubert, J. (1999). Effect of small variations in salinity on the rates of photosynthesis and respiration of the zooxanthellate coral *Stylophorapistillata*. *Marine Ecology Progress Series*, *181*, 309–314.

Fica, E., Diaz, E. R., & Valdivia, N. (2017). General spatial spectral variation in rocky intertidal communities from three biogeographical regions. *Journal of Biogeography, 44* (10), 2354–2364.

Franz, D. R., & Merrill, A. S. (1980). The origins and determinants of distribution of molluscan faunal groups on the shallow continental shelf of the Northwest Atlantic. *Malacologia, 19*, 227–248.

Gammal, J., Jarnstrom, M., Bernard, G., Norkko, J., & Norkko, A. (2019). Environmental context mediates biodiversity ecosystem functioning relationships in coastal soft sediment habitats. *Ecosystems, 22*, 137–151. Available from https://doi.org/10.1007/s10021-018-0258-9.

Gartner, A. F., Tuya, P. S., Lavery., & McMahon, K. (2013). Habitat preferences of macroinvertebrate fauna among seagrasses with varying structural forms. *Journal of Experimental Marine Biology and Ecology, 439*, 143–151.

Gattuso, J. P., Frankignoulle, M., & Wollast, R. (1998). Carbon and carbonate metabolism in coastal aquatic ecosystems. *Annual Review of Ecological Systems, 29*, 405–434.

Gattuso, J. P., Gentili, B., Duarte, C. M., Kleypas, J. A., Middelburg, J. J., & Antoine, D. (2006). Light availability in the coastal ocean: Impact on the distribution of benthic photosynthetic organisms and contribution to primary production. *Biogeosciences, 3*, 489–513.

Gelin, A., Crivelli, A. J., Rosecchi, E., & Kerambrun, P. (2001). Can salinity changes affect reproductive success in the brown shrimp Crangoncrangon? *Journal of Crustacean Biology, 21*(4), 905–911. Available from https://doi.org/10.1163/20021975-99990182.

Gilbert, S. F., McDonald, E., Boyle, N., Buttino, N., Gyi, L., Mai, M., Prakash, N., & Robinson, J. (2010). Symbiosis as a source of selectable epigenetic variation: Taking the heat for the big guy. *Philosophical Transactions of the Royal Society B: Biological Sciences, 365*(1540), 671–678.

Glasby, T. M. (1999). Interactive effects of shading and proximity to the seafloor on the development of subtidal epibiotic assemblages. *Marine Ecology Progress Series, 190*, 113–124.

Glynn, P. W. (1993). Coral reef bleaching: Ecological perspectives. *Coral Reefs, 12*(1), 1–17.

Glynn, P. W., & Enochs, I. C. (2011). An ecosystem in transition. In Z. Dubinsky, & N. Stambler (Eds.), *Coral Reefs* (pp. 273–325). Springer.

Gogina, M., & Zettler, M. L. (2010). Diversity and distribution of benthic macrofauna in the Baltic Sea: Data inventory and its use for species distribution modelling and prediction. *Journal of Sea Research, 64*(3), 313–321. Available from https://doi.org/10.1016/j.seares.2010.04.005.

Gray, J., & Elliott, M. (2009). *Ecology of marine sediments* (2nd ed.). Oxford University Press.

Guichard, F., Bourget, E., & Robert, J. L. (2001). Scaling the influence of topographic heterogeneity on an intertidal benthic communities: Alternate trajectories mediated by hydrodinamics and shading. *Marine Ecology Progress Series, 217*, 27–41.

Hagberg, J., & Tunberg, B. (2000). Studies on the covariation between physical factors and the long-term variation of the marine soft bottom macrofauna in western Sweden. *Estuarine Coastal and Shelf Science, 50*, 373–385. Available from https://doi.org/10.1006/ecss.1999.0578.

Hjorth, M., Vester, J., Henriksen, P., Forbes, V., & Dahllöf, I. (2007). Functional and structural responses of marine plankton food web to pyrene contamination. *Marine Ecology Progress Series, 338*, 21–31.

Hale, S. S. (2010). Biogeographical patterns of marine benthic macroinvertebrates along the Atlantic coast of the northeastern USA. *Estuaries and Coasts, 33*(5), 1039–1053. Available from https://doi.org/10.1007/s12237-010-9332-z.

Hale, S. S., Buffum, H. W., Kiddon, J. A., & Hughes, M. M. (2017). Subtidal benthic invertebrates shifting northward along the US Atlantic Coast. *Estuaries and Coasts*, *40*, 1744–1756. Available from https://doi.org/10.1007/s12237-017-0236-z.

Hale, S. S., Cicchetti, G., & Deacutis, C. F. (2016). Eutrophication and hypoxia diminish ecosystem functions of benthic communities in a New England estuary. *Frontiers in Marine Science*, *3*, 249. Available from https://doi.org/10.3389/fmars.2016.00249.

Harley, C. D., Denny, M. W., Mach, K. J., & Miller, L. P. (2009). Thermal stress and morphological adaptations in limpets. *Functional Ecology*, *23*, 292–301.

Harley, C. D., Randall Hughes, A., Hultgren, K. M., Miner, B. G., Sorte, C. J., Thornber, C. S., et al. (2006). The impacts of climate change in coastal marine systems. *Ecology Letters*, *9*(2), 228–241. Available from https://doi.org/10.1111/j.1461-0248.2005.00871.x.

Harley, C. D. G., & Helmuth, B. S. T. (2003). Local- and regional-scale effects of wave exposure, thermal stress, and absolute vs. effective shore level on patterns of intertidal zonation. *Limnology and Oceanography*, *48*, 1498–1508.

Hawkins, S. J., & Hartnoll, R. G. (1985). Factors determining the upper limits of intertidal canopy-forming algae. *Marine Ecology Progress Series*, *20*, 265–271.

Heck, K. L., & Valentine, J. F. (2006). Plant–herbivore interactions in seagrass meadows. *Journal of Experimental Marine Biology and Ecology*, *330*, 420–436.

Hemminga, M. A., & Duarte, C. M. (2000). *Seagrass ecology* (p. 298). Cambridge University Press.

Herkul, K., Kotta, J., & Parnoja, M. (2011). Effect of physical disturbance on the soft sediment benthic macrophyte and invertebrate community in the northern Baltic Sea. *Boreal Environment Research*, *16*(A), 209–219.

Herkul, K., Torn, K., Suursaar, U., Alari, V., & Peterson, A. (2016). Variability of benthic communities in relation to hydrodynamic conditions in the north-eastern Baltic Sea. *Journal of Coastal Research*, *75*, 867–871. Available from https://doi.org/10.2112/SI75-174.1.

Heron, S. F., Maynard, J. A., Van Hooidonk, R., & Eakin, C. M. (2016). Warming trends and bleaching stress of the world's coral reefs 1985–2012. *Scientific Reports*, *6*, 38402. Available from https://doi.org/10.1038/srep38402.

Hiddink, J. G., Burrows, M. T., & Garcia, J. G. (2015). Temperature tracking by North Sea benthic invertebrates in response to climate change. *Global Change Biology*, *21*, 117–129. Available from https://doi.org/10.1111/gcb.12726.

Hillebrand, H., Blasius, B., Borer, E., Elizabeth, T., et al. (2018). Biodiversity change is uncoupled from species richness trends: Consequences for conservation and monitoring. *Journal of Applied Ecology*, *55*, 169–184. Available from https://doi.org/10.1111/1365-2664.12959.

Hily, C., Le Loc'h, F., Grall, J., & Glemarec, M. (2008). Soft bottom macrobenthic communities of North Biscay revisited: Long-term evolution under fisheries-climate forcing. *Estuarine, Coastal and Shelf Science*, *78*(2), 413–425.

Hinz, H., Capasso, E., Lilley, M., Frost, M., & Jenkins, S. R. (2011). Temporal differences across a bio-geographical boundary reveal slow response of sub-littoral benthos to climate change. *Marine Ecology Progress Series*, *423*, 69–82.

Hiscock, K., Southward, A., Tittley, I., & Hawkins, S. (2004). Effects of changing temperature on benthic marine life in Britain and Ireland. *Aquatic Conservation: Marine and Freshwater Ecosystems*, *14*(4), 333–362. Available from https://doi.org/10.1002/aqc.628.

Hochachka, P. W., & Somero, G. N. (2002). *Biochemical adaptation: Mechanism and process in physiological evolution*. Oxford University Press.

Van Hoey, G., Degraer, S., & Vincx, M. (2004). Macrobenthic community structure of soft-bottom sediments at the Belgian Continental Shelf. *Estuarine, Coastal and Shelf Science*, *59*(4), 599–613.

Hope, J. A., Hewitt, J. E., Pilditch, C. A., Savage, C., & Thrush, S. F. (2020). Effect of nutrient enrichment and turbidity on interactions between microphytobenthos and a key bivalve: Implications for higher trophic levels. *Frontiers in Marine Science*, 7, 695. Available from https://doi.org/10.3389/fmars.2020.00695.

Hughes, T. P., Baird, A. H., Bellwood, D. R., Card, M., Connolly, S. R., Folke, C., Grosberg, R., et al. (2003). Climate change, human impacts, and the resilience of coral reefs. *Science (New York, N.Y.)*, *301*, 929–933. Available from https://doi.org/10.1126/science.1085046.

Hughes, T. P., Rodrigues, M. J., Bellwood, D. R., Ceccarelli, D., Hoegh-Guldberg, O., McCook, L., et al. (2007). Phase shifts, herbivory, and the resilience of coral reefs to climate change. *Current Biology*, *17*, 360–365. Available from https://doi.org/10.1016/j.cub.2006.12.049.

Hutchings, P. (1998). Biodiversity and functioning of polychaetes in benthic sediments. *Biodiversity and Conservation*, 7(9), 1133–1145.

Irving, A., & Connell, S. (2002). Sedimentation and light penetration interact to maintain heterogeneiry of subtidal habitats: Algal vs invertebrate dominated assemblages. *Marine Ecology Progress Series*, *245*, 83–91.

Jackson, J. B. C. (1977). Competition on marine hard substrata: The adaptive significance of solitary and colonial strategies. *The American Naturalist*, *111*(980), 743–767.

Jackson, J. B. C., Kirby, M. X., Berger, W. H., Bjorndal, K. A., Botsford, L. W., Bourque, B. J., et al. (2001). Historical overfishing and the recent collapse of coastal ecosystems. *Science (New York, N.Y.)*, *293*, 629–638. Available from https://doi.org/10.1126/science.1059199.

Jedruch, A., Bełdowska, M., & Graca, B. (2019). Seasonal variation in accumulation of mercury in the benthic macrofauna in a temperate coastal zone (Gulf of Gdansk). *Ecotoxicology and Environmental Safety*, *164*, 305–316. Available from https://doi.org/10.1016/j.ecoenv.2018.08.040.

Jorda, G., Marba, N., & Duarte, C. M. (2012). Mediterranean seagrass vulnerable to regional climate warming. *Nature Climate Change*, *2*, 821–824.

Kang, Y. H., Kim, S., Choi, S. K., Moon, K., Choi, H. G., Ko, Y. W., Hawes, I., Kim, S. H., et al. (2019). Composition and structure of the marine benthic community in Terra Nova Bay, Antarctica: Responses of the benthic assemblage to disturbances. *PLoS One*, *14*(12), e0225551.

Kautsky, H., Martin, G., Makinen, A., Borgiel, M., Vahteri, P., & Rissanen, J. (1999). Structure of phytobenthic and associated animal communities in the Gulf of Riga. *Hydrobiologia*, *393*, 191–200.

Khalaman, V. V., Golubovskaya, N. S., Komendantov, A. Y., et al. (2018). Effect of the spatial orientation of a substrate on the formation of early fouling communities in the White Sea. *Biology Bulletin*, *45*, 82–90. Available from https://doi.org/10.1134/S1062359018010065.

Khan, A., Manokaran, S., & Lyla, S. (2017). Changes in macrobenthic community structure from estuary to continental slope in the south-east coast of India. *Journal of the Marine Biological Association of the United Kingdom*, *97*(1), 161–180.

Khedhri, I., Djabou, H., & Afli, A. (2015). Trophic and functional organization of the benthic macrofauna in the lagoon of Boughrara–Tunisia (SW Mediterranean Sea). *Journal of the Marine Biological Association of the United Kingdom*, *95*(4), 647–659. Available from https://doi.org/10.1017/S0025315414001544.

Kirby, R. R., Beaugrand, G., & Lindley, J. A. (2008). Climate-induced effects on the meroplankton and the benthic-pelagic ecology of the North Sea. *Limnology and Oceanography*, *53*, 1805–1815.

Kon, K., Kurokura, H., & Tongnunui, P. (2010). Effects of the physical structure of mangrove vegetation on a benthic faunal community. *Journal of Experimental Marine Biology and Ecology*, *383*, 171–180.

Kristensen, E. (2008). Mangrove crabs as ecosystem engineers; with emphasis on sediment processes. *Journal of Sea Research, 59*, 30–43. Available from https://doi.org/10.1016/j.seares.2007.05.004.

Kroeker, K. J., Sanford, E., Rose, J. M., Blanchette, C. A., Chan, F., Chavez, F. P., Gaylord, B., et al. (2016). Interacting environmental mosaics drive geographic variation in mussel performance and predation vulnerability. *Ecology Letters, 19*, 771–779.

Kuanui, P., Chavanich, S., Viyakarn, V., Omori, M., Fujita, T., & Lin, C. (2020). Effect of light intensity on survival and photosynthetic efficiency of cultured corals of different ages. *Estuarine, Coastal and Shelf Science, 235*, 106515. Available from https://doi.org/10.1016/j.ecss.2019.106515.

Kumar, P. S., & Khan, A. B. (2013). The distribution and diversity of benthic macroinvertebrate fauna in Pondicherry mangroves, India. *Aquatic Biosystems, 9*(1), 15. Available from https://doi.org/10.1186/2046-9063-9-15.

Labrune, C., Gremare, A., Amouroux, J. M., Sarda, R., Gil, J., & Taboada, S. (2008). Structure and diversity of shallow soft-bottom benthic macrofauna in the Gulf of Lions (NW Mediterranean). *Helgoland Marine Research, 62*(3), 201–214.

Lee, S. Y. (1998). Ecological role of grapsid crabs in mangrove ecosystems: a review. *Marine and Freshwater Research, 49*, 335–343. Available from https://doi.org/10.1071/MF97179.

Lee, S. Y., Fong, C. W., & Wu, R. S. S. (2001). The effects of seagrass (*Zostera japonica*) canopy structure on associated fauna: A study using artificial seagrass units and sampling of natural beds. *Journal of Experimental Marine Biology and Ecology, 259*(1), 23–50. Available from https://doi.org/10.1016/S0022-0981(01)00221-0.

Leopardas, V., Uy, W., & Nakaoka, M. (2014). Benthic macrofaunal assemblages in multispecific seagrass meadows of the southern Philippines: Variation among vegetation dominated by different seagrass species. *Journal of Experimental Marine Biology and Ecology, 457*, 71–80.

Levin, L. A., Boesch, D. F., Covich, A., Dahm, C., Erseus, C., Ewel, K. C., Kneib, R. T., et al. (2001). The function of marine critical transition zones and the importance of sediment biodiversity. *Ecosystems, 4*, 430–451. Available from https://doi.org/10.1007/s10021-001-0021-4.

Levin, P., & Poe, M. (2017). *Conservation for the Anthropocene Ocean: Interdisciplinary science in support of nature and people* (p. 530). Academic Press.

Levinton, J. S. (2009). *Marine biology: Function, biodiversity, ecology* (3rd ed., p. 588). Oxford University Press, Inc.

Lin, J., Huang, Y., Arbi, U. Y., Lin, H., Azkab, M. H., Wang, J., He, X., Mou, J., Liu, K., & Zhang, S. (2018). An ecological survey of the abundance and diversity of benthic macrofauna in Indonesian multispecific seagrass beds. *Acta Oceanologica Sinica, 37*(6), 82–89.

Lindeyer, F., & Gittenberger, A. (2011). Ascidians in the succession of marine fouling communities. *Aquatic Invasions, 6*(4), 421–434. Available from https://doi.org/10.3391/ai.2011.6.4.07.

Lirman, D. (2001). Competition between macroalgae and corals: Effects of herbivore exclusion and increased algal biomass on coral survivorship and growth. *Coral Reefs, 19*, 392–399. Available from https://doi.org/10.1007/s003380000125.

Lirman, D., & Cropper, W. P. (2003). The influence of salinity on seagrass growth, survivorship, and distribution within Biscayne Bay, Florida: Field, experimental, and modeling studies. *Estuaries, 26*, 131–141. Available from https://doi.org/10.1007/BF02691700.

Little, C., & Kiching, J. A. (1996). *The biology of rocky shores*. Oxford University Press.

Liu, P. J., Lin, S. M., Fan, T. Y., Meng, P. J., Shao, K. T., & Lina, H. J. (2009). Rates of overgrowth by macroalgae and attack by sea anemones are greater for live coral than dead coral under conditions of nutrient enrichment. *Limnology and Oceanography 54*(4), 1167–1175.

Llanso, R. J., Scott, L. C., Dauer, D. M. M., Hyland, J. L., & Russell, D. E. (2002). An estuarine benthic index of biotic integrity for the mid-Atlantic region of the United States. II. Index development. *Estuaries, 25*, 1231–1242. Available from https://doi.org/10.1007/BF02692220.

Londono-Cruz, E., De Mesa-Agudelo, L. L. A., Arias-Galvez, F., Herrera-Paz, D. L., et al. (2014). Distribution of macroinvertebrates on intertidal rocky shores in Gorgona Island, Colombia (Tropical Eastern Pacific). *Revista de Biologia Tropical, 62*, 189–198.

Lubchenco, J. (1980). Algal zonation in the New England rocky intertidal community: An experimental analysis. *Ecology, 61*(2), 333–344.

Lundquist, C. J., Jones, T. C., Parkes, S. M., & Bulmer, R. H. (2018). Changes in benthic community structure and sediment characteristics after natural recolonisation of the seagrass *Zosteramuelleri*. *Scientific Reports, 8*(1), 1–9. Available from https://doi.org/10.1038/s41598-018-31398-2.

Macreadie, P. I., Jarvis, J., Trevathan-Tackett, S. M., & Bellgrove, A. (2017). Seagrasses and macroalgae: Importance, vulnerability and impacts. In (1st ed.B. F. Phillips, & M. Perez-Ramirez (Eds.), Climate change impacts on fisheries and aquaculture: A global analysis (IIWiley-Blackwell.

Magel, J. M., Burns, J. H., Gates, R. D., & Baum, J. K. (2019). Effects of bleaching-associated mass coral mortality on reef structural complexity across a gradient of local disturbance. *Scientific Reports, 9*(1), 1–12. Available from https://doi.org/10.1038/s41598-018-37713-1.

Main, K. L., & Goodbody-Gringley, G. (2010). Effect of light on the growth of four corals species in land-based nursery systems Book of Abstract, Linking Science to Management: A Conference and Workshop on the Florida Keys Marine Ecosystem, 19–22 October 2010, Duck Key, Florida, 109 pp.

Maldonado, M., Aguilar, R., Bannister, R. J., et al. (2015). Sponge grounds as key marine habitats: A synthetic review of types, structure, functional roles, and conservation concerns. In S. Rossi, L. Bramanti, A. Gori, et al. (Eds.), *Marine animal forests: The ecology of benthic biodiversity hotspots* (1st ed., pp. 1–39). Springer.

Maldonado, M., Durfort, M., McCarthy, D. A., & Young, C. M. (2003). The cellular basis of photobehavior in the tufted parenchymella larva of demosponges. *Marine Biology, 143*, 427–441.

Mariano, D. L. S., & Barros, F. (2015). Intertidal benthic macrofaunal assemblages: changes in structure along entire tropical estuarine salinity gradients. *Journal of the Marine Biological Association of the United Kingdom, 95*(1), 5–15.

Matthews, T. G., & Fairweather, P. G. (2004). Effect of lowered salinity on the survival, condition and reburial of *Soletellina alba* (Lamarck, 1818) (Bivalvia: Psammobiidae). *Austral Ecology, 29*, 250–257. Available from https://doi.org/10.1111/j.1442-9993.2004.01345.x.

Mazzuco, A. C. D. A., Stelzer, P. S., & Bernardino, A. F. (2020). Substrate rugosity and temperature matters: Patterns of benthic diversity at tropical intertidal reefs in the SW Atlantic. *PeerJ, 8*, e8289. Available from https://doi.org/10.7717/peerj.8289.

McCloskey, R. M., & Unsworth, R. K. (2015). Decreasing seagrass density negatively influences associated fauna. *PeerJ, 3*, e1053.

Menge, B. A., & Sutherland, W. J. (1987). Community regulation: Variation in disturbance, competition, and predation in relation to environmental stress and recruitment. *American Naturalist, 130*, 730–757. Available from https://doi.org/10.1086/284741.

Middelburg, J. J., & Levin, L. A. (2009). Coastal hypoxia and sediment biogeochemistry. *Biogeosciences, 6*, 1273–1293. Available from https://doi.org/10.5194/bg-6-1273-2009.

Miller, D. C., Muir, C. L., & Hauser, O. A. (2002). Detrimental effects of sedimentation on marine benthos: What can be learned from natural processes and rates? *Ecological Engineering, 19*(3), 211–232.

Miller, R. J., & Etter, R. J. (2008). Shading facilitates sessile invertebrate dominance in the rocky subtidal Gulf of Maine. *Ecology, 89*, 452–462.

Milne, R., & Griffiths, C. (2014). Invertebrate biodiversity associated with algal turfs on a coral-dominated reef. *Marine Biodiversity*, *44*, 181–188. Available from https://doi.org/10.1007/s12526-013-0199-7.

Mittelbach, G. G., Schemske, D. W., Cornell, H. V., Allen, A. P., Brown, J. M., et al. (2007). Evolution and the latitudinal diversity gradient: Speciation, extinction and biogeography. *Ecology Letters*, *10*, 315–331.

Moles, A. T., Bonser, S. P., Poore, A. G. B., Wallis, I. R., & Foley, W. J. (2011). Assessing the evidence for latitudinal gradients in plant defence and herbivory. *Functional Ecology*, *25*, 380–388.

Montague, C. L., & Ley, J. A. (1993). A possible effect of salinity fluctuation on abundance of benthic vegetation and associated fauna in northeastern Florida Bay. *Estuaries*, *16*(4), 703–717. Available from https://doi.org/10.2307/1352429.

Mucha, A. P., Vasconcelos, M. T., & Bordalo, A. A. (2004). Vertical distribution of the macrobenthic community and its relationships to trace metals and natural sediment characteristics in the lower Douro estuary, Portugal. *Estuarine, Coastal and Shelf Science*, *59*(4), 663–673.

Muniz, P., Pires-Vanin, A. M. S., & Venturini, N. (2013). Vertical distribution patterns of macrofauna in a subtropical near-shore coastal area affected by urban sewage. *Marine Ecology*, *34*(2), 233–250. Available from https://doi.org/10.1111/maec.12010.

Murugesan, P., Sarathy, P. P., Muthuvelu, S., & Mahadevan, G. (2018). Diversity and distribution of polychaetes in mangroves of east coast of India. *Mangrove Ecosystem Ecology and Function, Edited by Sahadev Sharma, IntechOpen*. Available from https://doi.org/10.5772/intechopen.78332. https://www.intechopen.com/books/mangrove-ecosystem-ecology-and-function/diversity-and-distribution-of-polychaetes-in-mangroves-of-east-coast-of-indiap.

Musale, A. S., Desai, D. V., Sawant, S. S., Venkat, K., & Anil, A. C. (2015). Distribution and abundance of benthic macroorganisms in and around Visakhapatnam Harbour on the east coast of India. *Journal of the Marine Biological Association of the United Kingdom*, *95*(2), 215–231. Available from https://doi.org/10.1017/S0025315414001490.

Nagelkerken, I., Blaber, S. J. M., Bouillon, S., Green, P., Haywood, M., Kirton, L. G., Meynecke, J. O., Pawlik, J., Penrose, H. M., Sasekumar, A., & Somerfield, P. J. (2008). The habitat function of mangroves for terrestrial and marine fauna: a review. *Aquatic Botany*, *89*, 155–185.

Neves, R. A. F., Carlos, E., Leandro, P., Paulo, P., Rodolfo., & Jean, V. (2013). Factors influencing spatial patterns of molluscs in a eutrophic tropical bay. *Journal of the Marine Biological Association of the United Kingdom*, *93*(3), 577–589. Available from https://doi.org/10.1017/S0025315412001105.

Orth, R. J., Heck, K. L., & van Montfrans, J. (1984). Faunal communities in seagrass beds: A review of the influence of plant structure and prey characteristics on predator-prey relationships. *Estuaries*, 7(4), 339–350.

Osinga, M., Schutter, B., Griffioen, R. H., Wijffels, J. A. J., Verreth, S., Shafir, S., Henard, M. et al. (2011). The biology and economics of coral growth. *Marine Biotechnology*, *13*, 658–671. Available from https://doi.org/10.1007/s10126-011-9382-7.

Osman, R. W., & Whitlatch, R. B. (1995). The influence of resident adults on recruitment—A comparison to settlement. *Journal of Experimental Marine Biology and Ecology*, *190*, 169–198.

Pabis, K., Sobczyk, R., Sicinski, J., Ensrud, T., & Serigstadt, B. (2020). Natural and anthropogenic factors influencing abundance of the benthic macrofauna along the shelf and slope of the Gulf of Guinea, a large marine ecosystem off West Africa. *Oceanologia*, *62*(1), 83–100.

Pacheco, A. S., Laudien, J., Thiel, M., Heilmayer, O., & Oliva, M. (2010). Hard-bottom succession of subtidal epibenthic communities colonizing hidden and exposed surfaces off northern Chile. *Scientia Marina, 74*, 147–154. Available from https://doi.org/10.3989/scimar.2010.74n1147.

Paine, R. T. (1974). Intertidal community structure: Experimental studies on the relationship between a dominant competitor and its principal predator. *Oecologia, 15*, 93–120.

Pandolfi, J. M., Bradbury, R. H., Sala, E., Hughes, T. P., Bjorndal, K. A., Cooke, R. G., et al. (2003). Global trajectories of the long-term decline of coral reef ecosystems. *Science (New York, N.Y.), 301*, 955–958. Available from https://doi.org/10.1126/science.1085706.

Pandolfi, J. M. (2011). Historical ecology of coral reefs. In D. Hopley (Ed.), *Encyclopedia of modern coral reefs. Encyclopedia of earth sciences series*. Springer. Available from https://doi.org/10.1007/978-90-481-2639-2_94.

Paterson, D. M., & Black, K. S. (1999). Water flow, sediment dynamics and benthic biology. *Advances in Ecology Research, 29*, 155–194.

Pearman, J. K., Leray, M., Villalobos, R., Machida, R. J., Berumen, M. L., Knowlton, N., & Carvalho, S. (2018). Cross-shelf investigation of coral reef cryptic benthic organisms reveals diversity patterns of the hidden majority. *Scientific Reports, 8*(1), 1–17. Available from https://doi.org/10.1038/s41598-018-26332-5.

Pearson, T. H., & Rosenberg, R. (1978). A macrobenthic succession in relation to organic enrichment and pollution of the marine environment. *Oceanography and Marine Biology: An Annual Review, 16*, 229–311.

Petersen, D. S., Gorb, S. N., & Heepe, L. (2020). The influence of material and roughness on the settlement and the adhesive strength of the barnacle *Balanus improvisus* in the Baltic Sea. *Frontiers in Marine Science, 6*, 664. Available from https://doi.org/10.3389/fmars.2020.00664.

Pfaff, M. C., Branch, G. M., Wieters, E. A., Branch, R. A., & Broitman, B. R. (2011). Upwelling intensity and wave exposure determine recruitment of intertidal mussels and barnacles in the southern Benguela upwelling region. *Marine Ecology Progress Series, 425*, 141–152. Available from https://doi.org/10.3354/meps09003.

Pianka, E. R. (1996). Latitudinal gradients in species diversity: A review of concepts. *American Naturalist, 100*, 33–46.

Poore, A. G., Campbell, A. H., Coleman, R. A., Edgar, G. J., Jormalainen, V., Reynolds, P. L., et al. (2012). Global patterns in the impact of marine herbivores on benthic primary producers. *Ecology Letters, 15*(8), 912–922. Available from https://doi.org/10.1111/j.1461-0248.2012.01804.x.

Post, A. L., Beaman, R. J., O'Brien, P. E., Eleaume, M., & Riddle, M. J. (2011). Community structure and benthic habitats across the George V Shelf, East Antarctica: Trends through space and time. *Deep Sea Research Part II, 58*, 105–118.

Price, D. M., Robert, K., Callaway, A., et al. (2019). Using 3D photogrammetry from ROV video to quantify cold-water coral reef structural complexity and investigate its influence on biodiversity and community assemblage. *Coral Reefs, 38*, 1007–1021. Available from https://doi.org/10.1007/s00338-019-01827-3.

Rattray, A., Ierodiaconou, D., & Womersley, T. (2015). Wave exposure as a predictor of benthic habitat distribution on high energy temperate reefs. *Frontiers in Marine Science, 2*, 8. Available from https://doi.org/10.3389/fmars.2015.00008.

Reiss, H., Meybohm, K., & Kröncke, I. (2006). Cold winter effects on benthic macrofauna communities in near-and offshore regions of the North Sea. *Helgoland Marine Research, 60*, 224–238.

Reiss, J., Bridle, J. R., Montoya, J. M., & Woodward, G. (2009). Emerging horizons in biodiversity and ecosystem functioning research. *Trends in Ecology & Evolution, 24*, 505–514. Available from https://doi.org/10.1016/j.tree.2009.03.018.

Rex, M. A., Crame, J. A., Stuart, C. T., & Clarke, A. (2005). Large-scale biogeographic patterns in marine mollusks: A confluence of history and productivity. *Ecology, 86*, 2288–2297.

Richardson, L. E., Graham, N. A., & Hoey, A. S. (2020). Coral species composition drives key ecosystem function on coral reefs. *Proceedings of the Royal Society B*, *287*(1921), 20192214. Available from https://doi.org/10.1098/rspb.2019.2214.

Rickert, E., Karsten, U., Pohnert, G., & Wahl, M. (2015). Seasonal fluctuations in chemical defenses against macrofouling in *Fucus vesiculosus* and *Fucus serratus* from the Baltic Sea. *Biofouling*, *31*(4), 363–377. Available from https://doi.org/10.1080/08927014.2015.1041020.

Roberts, C. M., McClean, C. J., Veron, J. E. N., Hawkins, J. P., Allen, G. R., McAllister, D. E., et al. (2002). Marine biodiversity hotspots and conservation priorities for tropical reefs. *Science (New York, N.Y.)*, *295*, 1280–1284.

Rosa, L. C., & Bemvenuti, C. E. (2006). Seasonal stratification of the estuarine macroinfauna of the Patos lagoon estuary, Southern Brazil. *Thalassas*, *22*, 17–23.

Rousi, H., Korpinen, S., & Bonsdorff, E. (2019). Brackish-Water benthic fauna under fluctuating environmental conditions: The role of eutrophication, hypoxia, and global change. *Frontiers in Marine Science*, *6*, 464. Available from https://doi.org/10.3389/fmars.2019.00464.

Rousi, H., Peltonen, H., Mattila, J., Back, S., & Bonsdorff, E. (2011). Impacts of physical environmental characteristics on the distribution of benthic fauna in the northern Baltic Sea. *Boreal Environment Research*, *16*, 521–533.

Satheesh, S., Siddik, A. A., Ba-Akdah, M. A., & Al-Sofyani, A. A. (2017). An introduction to the ecological significance of seaweeds on coastal ecosystems. In E. Nabti (Ed.), *Biotechnological applications of seaweeds* (pp. 1–14). Nova Science Publishers.

Satyam, K., & Thiruchitrambalam, G. (2018). Habitat ecology and diversity of rocky shore fauna. In C. Sivaperuman, A. Velmurugan, A. Kumar Singh, & I. Jaisankar (Eds.), *Biodiversity and climate change adaptation in tropical islands* (pp. 187–215). Academic Press. Available from http://doi.org/10.1016/B978-0-12-813064-3.00007-7.

Saunders, R., & Connell, S. (2001). Interactive effects of shade and surface orientation on the recruitment of spirorbid polychaetes. *Australian Ecology*, *26*, 109–115.

Schiel, D. R., Steinbeck, J. R., & Foster, M. S. (2004). Ten years of induced ocean warming causes comprehensive changes in marine benthic communities. *Ecology*, *85*, 1833–1839.

Schories, D., Bergan, A. B., Barletta, M., Krumme, U., Mehlig, U., & Rademaker, V. (2003). The keystone role of leaf-removing crabs in mangrove forests of North Brazil. *Wetlands Ecology and Management*, *11*, 243–255.

Schroeder, A. (2003). Community dynamics and development of soft bottom macrozoobenthos in the German Bight (North Sea) 1969–2000. Ph.D. thesis, University of Bremen, Germany.

Schuckel, U., Beck, M., & Kroncke, I. (2015). Macrofauna communities of tidal channels in Jade Bay (German Wadden Sea): Spatial patterns, relationships with environmental characteristics and comparative aspects. *Marine Biodiversity*, *45*, 841–855.

Schultz, S. T., Kruschel, C., & Bakran-Petricioli, T. (2009). Influence of seagrass meadows on predator–prey habitat segregation in an Adriatic lagoon. *Marine Ecology Progress Series*, *374*, 85–99.

Shinen, J. L., & Navarrete, S. A. (2014). Lottery coexistence on rocky shores: weak niche differentiation or equal competitors engaged in neutral dynamics? *The American Naturalist*, *183*, 342–362.

Short, F., Carruthers, T., Dennison, W., & Waycott, M. (2007). Global seagrass distribution and diversity: A bioregional model. *Journal of Experimental Marine Biology and Ecology*, *350*, 3–20. Available from https://doi.org/10.1016/j.jembe.2007.06.012.

Shou, L., Huang, Y., Zeng, J., Gao, A., Liao, Y., & Chen, Q. (2009). Seasonal changes of macrobenthos distribution and diversity in Zhoushan sea area. *Aquatic Ecosystem Health & Management*, *12*(1), 110–115. Available from https://doi.org/10.1080/14634980802697902.

Sink, K. J., Branch, G. M., & Harris, J. M. (2005). Biogeographic patterns in rocky intertidal communities in KwaZulu-Natal, South Africa. *African Journal of Marine Science*, *27*(1), 81–96. Available from https://doi.org/10.2989/18142320509504070.

Sivadas, S. K., & Ingole, B. S. (2016). Biodiversity and biogeography pattern of benthic communities in the coastal basins of India. *Marine Biology Research*, *12*(8), 797–816. Available from https://doi.org/10.1080/17451000.2016.1203949.

Skinner, L., & Coutinho, R. (2002). Preliminary results on settlement of the barnacles *Tetraclita stalactifera* and *Chthamalus bisinuatus* on a Brazilian tropical rocky shore under upwelling conditions. *Invertebrate Reproduction and Development*, *41*, 151–156. Available from https://doi.org/10.1080/07924259.2002.9652746.

Smith, D. (2013). *Ecology of the New Zealand rocky shore community* (p. 61). New Zealand Marine Studies Centre, University of Otago.

Smyth, K., & Elliott, M. (2016). *Effects of changing salinity on the ecology of the marine environment. Stressors in the marine environment.* Open University Press. Available from https://doi.org/10.1093/acprof:oso/9780198718826.003.0009.

Snelgrove, P. V. (1998). The biodiversity of macrofaunal organisms in marine sediments. *Biodiversity and Conservation*, 7(9), 1123–1132.

Snelgrove, P. V. (1999). Getting to the bottom of marine biodiversity: Sedimentary habitats: Ocean bottoms are the most widespread habitat on earth and support high biodiversity and key ecosystem services. *Bioscience*, *49*(2), 129–138. Available from https://doi.org/10.2307/1313538.

Sokołowski, A., Wołowicz, M., Asmus, H., Asmus, R., Carlier, A., Gasiunaite, Z., Grémare, A., Hummel, H., Lesutiené, J., Razinkovas, A., & Renaud, P. E. (2012). Is benthic food web structure related to diversity of marine macrobenthic communities? *Estuarine, Coastal and Shelf Science*, *108*, 76–86.

Solan, M., Aspden, R. J., & Paterson, D. M. (Eds.), (2012). *Marine biodiversity and ecosystem functioning: Frameworks, methodologies, and integration.* Oxford University Press.

Starko, S., Bailey, L. A., Creviston, E., James, K. A., Warren, A., Brophy, M. K., Danasel, A., et al. (2019). Environmental heterogeneity mediates scale-dependent declines in kelp diversity on intertidal rocky shores. *PLoS One*, *14*(3), e0213191. Available from https://doi.org/10.1371/journal.pone.0213191.

Steen, H. (2004). Effects of reduced salinity on reproduction and germling development in *Sargassum muticum* (*Phaeophyceae, Fucales*). *European Journal of Phycology*, *39*(3), 293–299. Available from https://doi.org/10.1080/09670260410001712581.

Steinberg, P. D., DeNys, R., & Kjelleberg, S. (1998). Chemical inhibition of epibiota by Australian seaweeds. *Biofouling*, *12*, 227–244.

Steneck, R. S. (2013). Sea Urchins as drivers of shallow benthic marine community structure. Sea Urchins. *Biology and Ecology*, *38*, 195.

Strasser, M., Dekker, R., Essink, K., Günther, C. P., Jaklin, S., Kroncke, I., Madsen, P. B., Michaelis, H., & Vedel, G. (2003). How predictable is high bivalve recruitment in the Wadden Sea after severe winter? *Journal of Sea Research*, *49*, 47–57.

Susan, V. D., Kumar, P. S., & Pillai, N. G. K. (2014). Biodiversity and seasonal variation of benthic macrofauna in Minicoy Island, Lakshadweep, India. *Acta Oceanologica Sinica*, *33*(10), 58–73.

Svane, I., & Dolmer, P. (1995). Perception of light at settlement: A comparative study of two invertebrate larvae, a scyphozoan planula and a simple ascidian tadpole. *Journal of Experimental Marine Biology and Ecology*, *187*, 51–61.

Terlizzi, A., Conte, E., Zupo, V., & Mazzella, L. (2000). Biological succession on silicone fouling-release surfaces: Long term exposure tests in the harbour of Ischia, Italy. *Biofouling*, *15*, 327–342.

Teso, V., Urteaga, D., & Pastorino, G. (2019). Assemblages of certain benthic molluscs along the southwestern Atlantic: From subtidal to deep sea. *BMC Ecology*, *19*(1), 49. Available from https://doi.org/10.1186/s12898-019-0263-7.

Thomson, D. P., Babcock, R. C., Vanderklift, M. A., Symonds, G., & Gunson, J. R. (2012). Evidence for persistent patch structure on temperate reefs and multiple hypotheses for their creation and maintenance. *Estuarine, Coastal and Shelf Science*, *96*, 105–113. Available from https://doi.org/10.1016/j.ecss.2011.10.014.

Thrush, S. F., & Dayton, P. K. (2002). Disturbance to marine benthic habitats by trawling and dredging: Implications for marine biodiversity. *Annual Review of Ecology and Systematics*, *33*(1), 449–473.

Tittensor, D. P., Mora, C., Jetz, W., Lotze, H. K., Ricard, D., Berghe, E. V., et al. (2010). Global patterns and predictors of marine biodiversity across taxa. *Nature*, *466*, 1098–1101.

Todd, C. D. (1998). Larval supply and recruitment of benthic invertebrates: Do larvae always disperse as much as we believe? In S. Baden, L. Phil, R. Rosenberg, J. O. Stromberg, I. Svane, & P. Tiselius (Eds.), *Recruitment, colonization and physical-chemical forcing in marine biological systems. Developments in hydrobiology* (132). Springer. Available from https://doi.org/10.1007/978-94-017-2864-5_1.

Tomanek, L., & Helmuth, B. (2002). Physiological ecology of rocky intertidal organisms: A synergy of concepts. *Integrative and Comparative Biology*, *42*(4), 771–775.

Touhami, F., Bazairi, H., Badaoui, B., & Benhoussa, A. (2018). Vertical distribution of benthic macrofauna in intertidal habitats frequented by shorebirds at MerjaZerga lagoon. *Thalassas*, *34*(2), 255–265. Available from https://doi.org/10.1007/s41208-017-0059-5.

Turner, S. J., Hewitt, J. E., Wilkinson, M. R., Morrisey, D. J., Thrush, S. F., Cummings, V. J., & Funnell, G. (1999). Seagrass patches and landscapes: The influence of wind-wave dynamics and hierarchical arrangements of spatial structure on macrofaunal seagrass communities. *Estuaries*, *22*(4), 1016–1032.

Underwood, A. J. (1981). Structure of a rocky intertidal community in New South Wales: Patterns of vertical distribution and seasonal changes. *Journal of Experimental Marine Biology and Ecology*, *51*, 57–85.

Underwood, A. J. (1985). Physical factors and biological interactions: The necessity and nature of ecological experiments. In P. G. Moore, & R. Seed (Eds.), *The ecology of rocky coasts* (pp. 372–390). Hodder and Stoughton.

Varfolomeeva, M. A. (2013). *Biotic interactions, structure, and long-term changes in marine benthic assemblages. Saint Petersburg State University Studies in biology* (p. 34). St. Petersburg University Press.

Vuorinen, I., Hanninen, J., Rajasilta, M., Laine, P., Eklund, J., Montesino-Pouzols, F., et al. (2015). Scenario simulations of future salinity and ecological consequences in the Baltic Sea and adjacent North Sea areas—Implications for environmental monitoring. *Ecological Indicators*, *50*, 196–205. Available from https://doi.org/10.1016/j.ecolind.2014.10.019.

Wabnitz, C. C., Lam, V. W., Reygondeau, G., Teh, L. C., Al-Abdulrazzak, D., Khalfallah, M., et al. (2018). Climate change impacts on marine biodiversity, fisheries and society in the Arabian Gulf. *PLoS One*, *13*(5), e0194537. Available from https://doi.org/10.1371/journal.pone.0194537.

Van der Wal, D., Ysebaert, T., & Herman, P. M. (2017). Response of intertidal benthic macrofauna to migrating megaripples and hydrodynamics. *Marine Ecology Progress Series*, *585*, 17–30.

Ward, T., & Hutchings, P. A. (1996). Effects of trace metals on infaunal species composition in polluted intertidal and subtidal marine sediments near a lead smelter, Spencer Gulf, South Australia. *Marine Ecology Progress Series*, *135*, 123–135.

Warwick, R. M. (1986). A new method for detecting pollution effects on marine macrobenthic communities. *Marine Biology*, *92*(4), 557–562.

Wendt, D. E., & Woollacott, R. M. (1999). Ontogenies of phototactic behavior and metamorphic competence in larvae of three species of *Bugula* (Bryozoa). *Invertebrate Biology*, *118*(1), 75–84.

Wernberg, T., Russell, B. D., Thomsen, M. S., & Connell, S. D. (2014). Marine biodiversity and climate change. In B. Freedman (Ed.), *Global environmental change. Handbook of global environmental pollution* (1). Springer. Available from https://doi.org/10.1007/978-94-007-5784-4_80.

Van Wesenbeeck, B. K., Van de Koppel, J., Herman, P. M. J., Bakker, J. P., & Bouma, T. J. (2007). Biomechanical warfare in ecology: Negative interactions between species by habitat modification. *Oikos, 116*, 742–750.

Wethey, D. S., Woodin, S. A., Hilbish, T. J., Jones, S. J., Lima, F. P., & Brannock, P. M. (2011). Response of intertidal populations to climate: Effects of extreme events vs long term change. *Journal of Experimental Marine Biology and Ecology, 400*, 132–144. Available from https://doi.org/10.1016/j.jembe.2011.02.008.

Wieking, G., & Kroncke, I. (2001). Decadal changes in macrofauna communities on the Dogger Bank caused by large-scale climate variability. *Senckenbergmarit, 31*, 125–141.

Willows, R. L., Widdows, J., & Wood, R. G. (1998). Influence of an infaunal bivalve on the erosion of an intertidal cohesive sediment: A flume and modelling study. *Limnology and Oceanography, 43*, 1332–1342.

Witman, J. D., Etter, R. J., & Smith, F. (2004). The relationship between regional and local species diversity in marine benthic communities: A global perspective. *Proceedings of the National Academy of Sciences, 101*(44), 15664–15669. Available from https://doi.org/10.1073/pnas.0404300101.

Włodarska-Kowalczuk, M., Jankowska, E., Kotwicki, L., & Balazy, P. (2014). Evidence of season-dependency in vegetation effects on macrofauna in temperate seagrass meadows (Baltic Sea). *PLoS One, 9*, e100788.

Wolanksi, E., & Elliott, M. (2015). *Estuarine ecohydrology: An introduction* (2nd ed.). Elsevier Science.

Worm, B., Barbier, E. B., Beaumont, N., Duffy, J. E., Folke, C., Halpern, B. S., Jackson, J. B., et al. (2006). Impacts of biodiversity loss on ocean ecosystem services. *Science (New York, N.Y.), 314*(5800), 787–790. Available from https://doi.org/10.1126/science.1132294.

Worm, B., & Karez, R. (2002). *Competition, coexistence and diversity on rocky shores, . Competition and coexistence. Ecological studies (analysis and synthesis)* (161). Berlin, Heidelberg: Springer. Available from https://doi.org/10.1007/978-3-642-56166-5_6.

Wu, P. P. Y., Mengersen, K., McMahon, K., Kendrick, G. A., Chartrand, K., York, P. H., et al. (2017). Timing anthropogenic stressors to mitigate their impact on marine ecosystem resilience. *Nature Communications, 8*, 1263. Available from https://doi.org/10.1038/s41467-017-01306-9.

Yallop, M. L., Paterson, D. M., & Wellsbury, P. (2000). Interrelationships between rates of microbial production, exopolymer production, microbial biomass, and sediment stability in biofilms of intertidal sediments. *Microbial Ecology, 39*, 116–127.

CHAPTER 8

Spatio-temporal variations of benthic communities along the coast

K.U. Abdul Jaleel[1], Usha V. Parameswaran[2], Aiswarya Gopal[2], Seerangan Manokaran[3] and Thadickal V. Joydas[3]
[1]CSIR-National Institute of Oceanography, Regional Centre, Kochi, Kerala, India
[2]Centre for Marine Living Resources & Ecology, Ministry of Earth Sciences, Kochi, Kerala, India
[3]Center for Environment and Marine Studies, Research Institute, King Fahd University of Petroleum and Minerals (KFUPM), Dhahran, Saudi Arabia

8.1 Introduction

Benthic fauna plays a key role in maintaining ecological stability, especially in coastal ecosystems. Their distribution varies across different spatial and temporal scales (Horne & Schneider, 1995; Ysebaert et al., 2002). Identification of the relevant scales of such variability in space and time is a prerequisite to understanding factors and processes generating patterns in biotic and abiotic components of ecosystems (Levin, 1992). Quantitative description and observation of distribution patterns at several spatial and/or temporal scales is a fundamental step in the scientific process from which questions and hypotheses arise and lead to the development of appropriate predictive models (Levin, 1992; Thrush et al., 1999; Underwood et al., 2000).

Both large-scale and small-scale processes cause patterns in benthic communities. Many patterns observed in benthic communities were frequently interpreted in the light of large-scale processes such as physicochemical properties of water masses, oceanic fronts, and patterns in sediment distribution (Jumars, 1993; Paiva, 2001). At the same time, small-scale processes such as differential larval settlement and biotic and physical interactions in the bottom boundary layers could give rise to strong variations in species or community responses and are responsible for complex patterns of benthic patchiness (Eckman, 1979; Trueblood, 1991; Thrush, 1991; Li et al., 1997). Small-scale processes like bioturbation and microscale topographic features such as ripple marks cause

Ecology and Biodiversity of Benthos
DOI: https://doi.org/10.1016/B978-0-12-821161-8.00003-9

macrobenthic patches on a scale of meters and centimeters (Hogue & Miller, 1981; Van Blaricom, 1982; Hall et al., 1991).

Spatiotemporal patterns in benthic communities vary with habitats. For example, in intertidal habitats, tidal action is the principal physical factor directly or indirectly governing all the other factors, including temperature (Crowe et al., 2000). Zonation with distinct benthic assemblages occurs in rocky, sandy, and muddy intertidal habitats and the patterns in zonation vary with seasons (La Valle et al., 2020). In estuaries, spatial variation in benthic communities is primarily governed by gradients in salinity (Lu et al., 2008). Pronounced seasonal variations occur in estuaries altering the salinity gradients and temperature regime (Rosa & Bemvenuti, 2006). In subtidal habitats, several physicochemical factors, including temperature, salinity, depth, sediment texture and flux of organic matter, heterogeneity of habitats such as seagrass meadows, algal beds, and coral reefs cause variations in benthic communities (Gray, 1974). Overall, the interaction between coastal morphology, land-ocean exchanges, meteorological and tidal conditions create a highly complex and finely scaled network of distinct communities.

This chapter describes the major spatial and temporal variations in benthic communities in the three coastal ecosystem compartments, namely, estuaries, intertidal habitats, and subtidal habitats.

8.2 Methods of obtaining data

Data presented in this chapter are based on an extensive literature review. Macrofauna samples mentioned in these studies were collected using either grab sampler (0.02–0.1 m^2 bite area) or box corer (0.1–0.2 m^2 bite area) in the case of estuarine and subtidal benthic studies and visual and quadrat (of varying sizes) surveys for intertidal benthic studies. In most cases, meiobenthic samples were collected using a core sampler as subsamples from the grab/box corer/quadrate. Epifauna samples were collected using dredge (for example, a naturalist dredge with 1 m × 0.3 m frame, 1 cm mesh bag, towed at 2 knots for 10 min). In general, density is expressed in terms of individuals per m^2 and biomass in gram per m^2 in the case of macrofauna, per 10 cm^2 for meiofauna and per hauls for epifauna.

8.3 Spatial variations of benthic communities

8.3.1 Estuaries

Estuaries are semienclosed bodies of brackish water (salinity between (and 35), which receive input from one or more rivers or streams, an(

have a free connection to the open sea (Pritchard, 1967). In estuaries, the spatial distribution of the macrobenthos is related to changes in freshwater flow and salinity (Salen-Picard et al., 2002; Rozas et al., 2005), nutrient input and primary production (Heip et al., 1995), sediment particle size (Snelgrove & Butman, 1994; Mannino & Montagna, 1997; Kendall & Widdicombe, 1999), periodic sediment anoxia (Diaz & Rosenberg, 1995), and the presence of grasses and seaweed (Hyndes et al., 2005).

The pronounced spatial patterns along the salinity gradient indicate that salinity majorly influences the existence of distinct benthic assemblages corresponding to the different salinity zones (euhaline, polyhaline, mesohaline, and oligohaline) in estuaries (Lu et al., 2008; Mulik et al., 2020). In the estuarine systems of the west coast of India (tropical), the euhaline zone was characterized by species such as *Mesopodopsis orientalis, Corophium* sp., *Barnea birmanica, Cheiriphotis megacheles, Melita dentata, Magelona cincta, Trochochaeta* sp., *Paraprionospio cordifolia, Nephtys hystricis,* and *Paraprionospio cristata,* while the polymesohaline zone was characterized by *Streblospio gynobranchiata, N. hystricis, Dendroneries arborifera,* and *Limnodrilus hoffmeisteri,* and oligohaline zone by Chironomids, *Namalycastis ouanaryensis,* and *Aulophorus furcatus* (Mulik et al., 2020). The spatial distribution of benthic communities along the upper, lower, and mouth locations of the estuary reflected the combined effect of sediment texture and salinity in the Richibucto Estuary, New Brunswick, Canada (Lu et al., 2008).

Structural indices of the benthic community also vary spatially according to the environmental gradients in estuaries (Nanami et al., 2005; Sousa et al., 2007). For example, an increase in species richness from upstream to downstream was reported in Paraíba do Norte estuary (species richness 2–52) and a similar pattern was observed in Mamanguape estuary (species richness 4–45) in Brazil (Ysebaert et al., 1993; Ysebaert et al., 2003; Medeiros et al., 2016). In these two estuaries, higher salinity and significant variations in sediment texture in the downstream part of the estuaries led to more species cohabitating than where the sediments were muddy with low salinity. Similar to the trends of species richness, species diversity increased with an increase in salinity (2.35 in the upstream part to 4.12 in the downstream part) in the Canada estuary (Lu et al., 2008). Downstream-upstream variations in the number of species (26–19 in Vellar and 6–3 in Uppanar) and density (44–28 ind./0.0251 m^2 in Vellar and 39–31 ind./0.0251 m^2 in Uppanar) are reported from estuaries along the east coast of India, with higher values downstream than upstream (Ajmal Khan et al., 2014). Spatial variations can also occur due to tidal activity that transports larvae and juveniles from the open sea to the estuarine environment. During this transportation, most of the adult specialist species assemble into low salinity areas. However, these cannot reproduce in these environmental conditions; moreover, they cannot generate the high energy levels for osmoregulation necessary to maintain vital functions (Medeiros et al., 2016).

Studies on functional traits of benthic communities reported the prevalence of deposit feeders in the inner estuarine regions (Herman et al., 1999) because of the availability of plentiful food in the form of large quantities of allochthonous organic matter and nutrients available where fine sediments prevailing in the estuarine ecosystem (Dauwe et al., 1998; Ysebaert et al., 2003; Ourives et al., 2011). The abundance of carnivores is observed in the mouth of estuaries (Fauchald & Jumars, 1979; Ourives et al., 2011) due to the availability of diverse prey types in the sandy sediments. The greater penetration of oxygen in the sediment can support the physiological requirements of active moving predators (Fauchald & Jumars, 1979; Ourives et al., 2011).

Spatial variations in benthic communities in estuarine environments are also caused by human pressures, which may or may not alter the natural spatial variations in benthic communities in the estuary. There are four general categories of human impacts on estuarine systems, including enrichment with excessive levels of organic material, inorganic nutrients, or heat; physical alterations and introduction of toxic materials can lead to direct changes in benthic community structures (Baltz & Yańez-Arancibia, 2012). A gradient of increasing quality from the most degraded northern site (where there are discharges from industries such as oil refinery, a fertilizer plant, a rare-earth processing plant, a minerals and Rutile plant, and a zinc smelter plant) to the main estuary, downstream was observed in the Cochin Estuary (Menon et al., 2000; Feebarani et al., 2016; Rehitha et al., 2017). Ajmal Khan et al. (2014) also compared spatial variations of benthic ecological quality in two different estuaries (Vellar and Uppanar River). As a positive human impact, Vellar Estuary shows the higher ecological status (undisturbed nature and high/good ecological quality) near the artificially developed Pitchavaram mangroves (Ajmal Khan et al., 2014).

8.3.2 Intertidal habitats

Intertidal habitats, the area between the low tide and high tide mark, experience steep environmental gradients (wave exposure, air exposure, temperature, substrate heterogeneity, etc.) that promote the development of the distinct spatial structure of benthic assemblages (Crowe et al., 2000; Zamprogno et al., 2012; Schlacher & Thompson, 2013). Flora and fauna in intertidal areas are subject to harsh environmental conditions. During low tide, organisms are exposed to air and sunlight, while they are submerged in seawater during the high tide. As a result, the fauna of these habitats is typically distributed in vertical zones, based on the environmental conditions that favor their survival. According to Peterson (1991), the zonation and patchiness in intertidal areas is a result of the organism's requirement to be submerged for feeding or other physiological needs, their susceptibility to desiccation or inability to otherwise cope with prolonged exposure to air, as well as biotic factors like predation and competition. The region just above the high tide mark characterized by an occasional exposure to moisture due to breaking waves is known as the supratidal zone. Between the high tide and low tide mark lies the intertidal zone proper, which may be delimited into the upper, middle, and lower intertidal zones. Beyond the low tide mark, the region which is permanently submerged in seawater, and influenced by dissipating wave energy, is known as the subtidal zone. Although zonation is a common feature in most shores, the number of zones, their faunal composition, etc., vary widely with latitude and nature of substrates (Valdivia et al., 2011; Veiga et al., 2014; Catalán et al., 2020). The major spatial variations of benthic communities in rocky, sandy, and muddy shores are described further.

In rocky shores, benthic fauna is subjected to several stressors such as increased air exposure, wave action, and tidal fluctuations, which are functions of the topography of the shore (Zamprogno et al., 2012). The influence of wave action is particularly important in the mid and low-shore communities while desiccating factors (air exposure and rock temperature) are significant in the supratidal and upper shore (Bustamante et al., 1995). Similarly, photoperiod and rainfall also play a significant role in structuring benthic communities (Cruz-Motta et al., 2010; Chappuis et al., 2014). Biotic factors that influence species distribution include predation, competition for space, and adaptations to specific environmental factors. These factors lead to strong zonation of sessile and motile communities at particular heights on the shore, based on adaptations of individual species (Chappuis et al., 2014). The supratidal zone in rocky shores is subjected to intense wave action and is occupied by lichens, which trap moisture (fungi) and carry out photosynthesis (associated algae), along with periwinkles, rock louse (*Ligia* sp.), etc. The upper intertidal zone is characterized by sessile and motile shelled invertebrates like periwinkle (*Littorina* spp.), limpets (*Acmaea* spp.), stellate barnacles (*Chthamalus* spp.), and chitons. The mid-intertidal zone is dominated by acorn and gooseneck barnacles (e.g., *Balanus* spp. and *Pollicipes* spp.) along with mussels and hermit crabs, and seaweeds. In the lower intertidal zone harbor a diverse assemblage of brittle stars (e.g., *Ophiocoma* spp.), sea stars (*Asterias* sp.), crabs, sponges, sea slugs, etc. (Lipkin & Safriel, 1971). Rocky intertidal areas also harbor tide pools, which are isolated pockets of seawater left in rocky depressions by receding tides. These offer important refugia to motile fauna from temperature stress and desiccation during low tide. However, they are still susceptible to other forms of stress like oxygen deficiency and increased predation by specialized predators (e.g., wading birds) (Metaxas & Scheibling, 1993). Large-scale latitudinal trends in the abundance and species richness of intertidal rocky shore echinoderms have been reported, with higher values in low latitudes and lower values in higher latitudes (Hillebrand, 2004; Iken et al., 2010). The higher values of abundance and diversity of sea urchins in low latitudes have also been found previously (Guidetti, Dulčić, 2007), attributing to the prevalence of thermophilic species among sea urchins (Francour et al., 1994), latitudinal differences in recruitment success (Ebert, 1983; Tsujino et al., 2010), salinity tolerance (Vidolin et al., 2007), competitive and predatory interactions (McClanahan & Shafir, 1990; Guidetti & Mori, 2005), relief from predation due to overfishing (Sala et al., 1998), as well as adaptability of sea urchins to environmental stress (Starr et al., 1993; Pancucci et al., 1993).

Sandy intertidal habitats cover around two-thirds of the ice-free coastlines globally (Brown & McLachlan, 2010). Most sandy shore species are found in no other environment, their unique adaptations for life in these dynamic systems include mobility, burrowing ability, protective exoskeletons, rhythmic behavior, orientation mechanisms, and behavioral plasticity (Brown, 1996). On a global scale, an increase of species richness from temperate to tropical sandy shores has been documented, highlighting beach morphodynamics and temperature in structuring benthic marine communities in these ecosystems (Barboza & Defeo, 2015). In general sediment grain size and organic matter also play a key role in determining the distribution of fauna in sandy shores. The nature of substrates is governed by beach topography, waves, currents, tides, and winds, which lead to erosion and accretion (Pereira et al., 2018). Cyclical exposure to air direct sunlight and temperature, lead to changes in sediment moisture content, which influences vertical as well as horizontal distribution of meio and macrofauna. Repeated exposure and submergence also lead to the development of anoxic layers within sediments, which limits the vertical distribution of meiofauna during tidal fluctuations (Pereira et al., 2018). Horizontal variation in floral and faunal composition in intertidal habitats is determined by a combination of biotic interactions, abiotic factors, dispersal processes, and priority effects (i.e., competitive dominance given by early colonization) (Catalán et al., 2020). Along the vertical axis, a strong gradient in environmental conditions such as desiccation, thermal stress, as well as the availability of food and shelter lead to distinct zonation (Araújo et al., 2005 Valdivia et al., 2011; Catalán et al., 2020). In general, the supratidal zone (also known as upper shore) in sandy shores is characterized by air-breathing crustaceans (crabs) and talitrid amphipods. The intertidal zone (or mid-shore) is characterized by a diverse assemblage including nematodes, cironlanid isopods, glycerid, and spionid polychaetes, as well as bivalves. The fauna in the lower shore includes burrowing amphipods, polychaetes, mysids, and echinoderms (Ansari et al., 1984; James & Fairweather, 1996; Schlacher & Thompson, 2013) (Fig. 8.1). Along the coralline sandy shores in the northwest coast of India, amphipods demonstrated sharp zone-specific distributions, with species

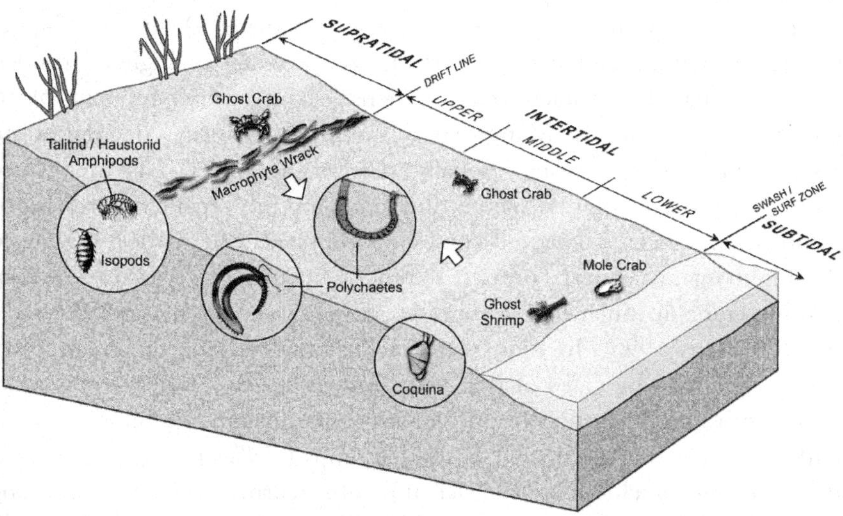

Figure 8.1 Distribution of sandy shore invertebrates in the three major zones. *From Michel, J., Fegley, S. R., Dahlin, J. A., & Wood, C. (2017). Oil spill response-related injuries on sand beaches: When shoreline treatment extends the impacts beyond the oil. Marine Ecology Progress Series, 576, 203–218 (Michel et al., 2017).*

Monoculodes sp., *Parhyale hawaiensis*, and *Urothoe brevicornis* occurring only at upper shore, *Ampelisca* spp., *Paracaprella* sp., *Peramphithoe* sp., etc., at mid-shore and *Apohyale prevostii*, *Cyproidea ornata*, *Idunella chilkensis*, etc., occurring only at the lower shore (Srinivas et al., 2020). In this study, the highest abundance and diversity were noted in the latter region, progressively declining toward the upper shore (Srinivas et al., 2020).

Muddy shores (also referred to as mudflats) where silt and clay are the major substrates are usually found in relatively sheltered depositional coastlines, including estuaries. Flooding by seawater (or brackish water) during high tide, followed by exposure to tropical sunlight, causes evaporation, which leads to a decrease in moisture content and an increase in salinity within sediments. The sediments also have very low oxygen content due to microbial degradation of the vast quantity of deposited organic matter. Despite the constantly and widely fluctuating temperature, moisture, and salinity as well as low-oxygen conditions prevailing in these swamps, they host a wide variety of biota, including benthic having specific adaptations, dominated by the mangrove plants (Giri et al., 2011). The distribution of benthic fauna in muddy shores is thus governed by salinity, fluctuations in temperature, dissolved oxygen, organic matter content, freshwater (or brackish water) influx, etc. (Smith & Able, 2003). In the tropical and subtropical belt, muddy shores are usually characterized by mangrove swamps (mangroves forests or mangals) in their upper reaches (supratidal and upper intertidal areas), while in higher latitudes, this zone may be characterized by dense stands of salt-tolerant plants known as salt marshes. Mangrove swamps or "mangals" occupy about 137,800 km^2 of the shores globally (Giri et al., 2011). As mentioned above, they occur in areas where fine sediment and organic matter deposition occurs, such as sheltered coastlines and banks of estuaries. Owing to the high organic matter content, sediments in mangrove swamps harbor benthic fauna at much higher densities than adjacent non-vegetated sediments (Edgar, 1990). In Indo-west Pacific mudflats, the upper intertidal area is inhabited by crabs of the genus *Macrophthalmus*, fiddler crabs, ocypodid crabs, mud-skippers, mud whelks, while the middle intertidal zone harbors burrowing fauna such as mud shrimps (*Callianasa* spp.), thalassinidian shrimps, brachiopods (*Lingula* sp.), tube worms (e.g., terebellids and onuphids), echiuroids, bivalves, and holothurians. Lower intertidal fauna is characterized by sand dollars, hemichordates, hermit crabs, and brittle stars (Dittmann, 2002). In the mudflats of Atharbanki mangrove waterways in Odisha, the benthic macrofauna was dominated by polychaetes, especially

Namalycastis sp. as well the amphipod *Victoriopisa chilkensis*. The highest diversity was associated with relatively undisturbed areas with neritic influences, while disturbed environments with human interferences harbored low diversity (Nayak et al., 2018).

8.3.3 Subtidal habitats

The subtidal zone, defined as the area beyond the influence of tidal fluctuations, is characterized by distinct environmental conditions compared to those in estuarine and intertidal habitats, being part of the open ocean and therefore directly influenced by its dynamics. The nature of the seabed (hard such as rocks and boulders or soft such as sand, silt, and clay) in subtidal habitats is determined by a set of factors including geomorphology, plate tectonics, proximity to river mouths and glaciers, which in turn determines the distribution of benthic community (Gray, 1974; Harkantra et al., 1980; Jayaraj et al., 2007). The nature of substrates and other physical factors influencing benthic communities tend to show large-scale spatial variations, for example, with depth (or distance from shore) and latitude. However, corresponding large-scale spatial patterns like latitudinal gradients in species richness (decrease in richness from the equator to poles), is poorly demonstrated in marine biota (Piacenza et al., 2015). The polar-equatorial cline is reported only in some benthic taxa like gastropods and bivalves (Sanders, 1968; Roy et al., 2000). Rather than latitudinal trends, increased species richness of benthic fauna appears to be associated with rocky outcrops, coral reefs, etc., where increased habitat complexity supports exceptional diversity (Huston, 1985; Sebens, 1991).

In higher latitudes, much of the coastal deposits are of glacial origin, and therefore comprise pebbles and coarse gravel (Barnes, 1999). Biogenic materials also make up a considerable portion of sediments in these regions, foraminiferan, and molluscan shells, sponge spicules, etc. The benthic fauna in these regions includes encrusting forms like sponges and bryozoans, sessile forms such as ascidians, barnacles, tube building polychaetes, etc., along with motile invertebrates like echinoderms, isopods, amphipods, etc. (Barnes et al., 1996, Barnes, 1999). In middle and lower latitudes, the factors determining the nature of bottom substrates are more heterogeneous. Regions with terrestrial discharge (runoff, rivers, etc.) as well as other erosional forces (waves and currents) are characterized by soft sediments, while other regions may be hard-bottomed. Soft-sediment habitats of the tropical and temperate regions harbor significant density and diversity of benthic infauna (Suchanek, 1995; Jayaraj et al., 2008; Joydas & Damodaran, 2009).

On a regional scale, an increase in depth is associated with decreasing light availability, temperature, productivity, etc., and a concurrent rise in pressure, as well as greater stability in salinity and temperature (Piacenza et al., 2015). Latitudinal and bathymetric variations in benthic community composition are widely reported off the west coast of India (Jayaraj et al., 2007, 2008; Joydas & Damodaran, 2009; Parameswaran et al., 2018). Nematodes were overwhelmingly dominant among the meiofauna along the west coast of India (88%), followed by harpacticoid copepods (8%), while macrofauna was dominated by polychaetes (57%), followed by crustaceans (28%) and epifauna by echinoderms (42%) and crustaceans (31%) (Joydas & Damodaran, 2009; Sajan et al., 2010; Parameswaran et al., 2018) (Fig. 8.2). In all three cases (meio, macro, and epifauna), a strong decreasing trend in abundance and diversity is reported with increasing water depth, attributed to shifting sediment nature and decline in oxygen availability. The southern region (up to ~8°N) is characterized by heterogeneous sandy sediments (coralline sand as well as siliceous sand) which

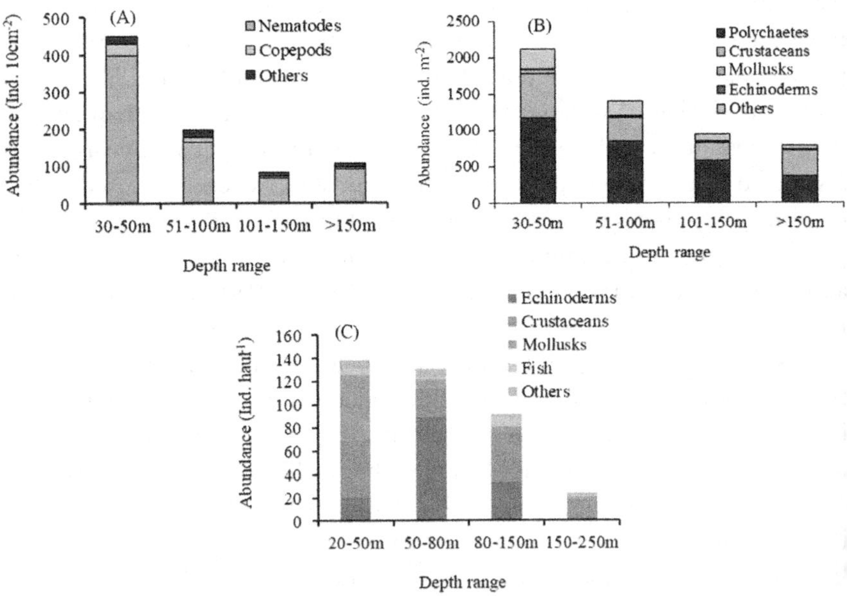

Figure 8.2 Bathymetric variations in abundance of (A) meiofauna, (B) macrofauna and (C) epifauna along the west coast of India. *(A) Joydas & Damodaran, 2009; (B) Sajan et al. 2010; (C) Parameswaran et al. 2018.*

harbor more diversified assemblages, with more complex food-webs encompassing diverse feeding modes (e.g., predators, filter feeders, deposit feeders, etc.), particularly among macrofaunal polychaetes, meiofaunal nematodes and epifaunal echinoderms (Joydas & Damodaran, 2009; Musale & Desai, 2011; Abdul Jaleel et al., 2015, 2021; Parameswaran et al., 2018; Rehitha et al., 2019; Gopal et al., 2020). By contrast, finer sediments, like silt and clay, frequently deposited around river mouths toward the north, harbor less diverse assemblages dominated by deposit-feeding forms sustained by the high organic matter content in finer sediments. This is attributed to bathymetric shifts in sediment characteristics and a decline in dissolved oxygen concentrations (Joydas & Damodaran, 2009; Sajan et al., 2010; Parameswaran et al., 2018; Naidu et al., 2018; Sivadas et al., 2020).

8.4 Temporal variations of benthic communities

8.4.1 Estuaries

In estuaries, seasonal patterns in benthos are strongly correlated with seasonal variations in physicochemical parameters of the water and sediments (Rosa & Bemvenuti, 2006). Winter–summer variations are well documented in benthic communities of estuaries. For example, a reduction in structural parameters of benthos during winter was reported from Lima estuary, NW of the Iberian Peninsula (Sousa et al., 2007). In this study, maximum values of winter–summer data were 17–29 for species richness, 2.37–2.51 for species diversity, 12–29 g for biomass, and 1448–8060 ind./m^2 for density. Similar seasonal variations in benthic community parameters were observed in other regions like the Chesapeake Bay, USA (Holland, 1985), Calcasieu lake estuary (Gaston et al., 1988), Rio Queule Estuary, Chile (Quijón & Jaramillo, 1993), the Mondego estuary, Western Portugal (Marques et al., 1993), Samborombon Bay, Argentina (Ieno & Bastida, 1998), Vellar estuary, India (Hemalatha et al., 2014), Cochin estuary, India (Feebarani et al., 2016), and Ulhas, Amba, and Savitri estuaries in India (Mulik et al., 2020). The changes in benthic community structures in winter are related to recruitment processes, which are influenced by a decrease in temperature and salinity. During summer, the reproductive activity of benthic animals increases due to the rise in temperature, resulting in dense recruitment in the estuarine regions (Bemvenuti, 1987, 1997; Tagliarolo et al., 2018; Rosa & Bemvenuti, 2006).

Precipitation alters salinity gradients, as well as erosional and depositional cycles, the latter in turn influencing the sediment composition in estuaries. These variations in properties of water and sediment affect the composition of benthic communities (Zajac & Whitlatch, 1982; Bonsdorff & Osterman, 1985; Lu et al., 2008; Shou et al., 2009). In general, in Portuguese estuaries (Canal de Mira and Ria de Aveiro), the dominance of amphipod species, *Corophium multisetosum*, has been reported during the rainy season, followed by a drop in density during summer (Queiroga, 1990; Cunha et al., 2000). On the other hand, the polychaete, *Streblospio shrubsolii*, which is an opportunistic species adapted to salinity changes, is abundant during all seasons, except spring due to natural (freshwater inflow) and anthropogenic (contaminants sources) stressors in the Mondego River estuary, Portugal (Chainho et al., 2006). Monsoon was reported to alter the quality of macrobenthos in Cochin Estuary differently in downstream and upstream locations (Feebarani et al., 2016). In the Cochin Estuary, in general, there is a gradient of increasing quality of macrobenthos from the most degraded northern site to the main estuary, downstream. Monsoon-induced reductions in the quality of macrobenthos are reported in the main estuary, while the degraded northern part showed improvement during the peak monsoon. This is because river flow caused dilution of pollutants and an increase in dissolved oxygen in the northern part, which improved the health of the benthic communities (Feebarani et al., 2016).

8.4.2 Intertidal habitats

In intertidal habitats, the distribution of benthic fauna can be influenced by a variety of abiotic (e.g., tidal fluctuations, wave exposure, topography, and substrate nature) and biotic (e.g., predation and competition) factors (Zamprogno et al., 2012). The physical factors (e.g., temperature) show significant temporal variations—from diurnal to seasonal time scales (La Valle et al., 2020). On a diurnal scale, intertidal fauna is subjected to substantial fluctuations in solar irradiation and consequently in temperature. Many intertidal taxa adapt to these variations by seeking shelter from intense sunlight, within rocks and crevices (e.g., echinoderms like sea cucumbers and brittle stars), within their shells (e.g., barnacles) or in deeper areas during the day; and emerging to feed during the night time (Williams & Morritt, 1995). The major implication of periodic tidal fluctuations is that intertidal organisms are alternately exposed to air and water, which increases the threat of desiccation. Osmoregulatory adaptations play a crucial role in the survival of fauna under these conditions (Oglesby, 1969).

The abiotic factors influencing benthic fauna vary widely globally. In the northern Indian Ocean, the influence of the Indian monsoon (seasonally reversing wind and weather patterns) causes significant variations in temperature, wave action, salinity (through freshwater inflow) in the nearshore areas (Nybakken & Bertness, 2004). Rocky intertidal communities in the Gulf of Oman were found to be heavily impacted by the increased wave action, erosion, and suspended sediment load during the southwest monsoon, which, apart from leading to dislodgement of algae and epifauna, also has a direct or indirect impact on the feeding, respiration, reproduction, and settlement of intertidal organisms (Ebrahimnezhad & Naderloo, 2019). This led to a very low abundance of diversity in decapod crustaceans during this season. The relatively stable conditions (lower water temperature, lesser wave stress, etc.) of the northeast monsoon supported relatively high density and diversity of decapods (Ebrahimnezhad & Naderloo, 2019). In some areas, an increase in nutrient levels due to coastal upwelling may cause an increased abundance of algae and plankton, which subsequently supports intertidal food webs (Reddin et al., 2015). Thermal stress is another important abiotic factor affecting seasonal variations in the benthic fauna of rocky shores. Heat kills are reported in the case of dominant populations like limpets, mussels, and barnacles in the tropics as well as in temperate regions (Harley, 2008; Wang et al., 2020).

In sandy shores, important factors causing community-level changes in benthic fauna include salinity, which varies as a result of precipitation, freshwater influx and groundwater seepage, as well as coastal morphodynamics, which influence grain size, sorting, pore water penetration, and organic matter content (Lercari & Defeo, 2006). In a Brazilian sandy beach abundance of infaunal macrobenthos is found to be positively influenced by seasonal recruitment peaks and species migration (between zones and vertically within sediments), as well as by the negative impact of predation, migration, and mortality (Neves et al., 2008). As a result, significant seasonal variations were observed in dominant species, with *Scolelepis gaucha* and *Mesodesma mactriodes* dominating in January–February, while *Donax haleyanus* and *Euzonus furciferus* dominated in May. The influence of coastal eutrophication and algal blooms on sandy shores has been studied to some extent (Quillien et al., 2018). The most significant impact was the massive development of opportunistic macroalgae, which washed ashore or floated above the sediments, which altered the microtidal sediment invertebrates as well as flatfishes in temperate waters of Europe (Pihl et al., 2005; Quillien et al., 2018). Although muddy shore is ecologically significant, temporal variations in benthic fauna in this habitat are poorly characterized. As these habitats are usually associated with depositional coastlines and terrestrial freshwater sources, they are subjected to seasonal pulses in sediments and organic matter (Brown & McLachlan, 2010). Microbial degradation of organic matter within sediments and poor oxygen penetration into the well-sorted sediments lead to anoxia and redox clines within sediments (Seidel et al., 2015). The depth and intensity of these redox clines vary with seasons, and organic matter fluxes (Brown & McLachlan, 2010). The implications of these variations in benthic fauna are poorly known.

Beyond diurnal and seasonal fluctuations, benthic fauna is also influenced by long-term changes in climate. In the Atlantic coast of Europe, the relative abundance of intertidal Lusitanian barnacles (*Chthamalus montagui* and *C. stellatus*) and the boreal species (*Semibalanus balanoides*) have been altered by long-term changes in temperature. These changes include the northward shift in the southern range limit of *S. balanoides* and the northward shift in the range of the *Chthamalus* spp. (Wethey & Woodin, 2008; Mieszkowska et al., 2006; Birchenough et al., 2015). The abundance of *S. balanoides* has also declined in recent years, in response to increases in temperature in recent decades (Poloczanska et al., 2008). Such long-term changes are poorly known from other parts of the world, especially the tropics and poles.

8.4.3 Subtidal

Benthic communities in subtidal habitats vary temporally due to variations in temperature, currents, upwelling, pelagic productivity, rainfall, and river runoff. In polar regions, temperature variations cause episodic or seasonal disturbance from sea ice, affecting benthic community structure (Clark et al., 2017). Processes like enclosure by ice foot, abrasion by ice scour and anchor ice, etc., regularly remove biota from the substratum almost completely (Barnes, 1999; Gutt, 2001). Frequent ice scours and recolonization in nearshore areas lead to well-marked zonation of benthic communities, particularly in the northern hemisphere, where the length of the nonglaciated coastline is greater (Gutt, 2001). Other factors tha impact benthic fauna in polar regions include temperature and seasonality in irradiation as well as organic flux. In temperate and tropical areas, seasonality in coastal circulation may lead to shifts in pelagic productivity regimes which in turn have consequences for benthic fauna. An increase in production due to seasonal processes like a current reversal, rainfall and river runoff, upwelling, etc., leads to increased biological production in the pelagic column and subsequent organic flux to the seafloor. Such seasonal organic matter "pulses" directly or indirectly alter benthic community structure.

High pelagic production sometimes also leads to rapid utilization of oxygen in subsurface waters, due to bacterial degradation of organic matter, which results in seasonal oxygen deficiencies (hypoxia/anoxia/dead zones) in coastal areas. Impingement of these oxygen-deficient waters on the seafloor has significant implications for benthic communities (Diaz & Rosenberg, 1995). Depending on the intensity and duration of the oxygen deficiency, benthic taxa show mortality, avoidance, migration, with some groups being more sensitive than others. Among the most vulnerable are the echinoderms and crustaceans, while some taxa among polychaetes and nematodes are relatively tolerant and adapted to hypoxia. In the eastern Arabian Sea, the seasonal process of monsoon-associated coastal upwelling leads to hypoxia/anoxia in coastal waters between June and September (Gupta et al., 2016). A shift in macrobenthic communities from being relatively abundant and diverse during the premonsoon to being dominated by one or a few species of opportunistic taxa, well adapted to thrive in organic-enriched, oxygen-deficient sediments during the summer monsoon have been reported (Abdul Jaleel et al., 2015, 2021; Naidu et al., 2018; Rehitha et al., 2019). This was found to be a direct result of the recruitment of dominant polychaete taxa in the region, coinciding with the highly productive monsoon season (Abdul Jaleel et al., 2015, 2021) (Fig. 8.3). Chief among these are deposit-feeding spionid polychaetes such as *Paraprionospio pinnata* and *Prionospio* spp. which have proliferous gills and relatively low surface area to volume ratios which enhance oxygen uptake, as well as long palps for feeding (Levin, 2003; Abdul Jaleel et al., 2015, 2021; Rehitha et al., 2019). In areas where the seasonal oxygen fluctuation is most intense (10–14°N), echinoderms (slow-growing, long-lived, and late-maturing) are not able to establish stable populations. Further south (7–10°N), where hypoxia does not occur or is less intense, echinoderms show high diversity and density (Parameswaran et al., 2018).

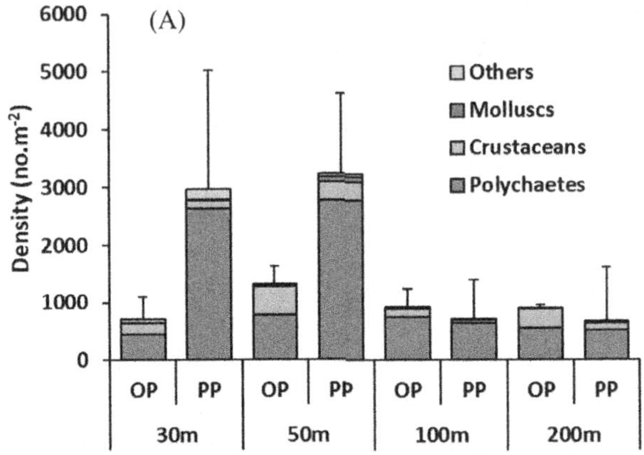

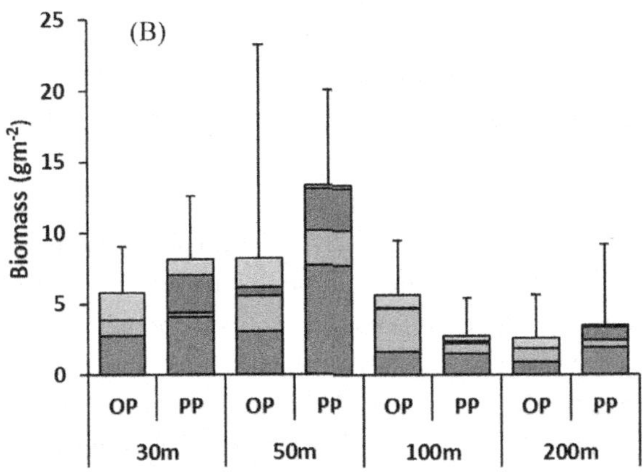

Figure 8.3 Mean density and biomass of macrofaunal groups at each depth stratum during the onset (OP) and peak (PP) of the southwest monsoon. (A) Mean density ($>500\ \mu m$) and (B) mean biomass ($>500\ \mu m$): OP and PP. Bar indicates the standard deviation of the total macrobenthic density (Abdul Jaleel et al., 2015).

Apart from natural, and seasonal variations, coastal benthos has also undergone substantial changes in recent decades, in response to long-term changes in marine environments associated with climate change as well as overexploitation of coastal ecosystems (e.g., Hily et al., 2008; Birchenough et al., 2015; Aguilar et al., 2017). Despite the perceived stability of subtidal habitats, relative to estuaries and intertidal areas, this zone is not free from the negative impacts of anthropogenic activities. Nearshore subtidal areas are the primary sinks for pollutants, sewage, and other land-derived waste, their further burial, dispersal, and dilution being determined by the physical dynamics of the sea. Heavy metal, microplastic, and organic pollutants have direct implications on soft-sediment benthos by causing death or sublethal stress (e.g., Bryan & Langston, 1992; Bour et al., 2018). Eutrophication and oxygen depletion in coastal areas has become common, especially off massively urbanized coastlines. Intensification of hypoxia in bottom waters leads to modification of benthic food-webs over time, for example, the replacement of predators and scavengers with opportunistic, rapidly colonizing taxa (Levin et al., 2009). Similarly, physical stressors like dredging, dumping of dredge spoils, anchoring and ballasting, also modify benthic communities by altering habitats, eliminating, clogging, or burying benthic fauna, thereby altering food webs (e.g., Harvey et al., 1998; Rehitha et al., 2017). Other disturbances to coastal benthos include oil and gas exploration, ship traffic, submarine cables, etc. Most of the aforementioned pressures do not act on coastal systems in isolation; instead, they usually occur in combinations to alter coastal benthic communities over the short and long term. Owing to the relative difficulty of collecting and analyzing benthos beyond the intertidal areas, very few datasets exist to assess the long-term changes in the benthos. A study in the Baltic Sea carried out over 30 years revealed substantial changes in benthic community structure in response to large-scale changes in salinity and temperature and increased fishing pressure (Olsson et al., 2013). The density and diversity of marine polychaetes decreased, and sensitive species were replaced by opportunistic species, tolerant to environmental perturbations, beginning in the early 1980s (Olsson et al., 2013). The exact mechanisms which lead to the aforementioned changes are not easy to pinpoint, owing to the combined effects of actual physiological stressors, physical barriers to dispersal, differences in life cycles, and anthropogenic changes to coastal habitats (Birchenough et al., 2015). A few examples are known of changes in the geographic distribution of species in response to temperature changes, such as the decapods *Diogenes pugilator*, *Goneplax rhomboides*, and *Liocarcinus vernalis*, which have extended their ranges from the Iberian Peninsula into the North Sea in recent decades (Birchenough et al., 2015). The changes in species distributions are also evident in the distribution of meroplanktonic larvae, with an increase in decapod and echinoderm larvae and a decrease in bivalve larvae in the North Sea from 1958 and 2005 (Kirby et al., 2008). Long-term, climate change-driven community shifts are also apparent off the coast of Antarctica, in the Weddell Sea, where macro and megabenthic biomass decreased by two-thirds between 1988 and

2014 (Pineda-Metz et al., 2020). Communities shifted from suspension feeders to deposit feeders, concurrent with a decrease in sea ice cover, which may have led to a decline in organic matter flux to the bottom (Pineda-Metz et al., 2020). There is an urgent need to build robust long-term datasets on coastal benthic faunal distribution from the tropics to the poles to be able to understand better the decadal-scale changes occurring as a result of Anthropocene climate change.

8.5 Conclusions

Benthic fauna is an important component of marine ecosystems, providing key services including secondary production and remineralization. Being sedentary or having only limited mobility, benthic communities are particularly vulnerable to variations in environmental and ecological factors. As a result, they exhibit distinct spatial and temporal distribution patterns on small and large scales. Coastal areas are naturally highly dynamic, with several distinct habitat types coexisting nearby (e.g., estuaries and intertidal habitats) and supporting high biodiversity. The abiotic factors structuring benthic communities include salinity, temperature, sediment characteristics, and oxygen availability, however, their relative importance varies among the different habitats. On a fine scale, biotic factors such as competition for food and space, predation, reproductive strategies, and life-history traits influence the distribution and abundance of individual species, in turn determining community structure. Moreover, coastal habitats are also the most impacted by anthropogenic pressures, from climate change-related warming and acidification to habitat degradation and pollution. Considering this, it is vital to understand the causes and effects of variations in benthic communities to develop effective ecosystem management strategies.

Acknowledgments

AJKU acknowledges the support of the Director, CSIR-National Institute of Oceanography, India as well as the Scientist-in-Charge CSIR-NIO Regional Centre Kochi. UVP and AG acknowledge the support of the Director, Centre for Marine Living Resources and Ecology, Ministry of Earth Sciences, India. SM and TVJ thank the Center for Environment & Marine Studies (CEMS) at the Research Institute of King Fahd University of Petroleum & Minerals, Dhahran, Saudi Arabia, for the support.

References

Abdul Jaleel, K. U., Parameswaran, U. V., Gopal, A., Khader, C., Ganesh, T., Sanjeevan, V. N., Shunmugaraj, T., Vijayan, A. K., & Gupta, G. V. M. (2015). Evaluation of changes in macrobenthic standing stock and polychaete community structure along the southeastern Arabian Sea shelf during the monsoon trawl-ban. *Continental Shelf Research*, *102*, 9–18.

Abdul Jaleel, K. U., Parameswaran, U. V., Gopal, A., Khader, C., Sanjeevan, V. N., Vijayan, A. K., & Gupta, G. V. M. (2021). Response of macrozoobenthic communities to summer monsoon upwelling and related hypoxia in the south eastern Arabian Sea shelf. *Marine Environmental Research*, *166*, 105278.

Aguilar, R., Perry, A. L., & López, J. (2017). Conservation and management of vulnerable marine benthic ecosystems. In S. Rossi, L. Bramanti, A. Gori, & C. Orejas (Eds.), *Marine animal forests*. Springer. Available from https://doi.org/10.1007/978-3-319-17001-5_34-1.

Ajmal Khan, S., Manokaran, S., & Lyla, P. S. (2014). Assessment of ecological quality of Vellar and Uppanar Estuaries (Southeast Coast of India) using benthos. *Indian Journal of Geo-Marine Sciences*, *43*, 0379–5136. Available from http://hdl.handle.net/123456789/34556.

Ansari, Z. A., Ingole, B. S., & Parulekar, A. H. (1984). Macrofauna and meiofauna of two sandy beaches at Mombassa, Kenya. *Indian Journal of Geo-Marine Sciences*, *13*(4), 187–189.

Araújo, R., Bárbara, I., Sousa-Pinto, I., & Quintino, V. (2005). Spatial variability of intertidal rocky shore assemblages in the northwest coast of Portugal. *Estuarine, Coastal and Shelf Science*, *64*(4), 658–670.

Baltz, D. M., & Yańez-Arancibia, A. (2012). Ecosystem-based management of coastal fisheries in the Gulf of Mexico: EnvironQ12 mental and anthropogenic impacts and essential habitat protection. In J. W. Day, & A. Yańez-Arancibia (Eds.), *The Gulf of Mexico ecosystem-based management*. A&M University Press, Chapter 19.

Barboza, F. R., & Defeo, O. (2015). Global diversity patterns in sandy beach macrofauna: A biogeographic analysis. *Scientific Reports*, *5*, 14515. Available from https://doi.org/10.1038/srep14515.

Barnes, D. (1999). The influence of ice on polar nearshore benthos. *Journal of the Marine Biological Association of the United Kingdom*, *79*(3), 401–407.

Barnes, D. K. A., Rothery, P., & Clarke, A. (1996). Colonisation and development in encrusting communities from the Antarctic intertidal and sublittoral. *Journal of Experimental Marine Biology and Ecology*, *196*, 251–265.

Bemvenuti, C. E. (1987). Predation effects on a benthic community in estuarine soft sediments. *Atlântica*, *9*, 5–32.

Bemvenuti, C. E. (1997). Benthic invertebrates. In U. Seeliger, C. Odebrecht, & J. P. Castello (Eds.), *Subtropical convergence environments: The coast and sea in the southwestern Atlantic* (pp. 43–46). Springer-Verlag.

Birchenough, S. N. R., Reiss, H., Degraer, S., Mieszkowska, N., Borja, Á., Buhl-Mortensen, L., Braeckman, U., Craeymeersch, J., De Mesel, I., Kerckhof, F., Kröncke, I., Parra, S., Rabaut, M., Schröder, A., Van Colen, C., Van Hoey, G., Vincx, M., & Wätjen, K. (2015). Climate change and marine benthos: A review of existing research and future directions in the North Atlantic. *Wiley Interdisciplinary Reviews: Climate Change*, *6*(2), 203–223. Available from https://doi.org/10.1002/wcc.330.

Bonsdorff, E., & Osterman, C. (1985). The establishment, successions and dynamics of a zoobenthic community: An experimental study. In P. E. Gibbs (Ed.), *Proceedings of 19th European Marine Biology Symposium, 1984 Sept 16–21* (pp. 287–297). Cambridge University Press.

Bour, A., Avio, C. G., Gorbi, S., Regoli, F., & Hylland, K. (2018). Presence of microplastics in benthic and epibenthic organisms: Influence of habitat, feeding mode and trophic level. *Environmental Pollution (Barking, Essex: 1987)*, *243*, 1217–1225.

Brown, A. C. (1996). Behavioural plasticity as a key factor in the survival and evolution of the macrofauna on exposed sandy beaches. *Revista Chilena de Historia Natural*, *69*, 469–474.

Brown, A. C., & McLachlan, A. (2010). *The ecology of sandy shores* (2nd ed., p. 392) Elsevier.

Bryan, G. W., & Langston, W. J. (1992). Bioavailability, accumulation and effects of heavy metals in sediments with special reference to United Kingdom estuaries: A review. *Environmental Pollution (Barking, Essex: 1987)*, *76*(2), 89–131.

Bustamante, R. H., Branch, G. M., & Eekhout, S. (1995). Maintenance of an exceptional intertidal grazer biomass in South Africa: Subsidy by sub tidal kelps. *Ecology*, *76*(7), 2314–2329.

Catalán, A. M., Valdivia, N., & Scrosati, R. A. (2020). Interhemispheric comparison of scale-dependent spatial variation in the structure of intertidal rocky-shore communities. *Ecosphere*, *11*(6), e03068.

Chainho, P., Costa, J. L., Chaves, M. L., Lane, M. F., Dauer, D. M., & Costa, M. J. (2006). *Seasonal and spatial patterns of distribution of subtidal benthic invertebrate communities in the Mondego River, Portugal—A poikilohaline estuary*. *Marine biodiversity* (pp. 59–74). Springer.

Chappuis, E., Terradas, M., Cefalì, M. E., Mariani, S., & Ballesteros, E. (2014). Vertical zonation is the main distribution pattern of littoral assemblages on rocky shores at a regional scale. *Estuarine, Coastal and Shelf Science*, *147*, 113–122.

Clark, G. F., Stark, J. S., Palmer, A. S., Riddle, M. J., & Johnston, E. L. (2017). The roles of sea-ice, light and sedimentation in structuring shallow Antarctic benthic communities. *PLoS One*, *12*(1), e0168391. Available from https://doi.org/10.1371/journal.pone.0168391.

Crowe, T. P., Thompson, R. C., Bray, S., & Hawkins, S. J. (2000). Impacts of anthropogenic stress on rocky intertidal communities. *Journal of Aquatic Ecosystem Stress and Recovery*, *7*(4), 273–297.

Cruz-Motta, J. J., Miloslavich, P., Palomo, G., Iken, K., Konar, B., Pohle, G., Trott, T., Benedetti-Cecchi, L., Herrera, C., Hernández, A., Sardi, A., Bueno, A., Castillo, J., Klein, E., Guerra-Castro, E., Gobin, J., Isabel Gómez, D., Riosmena-Rodríguez, R., Mead, A., ... Knowlton, A. (2010). Patterns of spatial variation of assemblages associated with intertidal rocky shores: A global perspective. *PLoS One*, *5*(12), e14354.

Cunha, M. R., Moreira, M. H., & Sorbe, J. C. (2000). The amphipod *Corophium multisetosum* (Corophiidae) in Ria de Aveiro (NW Portugal). II. Abundance, biomass and production. *Marine Biology*, *137*, 651–660.

Dauwe, B., Herman, P. M. J., & Heip, C. H. R. (1998). Community structure and bioturbation potential of macrofauna at four North Sea stations with contrasting food supply. *Marine Ecology Progress Series*, *173*, 67–83.

Diaz, R. J., & Rosenberg, R. (1995). Marine benthic hypoxia: A review of its ecological effects and the behavioural responses of benthic macrofauna. *Oceanography and Marine Biology Annual Review*, *33*, 245-03.

Dittmann, S. (2002). Benthic fauna in tropical tidal flats—A comparative perspective. *Wetlands Ecology and Management*, *10*(3), 189–195.

Ebert, T. A. (1983). Recruitment in echinoderms. In M. Jangoux, & J. M. Lawrence (Eds.), *Echinoderm studies I* (pp. 169–203). AA Balkema.

Ebrahimnezhad, S., & Naderloo, R. (2019). Quantitative analyses of decapod crustaceans at a rocky intertidal shore in the Gulf of Oman, Iran: Diversity, zonation and monsoonal fluctuations. *Journal of the Marine Biological Association of the United Kingdom*, *99*(2), 445–459.

Eckman, J. E. (1979). Small-scale patterns and processes in a soft substratum, intertidal community. *Journal of Marine Research*, *37*, 437–457.

Edgar, G. J. (1990). The use of the size structure of benthic macrofaunal communities to estimate faunal biomass and secondary production. *Journal of Experimental Marine Biology and Ecology*, *137*(3), 195–214.

Fauchald, K., & Jumars, P. A. (1979). The diet of worms: A study of polychaete feeding guilds. *Oceanography and Marine Biology Annual Review*, *17*(1), 193–284.

Feebarani, J., Joydas, T. V., Damodaran, R., & Borja, A. (2016). Benthic quality assessment in a naturally- and human-stressed tropical estuary. *Ecological Indicators*, *67*, 380–390. Available from https://doi.org/10.1016/j.ecolind.2016.03.005.

Francour, P., Boudouresque, C. F., Harmelin, J. G., Harmelin-Vivien, M. L., & Quignard, J. P. (1994). Are the Mediterranean waters becoming warmer? Information from biological indicators. *Marine Pollution Bulletin*, *28*, 523–536.

Gaston, G. R., Lee, D. L., & Nasci, J. C. (1988). *Estuarine macrobenthos in Calcasieu Lake, Louisiana: Community and trophic structure. Estuaries* (11, pp. 192–200).

Giri, C., Ochieng, E., Tieszen, L. L., Zhu, Z., Singh, A., Loveland, T., Masek, J., & Duke, N. (2011). Status and distribution of mangrove forests of the world using earth observation satellite data. *Global Ecology and Biogeography*, *20*(1), 154–159.

Gopal, A., Abdul Jaleel, K. U., Parameswaran, U. V., Sanjeevan, V. N., Saramma, A. V., Vijayan, A. K., Saravanane, N., Gupta, G. V. M., & Sudhakar, M. (2020). Distinctive community patterns with exceptional diversity of Polychaetes around a tectonically active Archipelago in the Tropical Indian Ocean. *Frontiers in Marine Science*, 7, 710. Available from https://doi.org/10.3389/fmars.2020.00710.

Gray, J. S. (1974). Animal-sediment relationships. *Oceanography and Marine Biology Annual Review*, *12*, 223–261.

Guidetti, P., & Dulčić, J. (2007). Relationships among predatory fish, sea urchins and barrens in Mediterranean rocky reefs across a latitudinal gradient. *Marine Environmental Research*, *63*, 168–184.

Guidetti, P., & Mori, M. (2005). Morpho-functional defences of Mediterranean sea urchins, Paracentrotus lividus and Arbacia lixula, against fish predators. *Marine Biology*, *147*, 797–802.

Gupta, G. V. M., Sudheesh, V., Sudharma, K. V., Saravanane, N., Dhanya, V., Dhanya, K. R., Dhanya, G., Lakshmi, G., Sudhakar, M., & Naqvi, S. W. A. (2016). Evolution to decay of upwelling and associated biogeochemistry over the south eastern Arabian Sea shelf. *Journal of Geophysical Research: Biogeosciences*, *121*(1), 159–175.

Gutt, J. (2001). On the direct impact of ice on marine benthic communities, a review. *Polar Biology*, *24*, 553–564. Available from https://doi.org/10.1007/s003000100262.

Hall, S. J., Basford, D. J., Robertson, M. R., Raffaelli, D. G., & Tuck, I. (1991). Patterns of recolonisation and the importance of pit-digging by the crab Cancer pagurus in a subtidal sand habitat. *Marine Ecology Progress Series*, *72*, 93–102.

Harkantra, S. N., Nair, A., Ansari, Z. A., & Parulekar, A. H. (1980). Benthos of the shelf region along the west coast of India. *Indian Journal of Geo-Marine Sciences*, *9*, 106–110.

Harley, C. D. (2008). Tidal dynamics, topographic orientation, and temperature-mediated mass mortalities on rocky shores. *Marine Ecology Progress Series*, *371*, 37–46.

Harvey, M., Gauthier, D., & Munro, J. (1998). Temporal changes in the composition and abundance of the macrobentic invertebrate communities at dredged material disposal sites in the Anse à Beaufils, Baie des Charleurs, eastern Canada. *Marine Pollution Bulletin*, *36*(1), 41–55.

Heip, C., Goosen, N. K., Herman, P. M. J., Kromkamp, J., Middelburg, J. J., & Soetaert, K. (1995). Production and consumption of biological particles in temperate tidal estuaries. *Oceanography and Marine Biology Annual Review*, *33*(1), 1–149.

Hemalatha, A., Ansari, K. G. M. T., Rajasekaran, R., & Fernando, O. J. (2014). Diversity of infaunal macrobenthic community in the intertidal zone of Vellar estuary (Southeast coast of India). *International Journal of Marine Science*, *4*(47), 1–11. Available from http://ijms.biopublisher.ca.

Herman, P. M. J., Middelburg, J. J., van de Koppel, J., & Heip, C. H. R. (1999). Ecology of estuarine macrobenthos. *Advances in Ecological Research*, *29*, 195–240.

Hillebrand, H. (2004). On the generality of the latitudinal diversity gradient. *The American Naturalist*, *163*, 192–211.

Hily, C., Le, Loc'h, F., Grall, J., & Glemarec, M. (2008). Soft bottom macrobenthic communities of North Biscay revisited: Long-term evolution under fisheries-climate forcing. *Estuarine, Coastal and Shelf Science*, *78*, 413–425.

Hogue, E. W., & Miller, C. B. (1981). Effects of sediment microtopography on small-scale spatial distributions of meiobenthic nematodes. *Journal of Experimental Marine Biology and Ecology*, *53*, 181–191.

Holland, A. F. (1985). Long-term variation of macrobenthos in a mesohaline region of Chesapeake Bay. *Estuaries*, *8*, 93–113.

Horne, J. K., & Schneider, D. C. (1995). Spatial variance in ecology. *Oikos*, *74*, 18–26.

Huston, M. A. (1985). Patterns of species diversity on coral reefs. *Annual Review of Ecology, Evolution, and Systematics*, *16*(1), 149–177.

Hyndes, G. A., Lavery, P. S., & Walker, D. I. (2005). Does transported seagrass provide an important trophic link in unvegetated, nearshore areas? *Estuarine, Coastal and Shelf Science*, *63*(1), 633–643.

Ieno, E. N., & Bastida, R. O. (1998). Spatial and temporal patterns in coastal macrobenthos of Samborombon Bay, Argentina: A case study of very low diversity. *Estuaries*, *21*, 690–699. Available from https://doi.org/10.2307/1353273.

Iken, K., Konar, B., Benedetti-Cecchi, L., Cruz-Motta, J. J., & Knowlton, A. (2010). Large-scale spatial distribution patterns of echinoderms in nearshore rocky habitats. *PLoS One*, *5*(11), e13845. Available from https://doi.org/10.1371/journal.pone.0013845.

James, R. J., & Fairweather, P. G. (1996). Spatial variation of intertidal macrofauna on a sandy ocean beach in Australia. *Estuarine, Coastal and Shelf Science*, *43*(1), 81–107.

Jayaraj, K. A., Jayalakshmi, K. V., & Saraladevi, K. (2007). Influence of environmental properties on macrobenthos in the northwest Indian shelf. *Environmental Monitoring and Assessment*, *127*(1-3), 459–475.

Jayaraj, K. A., Sheeba, P., Jacob, J., Revichandran, C., Arun, P. K., Praseeda, K. S., Nisha, P. A. & Rasheed, K. A. (2008). Response of infaunal macrobenthos to the sediment granulometry in a tropical continental margin-southwest coast of India. *Estuarine, Coastal and Shelf Science*, 77, 743–754.

Joydas, T. V., & Damodaran, R. (2009). Infaunal macrobenthos along the shelf waters of the west coast of India, Arabian Sea. *Indian Journal of Geo-Marine Sciences*, *38*, 191–204.

Jumars, P. A. (1993). *Concepts in biological oceanography: An interdisciplinary primer* (p. 358). Oxford University Press.

Kendall, M. A., & Widdicombe, S. (1999). Small scale patterns in the structure of macrofaunal assemblages of shallow soft sediments. *Journal of Experimental Marine Biology and Ecology*, *237*(1), 127–140.

Kirby, R. R., Beaugrand, G., & Lindley, J. A. (2008). Climate-induced effects on the meroplankton and the benthic-pelagic ecology of the North Sea. *Limnology and Oceanography*, *53*(5), 1805–1815.

La Valle, F. F., Schaefer, J. L. B., Cox, T. E., & Philippoff, J. (2020). Spatial and temporal patterns in Hawai'i's intertidal: Decadal changes in benthic community composition. *Hydrobiologia*, *847*(15), 3321–3336.

Lercari, D., & Defeo, O. (2006). Large-scale diversity and abundance trends in sandy beach macrofauna along full gradients of salinity and morphodynamics. *Estuarine, Coastal and Shelf Science*, *68*(1-2), 27–35.

Levin, L. A. (2003). Oxygen minimum zone benthos: Adaption and community response to hypoxia. *Oceanography and Marine Biology Annual Review*, *41*, 1–45.

Levin, L. A., Ekau, W., Gooday, A., Jorissen, F., Middelburg, J., Naqvi, S., Neira, C., Rabalais, N. N., & Zhang, J. (2009). Effects of natural and human-induced hypoxia on coastal benthos. *Biogeosciences Discussions*, *6*, 3563–3654.

Levin, S. A. (1992). The problem of pattern and scale in ecology. *Ecology*, *73*, 1943–1967.

Li, J., Vincx, M., Herman, P. M. J., & Heip, C. (1997). Monitoring meiobenthos using cm-, m- and km-scales in the Southern Bight of the North Sea. *Marine Environmental Research*, *43*, 265–278.

Lipkin, Y., & Safriel, U. (1971). Intertidal zonation on rocky shores at Mikhmoret (Mediterranean, Israel). *Journal of Ecology*, *59*(1), 1–30.

Lu, L., Grant, J., & Barrell, J. (2008). Macrofaunal spatial patterns in relationship to environmental variables in the Richibucto estuary, New Brunswick, Canada. *Estuaries and Coasts*, *31*, 994–1005. Available from https://doi.org/10.1007/s12237-008-9097-9.

Mannino, A., & Montagna, P. A. (1997). Small-scale spatial variation of macrobenthic community structure. *Estuaries*, *20*(1), 159–173.

Marques, J. C., Maranhão, P., & Pardal, M. A. (1993). Human impact assessment on the subtidal macrobenthic community structure in the Mondego Estuary (Western Portugal). *Estuarine, Coastal and Shelf Science*, *37*(4), 403–419. Available from https://doi.org/10.1006/ecss.1993.1064.

McClanahan, T. R., & Shafir, S. H. (1990). Causes and consequences of sea urchin abundance and diversity in Kenyan coral reef lagoons. *Oecologia*, *83*, 362–370.

Medeiros, C. R., Hepp, L. U., Patrício, J., & Molozzi, J. (2016). Tropical estuarine macrobenthic communities are structured by turnover rather than nestedness. *PLoS One*, *11*(9), e0161082. Available from https://doi.org/10.1371/journal.pone.0161082.

Menon, N. N., Balchand, A. N., & Menon, N. R. (2000). Hydrobiology of the Cochin backwater system—A review. *Hydrobiologia*, *430*, 149–183.

Metaxas, A., & Scheibling, R. E. (1993). Community structure and organization of tidepools. *Marine Ecology Progress Series*, *98*, 187–198.

Michel, J., Fegley, S. R., Dahlin, J. A., & Wood, C. (2017). Oil spill response-related injuries on sand beaches: When shoreline treatment extends the impacts beyond the oil. *Marine Ecology Progress Series*, *576*, 203–218.

Mieszkowska, N., Leaper, R., Moore, P., Kendall, M. A., Burrows, M. T., Lear, D., Poloczanska, E. S., Hiscock, K., Moschella, P., Thompson, R. C., Herbert, R. J. H., Laffoley, D., Baxter, J., Southward, A. J., Hawkins, S. J. (2006). Marine biodiversity and climate change: Assessing and predicting the influence of climatic change using intertidal rocky shore biota. Occasional Publication of the Marine Biological Association 20. Plymouth (UK), Marine Biological Association of the United Kingdom.

Mulik, J., Sukumaran, S., & Srinivas, T. (2020). Factors structuring spatio-temporal dynamics of macrobenthic communities of three differently modified tropical estuaries. *Marine Pollution Bulletin*, *150*, 110767.

Musale, A. S., & Desai, D. V. (2011). Distribution and abundance of macrobenthic polychaetes along the South Indian coast. *Environmental Monitoring and Assessment*, *178*(1-4), 423–436.

Naidu, S. A., Rao, G. D., Vardhan, K. V., Kumaraswami, M., Rao, V. R., & Ramu, K. (2018). Impact of natural and Anthropogenic disturbances on the benthic ecological quality of the coastal waters of Kochi, southwest coast of India. *International Journal of Environmental Research*, *12*(6), 861–875.

Nanami, A., Saitoa, H., Akitab, T., Motomatsuc, K., & Kuwaharaa, H. (2005). Spatial distribution and assemblage structure of macrobenthic invertebrates in a Brackish Lake in relation to environmental variables. *Estuarine, Coastal and Shelf Science, 63*, 167–176. Available from https://doi.org/10.1016/j.ecss.2004.11.004.

Nayak, G. N., Volvoikar, S. P., & Hoskatta, T. (2018). Changing depositional environment and factors controlling the growth of mudflat in a tropical estuary, west coast of India. *Environmental Earth Sciences*, 77(21), 741.

Neves, L. P. D., da Silva, P. D. S., & Bemvenuti, C. E. (2008). Temporal variability of benthic macrofauna on Cassino beach, southernmost Brazil. *Iheringia Série Zoologia, 98*(1), 36–44.

Nybakken, J. W., & Bertness, M. D. (2004). *Marine biology: An ecological approach* (p. 592) Pearson.

Oglesby, L. C. (1969). Salinity-stress and desiccation in intertidal worms. *American Zoologist, 9*(2), 319–331.

Olsson, J., Bergström, L., & Gårdmark, A. (2013). Top-down regulation, climate and multi-decadal changes in coastal zoobenthos communities in two Baltic Sea areas. *PLoS One, 8*(5), e64767. Available from https://doi.org/10.1371/journal.pone.0064767.

Ourives, T. M., Rizzo, A. E., & Boehs, G. (2011). Composition and spatial distribution of the benthic macrofauna in the Cachoeira River estuary, Ilhéus, Bahia, Brazil. *Revista de Biología Marina y Oceanografía, 46*(1), 17–25.

Paiva, P. C. (2001). Spatial and temporal variation of a near-shore benthic community in southern Brazil: Implications for the design of monitoring programs. *Estuarine, Coastal and Shelf Science, 52*, 423–433.

Pancucci, P., Panayotidis, P., & Zenetos, A. (1993). Morphological changes in sea urchin populations as a response to environmental stress. In J. C. Aldrich (Ed.), *Quantified phenotypic responses in morphology and physiology* (pp. 247–257). Ashford JAPAGA.

Parameswaran, U. V., Jaleel, K. A., Sanjeevan, V. N., Gopal, A., Vijayan, A. K., Gupta, G. V. M., & Sudhakar, M. (2018). Diversity and distribution of echinoderms in the South Eastern Arabian Sea shelf under the influence of seasonal hypoxia. *Progress in Oceanography, 165*, 189–204.

Pereira, T. J., Gingold, R., Villegas, A. D. M., & Rocha-Olivares, A. (2018). Patterns of spatial variation of meiofauna in sandy beaches of northwestern Mexico with contrasting levels of disturbance. *Thalassas: International Journal of Marine Science, 34*(1), 53–63.

Peterson, C. H. (1991). Intertidal zonation of marine invertebrates in sand and mud. *American Scientist, 79*(3), 236–249.

Piacenza, S. E., Barner, A. K., Benkwitt, C. E., Boersma, K. S., Cerny-Chipman, E. B., & Ingeman, K. E. (2015). Patterns and variation in benthic biodiversity in a large marine ecosystem. *PLoS One, 10*(8), e0135135. Available from https://doi.org/10.1371/journal.pone.0135135.

Pihl, L., Modin, J., & Wennhage, H. (2005). Relating plaice (*Pleuronectes platessa*) recruitment to deteriorating habitat quality: Efects of macroalgal blooms in coastal nursery grounds. *Canadian Journal of Fisheries and Aquatic Sciences, 62*(5), 1184–1193.

Pineda-Metz, S. E. A., Gerdes, D., & Richter, C. (2020). Benthic fauna declined on a whitening Antarctic continental shelf. *Nature Communication, 11*, 2226. Available from https://doi.org/10.1038/s41467-020-16093-z.

Poloczanska, E. S., Hawkins, S. J., Southward, A. J., & Burrows, M. T. (2008). Modeling the response of populations of competing species to climate change. *Ecology, 89*(11), 3138–3149.

Pritchard, D. W. (1967). What is an estuary: Physical viewpoint. In G. H. Lauff (Ed.), Estuaries. (Vol. 83, pp. 3–5). American Association for the Advancement of Science Publication.

Queiroga, H. (1990). *Corophium multisetosum (Amphipoda: Corophiidae) in Canal de Mira, Portugal: Some factors that affect its distribution. Marine Biology* (104, pp. 397–402).

Quijón, P., & Jaramillo, E. (1993). Temporal variability in the intertidal macroinfauna in the Queule River Estuary, South-central Chile. *Estuarine, Coastal and Shelf Science, 37*, 655–667.

Quillien, N., Nordström, M. C., Le Bris, H., Bonsdorff, E., & Grall, J. (2018). Green tides on inter-and subtidal sandy shores: Differential impacts on infauna and flatfish. *Journal of the Marine Biological Association of the United Kingdom, 98*(4), 699.

Reddin, C. J., Docmac, F., O'Connor, N. E., Bothwell, J. H., & Harrod, C. (2015). Coastal upwelling drives intertidal assemblage structure and trophic ecology. *PLoS One, 10*(7), e0130789.

Rehitha, T. V., Madhu, N. V., Vineetha, G., Vipindas, P. V., Resmi, P., & Revichandran, C. (2019). Spatio-temporal variability in macrobenthic communities and trophic structure of a tropical estuary and its adjacent coastal waters. *Environmental Monitoring and Assessment, 191*(6), 341.

Rehitha, T. V., Ullas, N., Vineetha, G., Benny, P. Y., Madhu, N. V., & Revichandran, C. (2017). Impact of maintenance dredging on macrobenthic community structure of a tropical estuary. *Ocean & Coastal Management, 144*, 71–82.

Rosa, L. C., & Bemvenuti, C. E. (2006). Temporal variability of the estuarine macrofauna of the Patos lagoon, Brazil. *Revista de Biologia Marina y Oceanografia, 41*(1), 1–9.

Roy, K., Jablonski, D., & Valentine, J. W. (2000). Dissecting latitudinal diversity gradients: Functional groups and clades of marine bivalves. *Proceedings of the Royal Society B: Biological Sciences, 267*, 293–299.

Rozas, L. P., Minello, T. J., Munuera-Fernández, I., Fry, B., & Wissel, B. (2005). Macrofauna distributions and habitat change following winter-spring release of freshwater into the Breton Sound estuary, Lousiana (USA). *Estuarine, Coastal and Shelf Science, 65*(1), 319–336.

Sajan, S., Joydas, T. V., & Damodaran, R. (2010). Meiofauna of the western continental shelf of India, Arabian Sea. *Estuarine, Coastal and Shelf Science, 86*(4), 665–674.

Sala, E., Boudouresque, C. F., & Harmelin-Vivien, M. (1998). Fishing, trophic cascades, and the structure of algal assemblages: Evaluation of an old but untested paradigm. *Oikos, 82*, 425–439.

Salen-Picard, C., Darnaude, A. M., Arlhac, D., & Harmelin–Vivien, M. L. (2002). Fluctuations of macrobenthic populations: A link between climate driven river run-–off and sole fishery yields in the Gulf of Lions. *Oecologia, 133*, 380–388.

Sanders, H. L. (1968). Marine benthic diversity: A comparative study. *The American Naturalist, 102*, 243–282.

Schlacher, T. A., & Thompson, L. (2013). Spatial structure on ocean-exposed sandy beaches: Faunal zonation metrics and their variability. *Marine Ecology Progress Series, 478*, 43–55.

Sebens, K. P. (1991). Habitat structure and community dynamics in marine benthic systems. In S. S. Bell, E. D. McCoy, & H. R. Mushinsky (Eds.), *Habitat structure. The physical arrangement of objects in space* (p. 438). Chapman & Hall.

Seidel, M., Beck, M., Greskowiak, J., Riedel, T., Waska, H., Suryaputra, I. N., Schnetger, B., Niggemann, J., Simon, M., & Dittmar, T. (2015). Benthic-pelagic coupling of nutrients and dissolved organic matter composition in an intertidal sandy beach. *Marine Chemistry, 176*, 150–163.

Shou, L., Huang, Y., Zeng, J., Gao, A., Liao, Y., & Chen, Q. (2009). Seasonal changes of macrobenthos distribution and diversity in Zhoushan sea area. *Aquatic Ecosystem Health & Management, 12*(1), 110–115.

Sivadas, S. K., Singh, D. P., & Saraswat, R. (2020). Functional and taxonomic (α and β) diversity patterns of macrobenthic communities along a depth gradient

(19–2639 m): A case study from the southern Indian continental margin. *Deep Sea Research Part I: Oceanographic Research Papers, 159*, 103250.

Smith, K. J., & Able, K. W. (2003). Dissolved oxygen dynamics in salt marsh pools and its potential impacts on fish assemblages. *Marine Ecology Progress Series, 258*, 223–232.

Snelgrove, P. V. R., & Butman, C. A. (1994). Animal-sediment relationships revisited: Cause vs effect. *Oceanography and Marine Biology Annual Review, 32*(1), 111–177.

Sousa, R., Antunes, C., & Guilhermino, L. (2007). Species composition and monthly variation of the Molluscan fauna in the freshwater subtidal area of the River Minho estuary. *Estuarine, Coastal and Shelf Science, 75*, 90–100.

Srinivas, T., Sukumaran, S., Neetu, S., & Ramesh Babu, K. (2020). Diversity and functional patterns of benthic amphipods in the coralline intertidal zones of a Marine National Park, India. *Frontiers in Marine Science*, 7, 589195. Available from https://doi.org/10.3389/fmars.2020.589195.

Starr, M., Himmelman, J., & Therriault, J. C. (1993). Environmental control of green sea urchin, *Strongylocentrotus droebachiensis*, spawning in the St. Lawrence Estuary. *Canadian Journal of Fisheries and Aquatic Sciences, 50*, 894–901.

Suchanek, T. H. (1995). Impacts of mercury on benthic invertebrate populations and communities within the aquatic ecosystem of Clear Lake, California. In D. B. Porcella, J. W. Huckabee, & B. Wheatley (Eds.), *Mercury as a global pollutant*. Springer. Available from https://doi.org/10.1007/978-94-011-0153-0-103.

Thrush, S. F. (1991). Spatial patterns in soft-bottom communities. *Trends in Ecology & Evolution, 6*, 75–79.

Thrush, S. F., Lawrie, S. M., Hewitt, J. E., & Cummings, V. J. (1999). The problem of scale: Uncertainties and implications for soft bottom marine communities and the assessment of human impacts. In J. S. Gray, W. G. Ambrose, & A. Szaniawska (Eds.), *Biogeochemical cycling and sediment ecology* (pp. 195–210). Kluwer Academic Publishers.

Trueblood, D. D. (1991). Spatial and temporal effects of terebellid polychaete tubes on soft-bottom community structure in Phosphorescent Bay, Puerto Rico. *Journal of Experimental Marine Biology and Ecology, 149*, 139–159.

Tsujino, M., Hori, M., Okuda, T., Nakaoka, M., & Yamamoto, T. (2010). Distance decay of community dynamics in rocky intertidal sessile assemblages evaluated by transition matrix models. *Population Ecology, 52*, 171–180.

Underwood, A. J., Chapman, M. G., & Connell, S. D. (2000). Observations in ecology: You can't make progress on processes without understanding the patterns. *Journal of Experimental Marine Biology and Ecology, 250*, 97–115.

Valdivia, N., Scrosati, R. A., Molis, M., & Knox, A. S. (2011). Variation in community structure across vertical intertidal stress gradients: How does it compare with horizontal variation at different scales? *PLoS One, 6*(8), e24062.

Van Blaricom, G. R. (1982). Experimental analysis of structural regulation in a marine sand community exposed to oceanic swell. *Ecological Monographs, 52*, 283–285.

Veiga, P., Rubal, M., & Sousa-Pinto, I. (2014). Structural complexity of macroalgae influences epifaunal assemblages associated with native and invasive species. *Marine Environmental Research, 101*, 115–123.

Vidolin, D., Santos-Gouvea, I. A., & Freire, C. (2007). Differences in ion regulation in the sea urchins *Lytechinus variegatus* and *Arbacia lixula* (Echinodermata: Echinoidea) *Journal of the Marine Biology Association of the United Kingdom, 87*, 769–775.

Wang, H. Y., Tsang, L. M., Lima, F. P., Seabra, R., Ganmanee, M., Williams, G. A., & Chan, B. K. (2020). Spatial variation in thermal stress experienced by barnacles on rocky shores: The interplay between geographic variation, tidal cycles and microhabitat temperatures. *Frontiers in Marine Science*, 7, 1–14.

Wethey, D. S., & Woodin, S. A. (2008). *Ecological hind casting of biogeographic responses to climate change in the European intertidal zone. Challenges to marine ecosystems* (pp. 139–151). Springer

Williams, G. A., & Morritt, D. (1995). Habitat partitioning and thermal tolerance in a tropical limpet, Cellana grata. *Marine Ecology Progress Series*, *124*, 89–103.

Ysebaert, T., Herman, P. M. J., Meire, P., Craeymeersch, J., Verbeek, H., & Heip, C. H. R. (2003). Large scale spatial patterns in estuaries: Estuarine macrobenthic communities in the Schelde estuary, NW Europe. *Estuarine, Coastal and Shelf Science*, *57*, 335–355.

Ysebaert, T., Meire, P., Herman, P. M. J., & Verbeek, H. (2002). Macrobenthic species response surfaces along estuarine gradients: Prediction by logistic regression. *Marine Ecology Progress Series*, *225*, 79–95.

Ysebaert, T., Meire, P., Maes, D., & Buijs, J. (1993). The benthic macrofauna along the estuarine gradient of the Schelde estuary. *Netherlands Journal of Aquatic Ecology*, *27*, 327–341.

Zajac, R. N., & Whitlatch, R. B. (1982). Responses of estuarine infauna to disturbance. II. Spatial and temporal variation of succession. *Marine Ecology Progress Series*, *10*, 15–27.

Zamprogno, G. C., Fernandes, F. C., & Fernandes, L. L. (2012). Temporal and spatial variation of rocky shores intertidal benthic communities in Southeast Brazil. *Iheringia Série Zoologia*, *102*(4), 375–383.

CHAPTER 9

Benthos and its interaction with marine and estuarine ecosystem

V. Sivapriya, K. Radhakrishnan and S.M. Hussain
Department of Geology, School of Earth and Atmospheric Sciences, University of Madras, Chennai, India

9.1 Introduction

Benthos is a term collectively used for organisms living in or on the bottom sediments of any aquatic system and includes both producers (phytobenthos) and consumers (zoobenthos). They are one of the structuring elements of the food web, which plays an important role in the system dynamics (Herman et al., 1999) and a key element of many marine and estuarine monitoring programs. Benthic animals can be defined based on the size or life habitat. Epibenthos and hyperbenthos are terms used to denote mobile benthic organisms that move freely between the sediments and the overlying water column. Endobenthos on the other hand denotes the organisms that dwell within the sediments; more specifically infaunal benthos burrow into the sediments while epifaunal benthos attaches themselves to hard substrates such as rocks, roots, or manmade structures. According to their size, benthic organisms are classified as microbenthos, meiobenthos, and macrobenthos. Microbenthos ($<32\ \mu m$) are usually protists and bacterial species which live in guilds simultaneously in the sediments (with a density of 10^9 cells/mL of sediments) and are metabolically dependent on each other. Meiobenthos (from $32\ \mu m$ to 1 mm) are usually dominated by nematodes and other dominant groups of Copepoda, Ostracoda, Turbellaria, and Foraminifera, with densities in the order of $10^6\ m^{-2}$ (Heip et al., 1995). These larger microbes, which include foraminiferans, turbellarians, and polychaetes, frequently dominate benthic food chains, filling the roles of nutrient recycler, decomposer, primary producer, and predator. Macrobenthos (>1 mm) comprises Mollusca, Polychaeta, Echinodermata, Crustacea, and other groups.

Ecology and Biodiversity of Benthos
DOI: https://doi.org/10.1016/B978-0-12-821161-8.00008-8

9.2 Benthos and their interaction with the ecosystem

The primary factors as such substrate, depth, temperature, dissolved oxygen, salinity, pH, food control, and habitat are all highly interconnected and possess a direct effect on the distribution and abundance of the organism (McIntire et al., 1996).

9.2.1 Response of the benthic organism to substrate changes

The substrate, being the physical habitat of the benthic community can influence and decide the assemblage that may establish. The mode of living has a direct relation with the substrate as infaunal benthos needs a softer substrate to enable easy borrowing while the epifaunal benthos needs a harder substrate for attachment. Sediment type is frequently related to the benthic communities and there is a long history of research pertaining to them. While many species are found over a wide range of sediment types, most display a clear affinity for a much narrower range. The relationship between the sediments and benthic communities may be due to the direct differences in the architecture and pore space of sediment particles or to the strong indirect effect of the chemical gradients within the sediments (Snelgrove & Butman, 1994).

Sediment type at a given location is considerably stable relative to the variability inherent in the water column and is often assigned to the category, that is, rock, sand, mud, etc. The nature of the sediments is also linked closely with the hydrographic dynamics and in particular by the relationship between energy and sediment particle size; the higher energy leads to the transportation of the large particles as well while the lower energy conditions are least transportive. This relationship is not so simple as the electrostatic attraction (cohesive force) between the particles also acts as a barrier to the fine (clayey and muddy) sediments from resuspension which requires slightly more energy. Sediments also serve as the repository of pollutants which makes sediment geochemistry acquire its higher importance in the study of risk assessment and environmental health (Table 9.1).

9.2.2 Habitat–biota–benthic relationships

The biota themselves provides stability or modification of the habitat Benthic communities often are found to cherish along with the vegetative ecosystems such as seagrass beds, mangroves, estuarine, reef systems, etc Extreme variation in salinity may be one of the major influential factors in determining plant biomass and animal densities. The fluctuation in salinity in any area could cause fluctuation in both the distribution and total

Table 9.1 Functional species trait categories which were expected to be most affected by pulsed sedimentation and the predicted mechanisms for effects on particular trait modalities.

Trait category	Possible mechanisms for any observed sedimentation effect
Relative size	Smaller animals may be more able to take advantage of refugia from sedimentation at microhabitat scales.
Body form	Spherical animals may be least able to avoid smothering, whereas cylindrical animals (e.g., worms) may be able to burrow into sediment.
Habitat preference	Near surface-dwelling taxa are likely to be more affected than deeper-burrowing taxa. Animals needing attachment sites are likely to be most affected by the coating of stable surfaces in unstable fine sediments.
Feeding habits	Filter-feeders are likely to be most affected by increased suspended sediment, which may clog their feeding structures. Grazers and deposit-feeders may be most affected by the smothering of their food sources. Predators and scavengers may benefit from decreased fitness of prey, although prey may be harder to find as a result of higher turbidity.
Dietary specificity	Dietary specialists are most likely to be affected by reductions in their specific food sources.
Mobility	More mobile animals are able to move away initially and move back quickly (likely to reduce densities of these animals initially, followed by rapid recovery as fine sediment is dispersed).

Source: After Reid, D.J., Chiaroni, L.D., Hewitt, J.E., Lohrer, D.M., Matthaei, C.D., Phillips, N.R., . . ., van Houte-Howes, K.S. (2011). Sedimentation effects on the benthos of streams and estuaries: A cross-ecosystem comparison. Marine and Freshwater Research, 62(10), 1201–1213 (Reid et al., 2011).

abundance of benthic vegetation thriving therein. Intolerance of sudden changes in salinity by some species of seagrass affects the benthic habitat and hence affects the benthic vegetation and the associated faunal assemblages (Montague & Ley, 1993).

9.2.3 Temperature and depth relationship with benthos

Water temperature plays a substantial role in the aquatic environment. Sunlight, atmospheric heat transfer, turbidity, the confluence of water bodies, and anthropogenic activities are the factors that affect the temperature of any water body. The depth of the water columns decides the amount of insolation it receives from the Sun, the primary source of heat. Based on this criterion, the water mass is divided into photic and aphotic zones; the photic zones are the upper region of the water mass that receives the maximum heat from the sun and the aphotic zone is the zone of deeper waters which is protected from the sun's heatwave. This implies that the depth and temperature are interconnected and that the depth is inversely proportional to the temperature.

Depth also has an influence on the type of sediment as the energy (which decides the sediment type) varies with the depth. The shallow nearshore regions and the mouth of the estuary are much agitated and high-energy environments while the regions of deeper ocean and lakes are calmer and low-energy environments. It plays a vital role in determining the amount of dissolved oxygen and the geochemistry of the water.

9.2.4 Salinity variation and their relationship with benthos

Salinity is simply the measure of dissolved salts in the water. Plants or animals are often sensitive to changes in salinity and the salinity level controls the local species composition. As salinity changes, the population of some species may decline or disappear while other species begin to thrive. However, frequent changes in salinity could be harsh on the organisms and may result in no species doing well. Interestingly, the predators and the parasites are frequently less tolerant to the variations in salinity than their prey or hosts. Hence the organism that adapts itself to the variability of the salinity escapes the predatory communities and may thus reduce the biological enemies.

The salinity of the water body affects the affluence and composition of the species assemblages. There exist different floral and faunal assemblages constrained to the waters of different salinity. The salinity-induced pH or alkalinity of the water affects the morphology, reproduction, and survival of the organisms. The freshwater assemblage would be different from the brackish water assemblage which in turn would differ from the marine assemblages. Interestingly, the assemblage found in any aquatic environment would serve as a proxy for determining the salinity of the water. This enables the study of a few meiofaunal communities which suggests them as an excellent proxy for paleosalinity and paleoenvironmental studies (Barbieri and Vaiani, 2018).

9.2.5 Relationships of benthos to changing dissolve oxygen (DO), total carbon (TC), and organic matter (OM)

Organic matter (OM) is decomposed aerobically by bacteria near the surface of the sediment wherein the oxygen is abundant. However, the consumption of oxygen is deprived along with the deeper layers of sediments and hence forms an anaerobic environment. The thickness of the oxygenated layers depends directly on the permeability of the sediments which in turn is determined by the grain size. Organisms have the tendency to evolve in various ways to cope with the lack of oxygen. Some anaerobes release hydrogen sulfide, ammonia, and other toxic reduced ions through metabolic processes. This biota, made up primarily of microorganisms, metabolizes sulfur. Infaunal organisms oxygenate their dwelling place by generating a respiratory current along with their burrow systems. The consequent "irrigation" of burrow systems can create oxygen and nutrient fluxes that stimulate the production of benthic producers (e.g., diatoms) (Kingsford & Battershill, 1998). Some meiofauna lives at the depth in the oxygenated halo around burrows of large macrofauna, some display body shapes that are longer and/or thinner to enhance diffusion or are long enough to extend into the oxic zone, and some appear to tolerate anaerobic conditions but migrate to the oxygenated surface for gas exchanges. Many factors influence the oxygen availability to the benthos and the faunal response to the reduced oxygen concentrations. The dissolved oxygen in combination with other factors such as organic carbon content, sediment size, and the station depth structure the composition of the benthos assemblage (Long et al., 2007). Benthic diversity and abundances are highly influenced by the dissolved oxygen conditions of the medium. The statistical analyses strongly indicated that many of the valued properties of the benthic assemblages (i.e., abundance and diversity) are considerably lower at stations with the lowest dissolve oxygen (DO) concentrations.

The grain size and $C_{org}\%$ are often linked, as the fineness of the sediment grain increases the surface area available for the OM to adhere, thus leading to better preservation, and vice versa (Burdige, 2007). The deficiency in food availability in the coarse-grained sediments restrains the proliferation of the benthic diversity as well as population.

9.2.6 Food control—benthic relationships

The marine ecosystem shares a cyclic flow of energy. A food web illustrates the energy transferred between the organisms and the hierarchy of organisms occupy different levels based on their prey and feeding habits. The feeding habits of the Benthos enable the grouping into functional groups such as deposivores, suspensivores, grazer, predators, and omnivores. The basic concepts of the food web evolved into two types in the aquatic environment, that is, the first one based on the plankton primary production and the second based on the detritus primary production. Benthic organisms are much dependent on the latter type. Nutrient levels impact the food chain depending on the number of grazers present and the number of trophic levels that exist. Grazer and predators form the backbone of the food web and influence the ecosystem in various ways. Grazers on inorganic nutrients such as nitrogen and phosphorus like primary producers exert a bottom-up force. The number of grazers in any food chain is regulated by the number of predators. Predators keep grazers from depleting resources and thus the role of predators is crucial to the survival of the ecosystem. Modern food webs have prioritized the importance of autotrophs and heterotrophs . Deposivores consume particulate matter and assimilate carbon from the detritus sediments, which is particularly a mixture of OM and sedimentary minerals. The term detritus also includes the remains of other living and dead biota and comprises feces, decomposers, and microorganisms such as algae, bacteria, and fungi that break down the dead materials and feces and are either unattached or attached to the sediments. Suspensivores on the other hand are filter feeders which extract their food from the materials suspended through the water column including phytoplanktons and zooplanktons.

It is concluded from decades of researches that the major found source of meiofaunal benthic communities is directly linked to the (DOM) concentration of the water. DOM is adsorbed by the amorphous particles and then consumed directly. DOM also indirectly keeps the food cycle intact by increasing the biomass productivity of the environment. The predatory pressure is also comparatively high along with the shallow water shelf environment, as the macro and megabenthic population is greater here (Ingole et al., 2010).

9.2.7 Relationship of the benthic organisms to biomass and secondary productivity changes

Benthic communities play significant roles in the flow of energy, in the cycling of nutrients, and trophic transfer. Another principal role of benthos is to mediate the transfer from lower trophic levels and thus the energy transferred depends a great deal on the input of energy to the benthos. Biomass is a biased measure of comparison where benthic organism may be many times larger than their planktonic counterparts, but productivity is similar.

A prominent contributing factor to the high spatial and temporal variability of the benthos is the ability of the juveniles and the adults of many soft-sediment species to disperse. This in turn is controlled by the hydrodynamic regime of the region. Juvenile bivalves may be suspended and deposited by the currents as an aid in finding the preferred habitat. For example, the hydrodynamic regime required for the transportation of certain bivalve (such as *Macoma balthica*) juveniles is similar to that which produces sediments with optimum particle size needed to promote the fitness of the species. The distribution and abundance of the meiofauna, which mostly lacks the dispersing larvae, are directed influenced by the hydrodynamic regime. The emergence of juveniles or adults into the water column could be passive (by erosion) or active (associated with mate finding, feeding, or avoidance of harsh environmental conditions). Besides, the abundance of meiofauna in sediments has been related to the flow of the water as such the highest densities are found at the slack tides (as some meiofauna have emerged during the times of high flow). Even the location of re-entry of emerged juveniles into the sediment is subject to the hydrodynamic constraints (e.g., pits and depressions are high in density with all small particles and meiofauna accumulation).

9.2.8 Human impact on benthic relationships

We have a long history of human exploitation of water bodies. Increasing anthropogenic activities such as fishing, pollution, aquaculture, freshwater withdrawal, reclamation or infilling of the water bodies, system management, etc., has resulted in the direct or indirect impact on the fauna such as removal or alteration of habitat, salinity changes, enhancement and breeding of selective species, cointroduction of exotics, and hence risking the species diversity and distribution. In general, studies have shown that crustaceans and echinoderms are intolerant of most kinds of sediment contamination while other smaller deposit feeders such as annelids and nematodes are most tolerant. Tolerance to contaminants may differ greatly with species; some species are very sensitive while others may be quite tolerant. As a result, pollution may bring about community change due to indirect effects, which causes the tolerant species to change in abundance because competitors, predators, or prey are affected by pollutants (Fleeger et al., 2003). Hence studying the effects of anthropogenic activities on the species takes an important stand in understanding the relationship each species has with the environment.

9.3 Relationship between benthos and marine environment

Marine ecology on the other hand is an open, dynamic system with varied water depth. The dynamicity of the marine system is remarkable as the oceans and seas of the world have changed significantly in the past 600 million years. The shape of the ocean has also been altered by the sea-level changes throughout history. Being a heterogeneous system, marine organisms are not evenly distributed throughout the oceans. Variations in the characteristics of the marine environment create different habitats and thus influence the type of organisms that inhabit them. The water depth, proximity to land, availability of light, the difference in salinity, dissolved oxygen, OM, and the complexity of the topography are all factors affecting marine habitats. The physical and chemical properties of the seawater vary in accordance with the depth, latitude, freshwater input, proximity to the land, and hydrological factors. These physical and chemical properties of the seawater have a greater effect on the distribution and abundance of the organism.

The size of the sediment affects the distribution of the marine benthos. As the marine realm shows remarkable variation in the sediment type which in turn depends on the combination of factors such as water depth, ocean currents, tidal forces, the topography of the sea bed, turbidity, etc., the species abundance, distribution and morphology vary in accordance with each zone. It is noteworthy that each realm has its assemblage of organisms that have evolved themselves to get adapted to the conditions of the system. It is found that the meiofaunal species distributions modify in sandy sediments due to the physical impositions on interstitial or burrowing lifestyles. Fine sediments may flocculate in marine environments (Xu et al., 2008) and prevail for a certain period before getting redistributed or buried by waves and tidal action (Cummings et al., 2003; Hewitt et al., 2003). Bergen et al. (2001) in their study along the continental shelf found that the benthic infaunal assemblages to be associated with the depth and sediment grain gradients but are not the only factors that determine the species distribution. They also concluded that the shallow assemblage was defined by the combination of depth and sediment gradient with hydrography as a controlling factor. However, in the shelf setting, multiple interactive factors such as the interaction of sedimentary, hydrodynamic, and biological processes and their variation in space and time are responsible for the meiofaunal turnover (Jorissen, 1987; Murray, 2006.

The influence of the river along the shelve region affects the spatial distribution pattern of the riverine input and hence strongly affects the meiofaunal assemblage (Goineau et al., 2015; Mojtahid et al., 2009). Snelgrove and Butman (1994) have suggested that the amount of organic material in the sediment along with the hydrodynamic environment are presumably primary driving forces along with the depth and sediment grain size as secondary correlates. As the hydrographic environment is difficult to measure especially during extreme events, the use of depth and sediment grain size surrogates the study. The majority of extant marine organisms are critically affected by dissolved oxygen and food (Wishner et al., 2018).

In the case of the marine environment, the greater the water depth, the less light can penetrate, and hence below a certain depth, there is no light. This area of darkness that occupies the greater part of the ocean is called the aphotic zone while the above-illuminated region is called the photic zone. The actual depth of these zones depends on the local meteorological conditions, water turbidity, and ocean surface. The marine benthic environment is also classed into different zones. The supralittoral is above the high tide mark and is usually not underwater. The littoral or the intertidal zone ranges from the high tide mark to the shallow offshore waters. The sublittoral is the region beyond the low tide mark and is mostly referred to as the substrates of the continental shelf reaching a depth between 150 and 300 m. Sediments of the continental shelf are generally originated from the land especially from the river runoff. Beyond the continental shelf is the bathyal zone, which occurs at the depth zone between 150 and 4000 m which includes the continental slope and rise. The abyssal zone which falls between the depth of 4000 and 6000 m forms a substantial portion of the oceans. The deepest region of the oceans with a depth greater than 6000 m is the hadal zone. Sediments of the deep sea/ocean originate primarily from the rain of dead marine organisms and their debris (Kingsford, 2018). Marine organisms have evolved a wide range of physiological and morphological features that enable them to live even in the harsh conditions of the sea.

It is important to understand the pelagic benthos balance in the marine system as it is highly complex. For example, pelagic planktons are the important food source for the benthos that dwell on the soft or rocky bottoms. Suspension feeders like sea anemones filter living and dead particles from the surrounding water while the detritus feeder grazes upon the particulate material which has accumulated from materials raining from above which comprises of the dead debris and feces of planktons, molts of crustaceans, marine snow from the pelagic environment. This fallout varies concerning the weather patterns and the seasonal cycles of productivity. This variation creates seasonality in the abiotic zone wherein there is little or no variation in temperature or light (Wetzel, 2002).

In a deeply stratified marine system, the benthos essentially receives the deposits of the pelagic export production while returning nutrients to the deeper layers of water. However, it may take many years before these nutrients can be utilized again by the photoautotrophs. The marine ecosystems closer to the shore contain the largest amounts of primary and secondary production. While in the open ocean, there is lesser production given the same amount of space. Food webs of the ocean bottom are highly dependent on the primary productivity of the photic zone above. It is found that the most productive areas in the open ocean are near the upwelling zones where the conditions are amicable for food webs to continue for a longer period. Benthic grazers regulate the effects of redundant algal blooms resulting from phosphorous pollution in marine ecosystems. Although it might be tranquil effect to predict the number of plants and algae that would be built due to the increased nutrient supply in a marine ecosystem, the net biomass produced relies on many factors including the trophic structure and profusion of the predators (Sabo & Gerber, 2007). It is important to remember that in ecosystems lacking nutrients, top-down control of primary production is more powerful while in nutrient-rich ecosystems, bottom-up control is much crucial (Wetzel, 2002).

9.4 Relationship between benthos and estuarine environment

Benthic organisms play a vital role in the marine food web while some are even fishery species. In this section, we discuss the factors that control their abundance and distribution and their interaction with the estuarine and marine environment quoting studies from works all over the world. The importance of the microphytobenthos in the aquatic environment is emphasized by Barranguet et al. (1997) and hence studies about the microfossil groups of Foraminifera and ostracods (meiofaunal community) are looked upon as examples for better understanding.

Estuaries are shallow, open, and dynamic systems that are highly variable environments as the variation in the physical/chemical factors such as temperature, salinity, depth, oxygen, OMs in the water column and sediments are high and stressful to the biota. The variable salinity regime of the estuaries has notably broad implications on the benthic communities, and this driver exerts a powerful influence on the biota and the behavior of the system. The lesser volume of water per m^2 of sediment surface, the presence of intertidal flats, and very shallow subtidal areas with a generally well-fixed nature of the water column are physical conditions that aggravate the exchange of matter and energy between the sediment system and water column.

The availability of light affects the distribution of the organism in any aquatic system. In the estuarine system due to the effectively limited light, any excess nutrient input cannot lead to primary productivity, and thus the role of the benthic system, microbial, floral, and faunal community is crucial.

One should be cautious while considering the pelagic and benthic environments in isolation from each other as they are both interlinked in various ways. Heip et al. (1995) explored the ecosystem with an attempt to understand benthos-pelagic coupling. They depicted that it is the relationship between the benthic and pelagic environment that essentially controls the functioning of the estuarine system and concluded that estuaries are net heterotrophic systems.

In the estuarine system, the pelagic-benthic links are not just quantitatively more important, but also qualitatively different as the benthos can be liable for direct grazing on live phytoplankton, nutrient recycling occurs within a growing season, and hence there is significance in the in situ benthic primary productivity.

Benthic communities in estuarine are sensitive to environmental changes and are often used to help characterize the environment. The bottom substrate along the estuary keeps changing, which plays an important factor in influencing the distribution of the biota. The sediments along the deeper estuarine would be undisturbed while those at the end members would be disturbed due to the turbulence and other hydrological forces. Depending on the slope, flow regime, and the local composition of the bed, fine sediments can be flushed downstream during high flows (Petticrew et al., 2007). These sediments are then discharged into the estuaries from the coastal catchments, where further ecological impacts may occur (Thrush et al., 2004). The disturbance exerted on the bottom sediment influences the communities in the estuaries (Lake, 1990; Wall et al., 2005) and numerous studies have shown that the benthic communities on these ecosystems are detrimentally affected by pulsed inputs of terrigenous sediment (Matthaei et al., 2006; Thrush et al., 2003; Townsend et al., 2008).

Remane (1933) hypothesized that because of the challenging physio-chemical conditions, macrofaunal marine species diversity declines with salinity and that the true brackish water species occurs at the salinity range <15 psu. This conceptual model formed the foundation for much of our understanding of the estuarine benthos and their distribution. Recent studies have emphasized the importance of salinity variation as it has control over species diversity. suggested that the rate and scale of salinity variation are significant as the increasing salinity variation diminishes diversity. Due to the strong control that the salinity exerts over the pool of biota in estuaries, species available for communities in the mid-estuary wherein salinity is most different and greater from that of the end members are most limited. The biological changes related to the salinity are not always linear, nor are they always correlated along the salinity gradient. As the estuaries have an environment of different salinity regimes, Josefson (2016) predicted that the metacommunity structure with species is sorted according to the tolerance ranges in the salinity gradient. Salinity and substrate gradient may affect and alter the relative abundance of infaunal organisms within the estuary (Wilson & Fleeger, 2012). Macrobenthic species decline in number from both end members (river and sea) into an estuary while the meiofaunal species experience a decrease in species diversity with increasing salinity. This causes the increase in meiofaunal biota relative to the macrofauna in low salinity conditions of the estuary.

The salinity variation of estuary may cause a reduction in the biological enemies and hence releasing low salinity tolerant species from the biological control. Estuarine areas of low salinity may act as refuges from the harmful biological interactions for many species and hence contributing to the higher abundance in the estuarine. Functional groupings of estuarine consumers are assigned in accordance with their trophic position. However, many estuarine consumers are much generalized in their feeding habits. For most benthic consumers, feeding type is categorized based on morphology or structure. The ability of most of the benthic species to take up DOM against the concentration gradient forms the primary source of food. Since the estuarine water is loaded with plenty of DOM, it forms a good food source that may be absorbed and consumed.

In the estuarine system, there is generally a greater transfer of primary productivity through the benthic system than the water column. This is one of the major driving forces for the significance of the benthic food webs in estuaries that contribute to the nursery function for fish and crustaceans, which include important fishery species. The structuring of the food webs, which may be only two steps long (e.g., primary production => benthos => consumers) may be a factor for the low diversity in estuaries. The shallowness and depositional cycle might cause the distinction between surficial sediments and water column suspended particles to blur. These causes may benthos such as benthic diatoms to be commonly found in the estuarine phytoplankton samples. The macrofaunal community is the major contributor to the estuarine food webs, while the meiofaunal community is significant as they serve as the food source to this macrofauna.

Many trends related to body size are apparent in the benthic community. The abundance of macrofauna in estuaries typically ranges from 500 to 10,000 ind/m^2, while biomass ranges from ~5 to 200 g/m^2. For meiofauna, the abundance typically ranges from 10^5 to 10^7 ind/m^2. Macrobenthos has the highest densities but sometimes lower biomass (Wilson & Fleeger, 2012). However, the turnover rate of the benthic population is related to the size of the individual, such that smaller individuals generally have faster turnover. While the abundance (number of species) and mass (individual size) control the biomass ($B = A^*M$), secondary productivity is also crucial in evaluating the importance of organisms in the estuarine system.

9.5 An attempt to understand the meiofaunal interaction with the ecosystem using Foraminifera and ostracods

Foraminifera is marine unicellular protists with a size range of 0.5–12 mm. They exhibit two modes of living—benthic and planktonic. Benthic Foraminifera includes a few thousand species worldwide, with a unique microhabitat preference (Lei et al., 2017, 2019). They possess a hard exoskeleton called "test" which is used in micropaleontological studies as a proxy in paleoclimatic reconstruction. They are very sensitive to changes in the environment and are proven highly useful in biostratigraphy, oil exploration, stratigraphic correlation, geochronological studies, and environmental and paleoenvironmental studies. They are useful bioindicators of dynamic estuarine conditions due to their widespread occurrence and sensitivity to environmental changes (Nash et al., 2010; Samir et al., 2003). The distribution of the Foraminifera along the coastal areas elucidates the relationship of the assemblage with the present-day habitats and environmental parameters (Nagendra et al., 2015) which in turn aids in monitoring the impact of anthropogenic activity in the present-day conditions (Mendes et al., 2004; Quilty & Hosie, 2006) and in interpreting the paleoecological conditions in

Quaternary estuarine deposits (Li et al., 2002). Being a marine organism, Foraminifera has evolved itself to a state wherein they happen to dwell in marginal margin environments like lagoons and estuaries.

Several studies have shown the distribution of benthic Foraminifera (meiofauna) in the estuarine bodies. Benthic foraminiferal distributions in the surface sediments were strongly influenced by salinity (Wu et al., 2015). Foraminiferal density tends to increase seawards implying that the increased salinity encourages the occurrence of the same (Liu et al., 2011). The relative abundance of three different kinds of tests (agglutinated, porcelaneous, and hyaline) is in accordance with the salinity of the water. Agglutinated Foraminifera develops their test by cementing smaller sediment grains and hence reach peak abundance in areas of low $CaCO_3$ and low salinity (Greiner, 1970). While calcareous Foraminifera requires a certain amount of dissolved $CaCO_3$ to build their test and is found to replace agglutinated forms in the littoral areas of higher salinity (Wu & Wang, 1989; Hayward, 1993). Consequently agglutinated Foraminifera are better indicators than calcareous Foraminifera of brackish water conditions and of poorly sorted sand-clay-silt deposits in the estuary. Shallow marine regions of salinity more than 25 ppm are more suitable for the calcareous Foraminifera to flourish. The littoral zones with not so high energy and a substrate dominated by sand—clay silt mixture are dominantly certainly by the calcareous Foraminifera (Wu et al., 2015). The rate of sedimentation is yet another factor that controls the abundance of Foraminifera in sediments (Li, 1985) for higher rates of sedimentation can dilute the Foraminiferal test within the sediments. The benthic Foraminifera community was found controlled by the OM content along the shelve region.

Kathal et al. 2017 introduced a new insight into the importance of spinosity in Foraminifera. They concluded that *Ammonia dentata* can adapt itself morphologically to the changing environment. The test of the organism becomes more robust with short and blunt spines in the shallow water while the test becomes less robust with long, slender spines in comparatively deeper waters (~30 m water depth). This enables the application of *A. dentata* in inferring the paleodepths provided its spines are well preserved.

Saraswat et al. (2017) studied the abundance and ecology of *Asterorotalia trispinosa*, a benthic Foraminifera endemic to the Bay of Bengal, and concluded that its relative abundance is not uniformly high to the shallow water regions but that the distribution and abundance along the continental shelf of the western coast of India are controlled by several parameters. They concluded that the abundance of *A. trispinosa* correlates positively with the temperature and negatively with the salinity and C_{org}%. They further found that the spatial distribution of the species was controlled by the terrigenous and freshwater influx and the resultant variations in the physicochemical conditions. The study depicted that the species prefers stenohaline warmer water found in the mouth of the outfall of rivers. Rao et al. (2013) studied the same species from the Muthupet lagoon and concluded that *A. trispinosa* can adapt itself to the estuarine and open marine mud dominated conditions provided that the salinity is as less as 29‰ and continuous rain of suspended fine particles along with sufficient DO and OM is available.

Kaithwar et al. (2020) have made an effort in studying the relationship of living benthic Foraminiferal assemblage with dissolved oxygen. The stressed environment such as the ODZ usually supports a few opportunistic faunas, thus reducing the diversity. Thus the low DO is suggested as severe stress for benthic Foraminifera (Jorissen et al., 1995; Kaminski et al., 1995) implying that the benthic foram diversity should decrease within the ODZ. The high diversity in the ODZ, thus advocates a combined effect of several factors. This implies that the coarse-grained sediments, higher energy conditions along lower C_{org}% reduce the foraminiferal diversity in the continental shelf of the SE Arabian Sea. Though the DO concentration decreases along the slope, the availability of plenty of OM coupled with the decreased predatory pressure leads to the proliferation of benthic foraminiferal diversity within the ODZ. They reported five assemblages of living benthic Foraminifera representing different bottom water conditions.

Ostracods are tiny bivalve microorganisms and one of the best-documented groups of microfossils due to the most characteristic features of their bodies and well calcified tiny bivalved carapace which fossilizes easily. Ostracods occur almost in all aquatic environments (Puri, 1966) that include freshwater (stream, lakes, rivers, and ponds), saline water from shallow sea to deep oceans, marginal marine (brackish water, estuaries, lagoons, salt, marshes, and mangroves). The ecology of the ostracods is often reflected in the shape and structure of their carapaces and hence makes them a useful indicator of paleoenvironment as such deep-sea ostracods are distinctly different from the shelf and marginal marine taxa.

Barbieri et al., 2019 has concluded that the primary controlling factor of the ostracod distribution is grain size and bathymetry. The distribution and abundance of Ostracoda show that the silty-sand substrate is congenial for the fauna to thrive (Sridhar et al., 1998). Sand with higher silt content is found to be a more congenial substrate for ostracod populations. The bottom water temperature shows an inverse proportion to the total population (Baskar et al., 2013). Organic carbon has a significant role in geochemical and biological cycles that accumulate in sediments. However, OM enhances the nutritive content of the water and substrate and thus favors the development of Ostracoda. The smooth and fragile forms occur at deeper depths; moderately calcified and pitted forms and fine to moderate reticulate and ridged forms in the medium depths; and conspicuously ornate forms in the nearshore where the depth is lower. A direct relationship exists between the ornamentation of the Ostracoda carapace and the nature of the substrate. The shells of ostracod show heterogeneity in ornamentation pattern few are smooth and devoid of any sculptures while others are intensely sculpted forms. Smoother forms are predominant in fine-grained muds, whereas highly ornamented forms are found in coarse or more calcareous sediments (Benson, 1961; Brasier, 1980; Hussain et al., 2002, 2004, 2007, 2010). The presence of coral reefs is attributed to the higher population of ostracods which gives an insight into the importance of the biota for the thriving of benthic species (Baskar et al., 2013).

The ostracod shell due to its bivalved form tends to get disarticulated after death due to bacterial actions and external forces. If the ostracod carapace is exposed for a longer period, chances are high for it to get disarticulated. When sedimentation is rapid, the shells are preserved intact as carapace (Oertli, 1971). High-energy conditions also cause the disarticulation of the carapace. Hence the rate of sedimentation and the energy condition of the region are two important governing factors for the carapace-open valve ratio. The study of the carapace-open valve ratio gives an insight into the rate of sedimentation or the hydrographic conditions of the region (Ganesan & Hussain, 2010). The color diversity of the ostracod carapace has been used to predict the ecology/paleoecology of the sediments. It also indicates the oxygenation of the environment as the white/milky white, pale yellow/light brown color were more indicators of a normal oxygenated environment while the brown/black/pyritized carapace was confined to the clayey sediments with granules of pyrite and carbonaceous matter (Honnappa & Venkatachalapathy, 1978). Figs. 9.1 and 9.2 show SEM images of a few estuarine and shallow marine ostracod species.

Hence benthic Foraminifera and Ostracoda are two important groups of meiofaunal species used as proxies for ecological indicators of the depositional dynamics of the continental shelves, due to their tiny size, abundance, and adaptations to different environmental conditions (Athersuch et al., 1989;

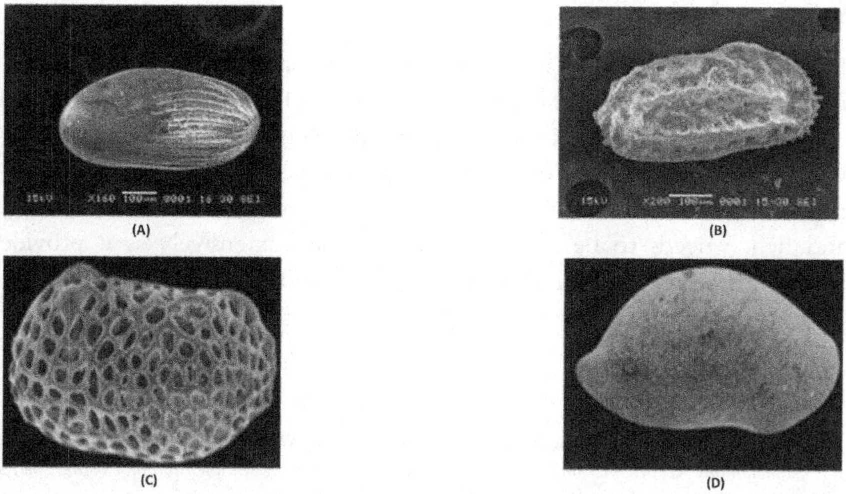

Figure 9.1 Shallow marine ostracods: (A) *Basslerites liebaui*, (B) *Stigmatocythere indica*, (C) *Lankacythere coralloides*, and (D) *Bairdoppilata (B) alcyonicola*.

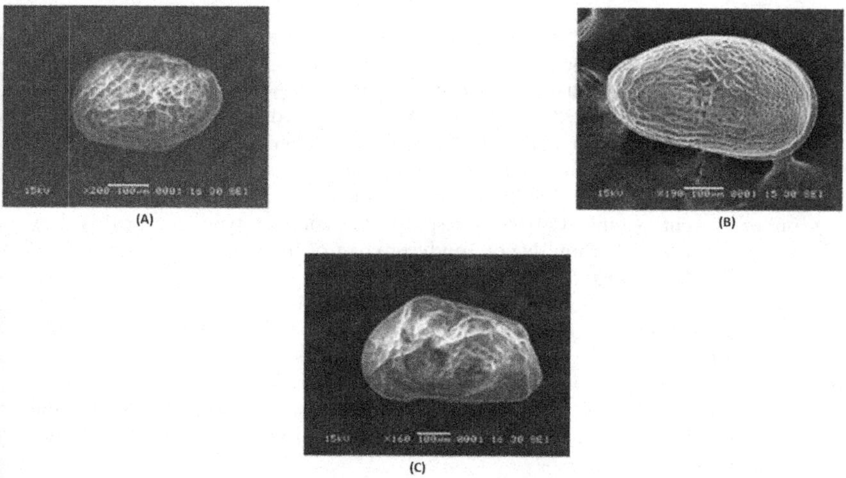

Figure 9.2 Brackish water ostracod species: (A) *Loxoconcha megapora indica*, (B) *Hemicytheridea paiki*, and (C) *Jankeijcythere mckenziei*.

Murray, 2006). The distribution and abundance of these organisms are highly influenced by external factors such as depth of the water column, substrate, temperature, $CaCO_3$, OM, salinity, DO, and geochemical and hydrographical parameters. They are found to actively interact with the environment and adapt themselves to the environmental factors thus having different assemblages pertaining to different conditions.

9.6 Conclusion

Benthic organisms are an important group comprising a diversity of organisms dwelling in or on the bottom sediments of any aquatic system. The type and number of species making up any benthic community are determined by a combination of physical and biological factors. Therefore benthic ecosystem is a sensitive index to ecological and environmental change and hence needs to be monitored and studied extensively as it provides important information on the health and wealth of the bottom sediments.

References

Athersuch, J., Horne, D. J., & Whittaker, J. E. (1989). *Marine and brackish water ostracods (superfamilies Cypridacea and Cytheracea): Keys and notes for the identification of the species. Synopses of British fauna* (Vol. 43). Leiden.

Barbieri, G., Rossi, V., Vaiani, S. C., & Horton, B. P. (2019). Benthic Ostracoda and Foraminifera from the North Adriatic Sea (Italy, Mediterranean Sea): A proxy for the depositional characterisation of river-influenced shelves. *Marine Micropaleontology, 153*, 101772.

Barbieri, G., & Vaiani, S. C. (2018). Benthic foraminifera or Ostracoda? Comparing the accuracy of palaeoenvironmental indicators from a Pleistocene lagoon of the Romagna coastal plain (Italy). *Journal of Micropalaeontology, 37*(1), 203–230.

Barranguet, C., Herman, P. M. J., & Sinke, J. J. (1997). Microphytobenthos biomass and community composition studied by pigment biomarkers: Importance and fate in the carbon cycle of a tidal flat. *Journal of Sea Research, 38*(1–2), 59–70.

Baskar, K., Sridhar, S. G. D., Sivakumar, T., & Hussain, S. M. (2013). Seasonal comparison of Recent benthic Ostracoda response to sediment type in the Palk bay, off Rameswaram Island, Tamil Nadu, southeast coast of India. *International Journal of Earth Sciences and Engineering, 6*(5), 1047–1059.

Benson, R. H. (1961). Ecology of ostracode assemblage. In R. C. Moore (Ed.), *Treatise on invertebrate paleontology, pt. Q, Arthropoda 3, Ostracoda* (pp. Q56–Q63). Geological Society of America.

Bergen, M., Weisberg, S. B., Smith, R. W., Cadien, D., Dalkey, A., Montagne, D., & Velarde, R. G. (2001). Relationship between depth, latitude and sediment and the structure of benthic infaunal assemblages on the mainland shelf of Southern California. *Marine Biology, 138*, 637–647.

Brasier, M. D. (1980). *Microfossils*. George Allen and Unwin Ltd.

Burdige, D. J. (2007). Preservation of organic matter in marine sediments: Controls, mechanisms, and an imbalance in sediment organic carbon budgets? *Chemical Reviews 107*(2), 467–485.

Cummings, V., Thrush, S., Hewitt, J., Norkko, A., & Pickmere, S. (2003). Terrestrial deposits on intertidal sandflats: Sediment characteristics as indicators of habitat suitability for recolonising macrofauna. *Marine Ecology Progress Series, 253*, 39–54.

Fleeger, J. W., Carman, K. R., & Nisbet, R. M. (2003). Indirect effects of contaminants in aquatic ecosystems. *Science of the Total Environment, 317*(1–3), 207–233.

Ganesan, P., & Hussain, S. M. (2010). Distribution of recent benthic Ostracoda in Tamiraparni Estuary, Punnaikayal, near Tuticorin, Southeast Coast of India—Implications on microenvironment. *Gondwana Geological Magazine, 25*, 103–114.

Goineau, A., Fontanier, C., Mojtahid, M., Fanget, A. S., Bassetti, M. A., Berné, S., & Jorissen, F. (2015). Live–dead comparison of benthic foraminiferal faunas from the Rhône prodelta (Gulf of Lions, NW Mediterranean): Development of a proxy for palaeoenvironmental reconstructions. *Marine Micropaleontology, 119*, 17–33.

Greiner, G. O. (1970). Distribution of major benthonic foraminiferal groups on the Gulf of Mexico continental shelf. *Micropaleontology, 16*(1), 83–101.

Hayward, B. W. (1993). Estuarine foraminifera, Helena Bay, Northland, New Zealand. *Tane, 34*, 79–88.

Heip, C. H. R., Goosen, N. K., Herman, P. M. J., Kromkamp, J. C., Middelburg, J. J., & Soetaert, K. E. R. (1995). Production and consumption of biological particles in temperate tidal estuaries. *Oceanography and Marine Biology: An Annual Review, 33*, 1–149.

Herman, P. M. J., Middelburg, J. J., Van de Koppel, J., & Heip, C. H. R. (1999). Ecology of estuarine macrobenthos. *Advances in Ecological Research, 29*, 195–240.

Hewitt, J. E., Cummings, V. J., Ellis, J. I., Funnell, G., Norkko, A., Talley, T. S., & Thrush, S. F. (2003). The role of waves in the colonisation of terrestrial sediments deposited in the marine environment. *Journal of Experimental Marine Biology and Ecology, 290*(1), 19–47.

HonnappaVenkatachalapathy, V. (1978). Some aspects of pyritised ostracode shells: a possible tool in petroleum sedimentology from the sediments of Mangalore Harbour area, Karnataka state, west coast of India. *Proceedings of VII Indian Colloquium on Micropaleontology and Stratigraphy*, 65–69.

Hussain, S. M., Ganesan, P., Ravi, G., Mohan, S. P., & Sridhar, S. G. D. (2007). Distribution of Ostracoda in marine and marginal marine habitats off Tamil Nadu and adjoining areas, southern east coast of India and Andaman Islands: Environmental implications. *Indian Journal of Geo-Marine Sciences, 36*(4), 369–377.

Hussain, S. M., Mohan, S. P., & Jonathan, M. P. (2010). Ostracoda as an aid in identifying 2004 tsunami sediments: A report from SE coast of India. *Natural Hazards, 55*(2), 513–522.

Hussain, S. M., Ravi, G., Mohan, S. P., & Rao, N. R. (2004). Recent benthic Ostracoda from the inner shelf off Chennai, south east coast of India—Implication of microenvironments. *Environmental Micropaleontology, Microbiology and Meiobenthology, 1*, 105–121.

Ingole, B. S., Sautya, S., Sivadas, S., Singh, R., & Nanajkar, M. (2010). Macrofaunal community structure in the western Indian continental margin including the oxygen minimum zone. *Marine Ecology, 31*(1), 148–166.

Jorissen, F. J. (1987). The distribution of benthic foraminifera in the Adriatic Sea. *Marine Micropaleontology, 12*, 21–48.

Jorissen, F. J., de Stigter, H. C., & Widmark, J. G. (1995). A conceptual model explaining benthic foraminiferal microhabitats. *Marine Micropaleontology, 26*(1–4), 3–15.

Josefson, A. B. (2016). Species sorting of benthic invertebrates in a salinity gradient—Importance of dispersal limitation. *PLoS One, 11*(12), e0168908.

Kathal, P. K., Nigam, R., & Talib, A. (2017). The Significance of Spinosity in Foraminifera–A New Perspective. *Micropaleontology and Its Applications*, 301.

Kaithwar, A., Singh, D. P., & Saraswat, R. (2020). A highly diverse living benthic foraminiferal assemblage in the oxygen deficient zone of the southeastern Arabian Sea. *Biodiversity and Conservation, 29*(14), 1–34.

Kaminski, M.A., Boersma, E., Tyszka, J., & Holbourn, A.E.L. (1995). Response of deep-water agglutinated foraminifera to dysoxic conditions in the California Borderland basins. Grzybowski Foundation.

Kingsford, M.J. (2018) Marine ecosystem—Britannica Online Encyclopedia. https://www.britannica.com/science/marine-ecosystem

Kingsford, M. J., & Battershill, C. N. (1998). *Subtidal habitats and benthic organisms of rocky reefs. Studying temperate marine environments: A handbook for ecologists* (pp. 84–114). Canterbury University Press.

Lake, P. S. (1990). Disturbing hard and soft bottom communities: A comparison of marine and freshwater environments. *Australian Journal of Ecology, 15*(4), 477–488.

Lei, Y., Li, C., Li, T., & Jian, Z. (2019). The effect of temperature and food concentration on ingestion rates of *Quinqueloculina seminula* on the diatom *Nitzschia closterium*. *Journal of Foraminiferal Research, 49*(1), 3–10.

Lei, Y., Li, T., Jian, Z., & Nigam, R. (2017). Taxonomy and distribution of benthic foraminifera in an intertidal zone of the Yellow Sea, PR China: Correlations with sediment temperature and salinity. *Marine Micropaleontology, 133*, 1–20.

Li, S. (1985). Distribution of the foraminiferal thanatocoenosis of Pearl River Estuary. *Marine Geology & Quaternary Geology, 5*(2), 83–101.

Li, S. L., Li, S. Q., Yang, W. D., Chen, Y. X., & Long, J. P. (2002). Oxygen and carbon isotopic record of foraminiferal crusts from HY126EA1 hole in the continental shelf of the East China Sea. *Acta Oceanologica Sinica, 24*, 81–87.

Liu, C., Wu, J., Yang, T., Fürsich, F. T., & Zhang, S. (2011). Rare earth element records from core sediments of southern Pearl River Delta: Geochemical response to palaeoenvironmental changes during the Late Quaternary. *Journal of Palaeogeography, 13*(2), 221–228.

Long, E. R., Dutch, M. E., Aasen, S., Welch, K. I., Partridge, V. A., & Shull, D. H. (2007). *Relationships between the composition of the benthos and sediment and water quality parameters in Hood Canal: Task IV-Hood Canal Dissolved Oxygen Program*. Washington State Department of Ecology Publication.

Matthaei, C. D., Weller, F., Kelly, D. W., & Townsend, C. R. (2006). Impacts of fine sediment addition to tussock, pasture, dairy and deer farming streams in New Zealand. *Freshwater Biology, 51*(11), 2154–2172.

McIntire, C. D., Gregory, S. V., Steinman, A. D., & Lamberti, G. A. (1996). *Modeling benthic algal communities: An example from stream ecology. Algal ecology: Freshwater benthic ecosystems* (pp. 669–704). Academic Press.

Mendes, I., Gonzalez, R., Dias, J. M. A., Lobo, F., & Martins, V. (2004). Factors influencing recent benthic foraminifera distribution on the Guadiana shelf (Southwestern Iberia). *Marine Micropaleontology, 51*(1–2), 171–192.

Mojtahid, M., Jorissen, F., Lansard, B., Fontanier, C., Bombled, B., & Rabouille, C. (2009). Spatial distribution of live benthic foraminifera in the Rhône prodelta: Faunal response to a continental–marine organic matter gradient. *Marine Micropaleontology, 70* (3–4), 177–200.

Montague, C. L., & Ley, J. A. (1993). A possible effect of salinity fluctuation on abundance of benthic vegetation and associated fauna in northeastern Florida Bay. *Estuaries, 16*(4), 703–717.

Murray, J. W. (2006). *Ecology and applications of benthic foraminifera*. Cambridge University Press.

Nagendra, R., Sathiyamoorthy, P., Reddy, A. N., & Ramachandran, A. (2015). Spatial distribution of benthic foraminifera in the Palar estuary, Tamil Nadu. *Journal of the Geological Society of India, 86*(3), 305–316.

Nash, G. J., Binnie, M. N., & Cann, J. H. (2010). Distribution of Foraminifera and ostracods in the Onkaparinga Estuary, South Australia. *Australian Journal of Earth Sciences, 57*(7), 901–910.

Oertli, H. J. (1971). The aspect of ostracode faunas—A possible new tool in petroleum sedimentology. *Bulletin du Centre de Recherche de Pau-SNPA, 5*, 137–151.

Petticrew, E. L., Krein, A., & Walling, D. E. (2007). Evaluating fine sediment mobilization and storage in a gravel-bed river using controlled reservoir releases. *Hydrological Processes, 21*(2), 198–210.

Puri, H.S. (1966). Ecology and distribution of recent Ostracoda, pp. 457–495. In: Proceedings of symposium on Crustacea, Part I, Marine Biological Association of India, Mandapam.

Quilty, P. G., & Hosie, G. (2006). Modern foraminifera, Swan River estuary, Western Australia: Distribution and controlling factors. *The Journal of Foraminiferal Research, 36* (4), 291–314.

Rao, N. R., Jayaprakash, M., & Velmurugan, P. M. (2013). The ecology of *Asterorotalia trispinosa*—New insights from Muthupet Lagoon, Southeast Coast of India. *The Journal of Foraminiferal Research, 43*(1), 14–20.

Reid, D. J., Chiaroni, L. D., Hewitt, J. E., Lohrer, D. M., Matthaei, C. D., Phillips, N. R., … … van Houte-Howes, K. S. (2011). Sedimentation effects on the benthos of streams and estuaries: A cross-ecosystem comparison. *Marine and Freshwater Research, 62*(10), 1201–1213.

Remane, A. (1933). Verteilung und organisation der benthonischen mikrofauna der KielerBucht.

Sabo, J. L., & Gerber, L. R. (2007). Predicting extinction risk in spite of predator–prey oscillations. *Ecological Applications, 17*(5), 1543–1554.

Samir, A. M., Abdou, H. F., Zazou, S. M., & El-Menhawey, W. H. (2003). Cluster analysis of recent benthic foraminifera from the northwestern Mediterranean coast of Egypt. *Revue de Micropaleontologie, 46*(2), 111–130.

Saraswat, R., Manasa, M., Suokhrie, T., Saalim, M. S., & Nigam, R. (2017). Abundance and ecology of endemic *Asterorotalia trispinosa* from the Western Bay of Bengal: Implications for its application as a Paleomonsoon proxy. *Acta Geologica Sinica—English Edition, 91*(6), 2268–2282.

Snelgrove, P. V. R., & Butman, C. A. (1994). Animal-sediment relationships revisited: Cause vs effect. *Oceanography and Marine Biology Annual Review, 32*, 111–177.

Sridhar, S. G. D., Hussain, S. M., Kumar, V., & Periakali, P. (1998). Benthic Ostracod responses to sediments in the Palk Bay, off Rameswaram, Southeast coast of India. *Journal Indian Association of Sedimentologists, 17*(2), 187–195.

Thrush, S. F., Hewitt, J. E., Cummings, V. J., Ellis, J. I., Hatton, C., Lohrer, A., & Norkko, A. J. F. I. E. (2004). Muddy waters: Elevating sediment input to coastal and estuarine habitats. *Frontiers in Ecology and the Environment, 2*(6), 299–306.

Thrush, S. F., Hewitt, J. E., Norkko, A., Cummings, V. J., & Funnell, G. A. (2003). Macrobenthic recovery processes following catastrophic sedimentation on estuarine sandflats. *Ecological Applications, 13*(5), 1433–1455.

Townsend, C. R., Uhlmann, S. S., & Matthaei, C. D. (2008). Individual and combined responses of stream ecosystems to multiple stressors. *Journal of Applied Ecology, 45*(6), 1810–1819.

Wall, D. H., Ayres, E., Behan-Pelletier, V., Covich, A. P., & Snelgrove, P. V. (2005). Soils, freshwater and marine sediments: The need for integrative landscape science. *Marine Ecology Progress Series, 304*, 302–307.

Wetzel, R. (2002). Grazers.

Wetzel, R. G. (2002). Lake and river ecosystems. Limnology.

Wilson, J. G., & Fleeger, J. W. (2012). *Estuarine benthos. Estuarine ecology* (pp. 303–325). Wiley Blackwell.

Wishner, K. F., Seibel, B. A., Roman, C., Deutsch, C., Outram, D., Shaw, C. T., & Riley, S. (2018). Ocean deoxygenation and zooplankton: Very small oxygen differences matter. *Science Advances, 4*(12), eaau5180.

Wu, N., & Wang, P. (1989). Distribution of arenaceous Foraminifera along China coast and its controlling factors. *Chinese Science Bulletin: English Edition, 22*, 1896–1901.

Wu, J., Liu, C., Fürsich, F. T., Yang, T., & Yin, J. (2015). Foraminifera as environmental indicators and quantitative salinity reconstructions in the Pearl River estuary, southern China. *The Journal of Foraminiferal Research, 45*(3), 205–219.

Xu, F., Wang, D. P., & Riemer, N. (2008). Modeling flocculation processes of fine-grained particles using a size-resolved method: Comparison with published laboratory experiments. *Continental Shelf Research, 28*(19), 2668–2677.

Further reading

Attrill, M. J. (2002). A testable linear model for diversity trends in estuaries. *Journal of Animal Ecology*, *71*(2), 262–269.

Friedrich, P. (1965). A Mexican cacicazgo. *Ethnology*, *4*(2), 190–209.

Golderman, G., & Connolly, B. (2002). AccessScience @ McGraw-Hill. *School Library Journal*, *48*(8), 44–46.

Heip, C., & Craeymeersch, J. A. (1995). Benthic community structures in the North Sea. *Helgoländer Meeresuntersuchungen*, *49*(1), 313–328.

Horne, D. J., Holmes, J. A., Rodriguez-Lazaro, J., & Viehberg, F. A. (2012). *Ostracoda as proxies for quaternary climate change: Overview and future prospects*, . *Developments in quaternary sciences* (Vol. 17, pp. 305–315). Elsevier.

Hussain, S. M. (1998). Recent benthic Ostracoda from the Gulf of Mannar, off Tuticorin, southeast coast of India. *Journal of the Palaeontological Society of India*, *43*, 1–22.

Kathal, P. K., Nigam, R., & Talib, A. (2017). *The significance of spinosity in foraminifera—A new perspective*. *Micropaleontology and its applications* (p. 301) Scientific Publishers.

Lutze, G. (1965). Zurforaminiferen-fauna der Ostsee. *Meyniana*, *15*, 75–142.

Nagarajan, R., Anandkumar, A., Hussain, S. M., Jonathan, M. P., Ramkumar, M., Eswaramoorthi, S., & Chua, H. B. (2019). *Geochemical characterization of beach sediments of Miri, NW Borneo, SE Asia: Implications on provenance, weathering intensity, and assessment of coastal environmental status*. *Coastal zone management* (pp. 279–330). Elsevier.

Rodriguez-Lazaro, J., & Ruiz-Muñoz, F. (2012). *A general introduction to ostracods: morphology, distribution, fossil record and applications*. *In*. *Developments in Quaternary Sciences* (Vol. 17, pp. 1–14). Elsevier.

Scott, D. B., Schafer, C. T., & Medioli, F. S. (1980). Eastern Canadian estuarine foraminifera; a framework for comparison. *The*. *Journal of Foraminiferal Research*, *10*(3), 205–234.

Wang, P., & Murray, J. W. (1983). The use of foraminifera as indicators of tidal effects in estuarine deposits. *Marine Geology*, *51*(3–4), 239–250.

Warwick, R. M., & Clarke, K. R. (1984). Species size distributions in marine benthic communities. *Oecologia*, *61*(1), 32–41.

CHAPTER 10

Benthic organisms as an ecological tool for monitoring coastal and marine ecosystem health

P.R. Jayachandran, S. Bijoy Nandan, M. Jima, J. Philomina and N.K. Vishnudattan
Department of Marine Biology, Microbiology and Biochemistry, School of Marine Sciences, Cochin University of Science and Technology, Kochi, India

10.1 Introduction

Coastal and marine ecosystems are historically important since the human settlements are concentrated along the coastal marine zone because of the available food resources, easiness in transportation, and numerous services they provide (Bijoy Nandan, 2008; McLusky and Elliott, 2004). However, these critical ecosystems are experiencing a variety of pressures from natural and anthropogenic causes (Dauvin, 2007; Rakocinski and Zapfe, 2005; Shannon and Weaver, 1949). The major anthropogenic factors influencing the coastal marine waters are land reclamation, dredging, overexploitation of resources, unmanaged tourism, and other anthropogenic causes like pollutant discharge from the point and nonpoint sources, the introduction of alien invasive species, ocean acidification, and climate change (Danovaro et al., 2016; Dauvin, 2007; Jayachandran and Bijoy Nandan, 2012). Coastal marine environments are also influenced by natural pressures like sudden salinity changes and highly turbid conditions (Asha et al., 2016; Jayachandran and Bijoy Nandan, 2012; Jayachandran et al., 2012). The environmental or ecological quality assessment concept is now widely used to assess such pressures on these dynamic ecosystems (Jayachandran and Bijoy Nandan, 2012; Shannon and Weaver, 1949; Simpson, 1949). The ecological quality assessment concept is developed based on the integrative approach that considers the information on biological, physicochemical, and pollution factors (Brillouin, 1956; Naidu et al., 2018; Shannon and Weaver, 1949). In addition to this, food web

Ecology and Biodiversity of Benthos
DOI: https://doi.org/10.1016/B978-0-12-821161-8.00004-0

dynamics, species diversity, and life-history traits are also considered in such analysis since they are potential biological indicators of ecological functioning and status (Borja et al., 2000; Pielou, 1966). Such an ecosystem-based approach is more successful than species level or chemical level assessment for classifying the coastal marine waters (Borja et al., 2009; Rakocinski and Zapfe, 2005). The anthropogenic impacts on coastal marine systems are extensively recognized, their quantification is difficult, due to its complex and unpredictable nature (Vitousek et al., 1997). The anthropogenic interventions in coastal marine waters primarily affect the community structure, such as abundance, biomass, and species diversity, or derived community variations such as diversity indices, abundance—biomass ratio, and evenness indices (Margalef, 1968; Rakocinski and Zapfe, 2005; Sivadas et al., 2016). Many studies have been conducted on these aspects to develop a suitable ecosystem health assessment tool that is capable to demonstrate the status of the ecosystem and are easily understandable for ecosystem managers to develop the management and conservation plan (Jørgensen et al., 2010). As a result, scientific interest in various indices employed in ecosystem health assessment has drastically increased for the last few decades (Hayes et al., 2008). Here we discuss the commonly used benthic indices in the coastal marine habitat for assessing ecological quality to develop suitable benthic quality criteria with special reference to the Indian scenario.

10.2 Importance of benthos in coastal monitoring programs

The benthic zone (bottom) of coastal marine ecosystems is relevant in maintaining the productivity, energy flow and nutrient recycling process (McLusky and Elliott, 2004). Organisms living in these zones are referred to as "benthos" which play a key role in the ecosystem processes (Coull, 1977; Jayachandran et al., 2019c; Lalli and Parsons, 1997). Based on the size, benthos can be classified as microfauna (>0.063 mm), meiofauna [0.063–1.0 (or 0.5) mm], macrofauna [>1.0 (or 0.5) mm] and megabenthos (>10.0 mm) (Mare, 1942). They are characterized by their sessile or sedentary nature, having a relatively long life span, susceptibility to any changes in the overlying water mass and sediment conditions. Several studies have also revealed their relatively rapid response to anthropogenic as well as natural disturbances and their role in energy transfer in the aquatic food web (Jayachandran et al., 2019b; Perus et al., 2007;

Simboura and Zenetos, 2002). In addition to this, they play a key role in chemical fluxes between the benthic-pelagic ecosystems through bioturbation and suspension-feeding process (Reiss and Kröncke, 2005). Since they are an extremely diverse group displaying a variety of lifestyle and tolerance levels to various ecological disturbance, which helps to classify them into different ecological groups (Jayachandran et al., 2019b; Pearson and Rosenberg, 1978a). Many recent studies considered functional aspects of benthos such as feeding guild (Jayachandran et al., 2019b; Margalef, 1968) and their response to organic matter accumulation in the bottom sediment (Borja et al., 2000; Jayachandran et al., 2019c). Abundance biomass comparison of benthic communities is also used to assess the ecological disturbances (Zhang et al., 2013). Macroinvertebrates are the major group of animal communities in the zoobenthos and they are highly sensitive to ecosystem disturbances. They respond rapidly to both natural and anthropogenic disturbances. Therefore, the use of benthic macroinvertebrates as bioindicators performing their functional role as both primary and secondary consumers in the food web possess certain advantages over other communities (Ajmal Khan et al., 2004; Aylagas et al., 2018; Borja et al., 2000; Jayachandran et al., 2019a, 2019c). For these reasons, macroinvertebrate-based indices have wide acceptance by scientific communities and ecosystem managers compared to other indices (McLusky and Elliott, 2004), and are frequently used in the coastal ecosystem health monitoring programs (Bald et al., 2005; Blanchet et al., 2008; Borja et al., 2008; Simboura and Zenetos, 2002). Even though such benthic indices are currently in use by scientists, researchers and policymakers to study and document the ecological health of the coastal and marine ecosystems, there are still several limitations to be resolved.

10.3 Marine benthic biotic indices and their relevance to coastal monitoring programs

Coastal marine ecosystems are characterized by high biological productivity and varying environmental conditions (McLusky and Elliott, 2004). Anthropogenic pressure from eutrophication, shipping (oil spills and invasive species), industrial activities, dredging and dumping, mariculture, fishing, tourism, ocean acidification, climate change, and other developmental activities impact the benthic diversity, at the gene to ecosystem level. Benthic ecosystems are also impacted by natural stressors, especially in the transitional waters, in which salinity influences the distribution of species along the different gradients (Anu et al., 2016; Don Xavier et al., 2019; McLusky and Elliott, 2004). The alien invasive species occurrence also affect the coastal ecology and species distribution, they compete with the native population and dominate in the community, they are adapted to live in a wide range of environmental conditions (Jayachandran et al., 2019a, 2018a, 2018b; Rakocinski and Zapfe, 2005). Even though benthic organisms are excellent indicators of ecosystem disturbances, their application remains difficult in transitional waters due to their dynamic nature and different geomorphologic features (Blanchet et al., 2008; Van Hoey et al., 2007). Therefore indices or indicators for the coastal and transitional waters should have the ability to simplify and differentiate the overall state of these ecosystems from diverse factors that affect the ecological quality (Bettencourt et al., 2004). Also, it should be easier to understand or be communicable to stakeholders and policymakers (Rakocinski and Zapfe, 2005). According to Dauvin (2007), any measure that allows the assessment and evaluation of a system's status are referred to as an indicator while one possible measure of a system's status is referred to as indices (Borja et al., 2011).

Benthic organisms are excellent indicators of biotic integrity due to their sedentary life with a relatively long life span, high level of tolerance to different stress levels, a key role in nutrient cycling, and significance in energy transfer to higher trophic levels (McLusky and Elliott, 2004). Benthic sentinel organisms like bivalves are widely recognized as excellent indicators of pollution (Farrington et al., 2016). The International Mussel Watch Project initiated by United Nations (UNESCO-IOC and UNEP) in 1986 under the leadership of Professor Edward D. Goldberg used this capability of mussel species to water-quality monitoring and now it is globally recognized (Farrington et al., 2016; Hawkins et al., 1986; Krishnakumar et al., 1990). In the first phase of the mussel watch program, it was implemented in South America, Central America, Mexico, and the Caribbean. Later in the second phase, it was launched as International Mussel Watch and implemented in the Asia Pacific Region and other related areas (Farrington et al., 2016). However, many studies have been pointed out the limitations of benthos in developing indices because they represent a static expression of an ecological state, not related to changes in ecological function, not specific to a diverse level of stressors, it is subjected to taxonomic changes across ecological gradients and it is labor intensive and less consistent with different biogeographic provinces (Rakocinski and Zapfe, 2005).

The concept of benthic indicator species is based on the presence and/or dominance of a species in a particular environment (Levin, 2000; Pearson and Rosenberg, 1978b). In an undisturbed benthic community, k-selected benthic species dominate, they are characterized by slow-growth rate, relatively large size, and having late maturity while r-selected benthic species dominate in a disturbed ecosystem which is characterized by fast-growth rate, relatively small size, and opportunistic nature (Zhang et al., 2013) (Fig. 10.1). Based on this, indicator species can be classified as positive indicators (pollution tolerant species) and negative indicator species (sensitive to disturbance) (Rygg, 1985). The species-level macroinvertebrate indicators are widely used to understand the habitat quality such as polychaete species *Capitella capitata* (Capitellidae), *Chaetozone setosa* (Cirratulidae), and a group of spionid species like *Prionospio cirrifera* is considered as a positive indicator, they proliferate in organic matter enriched sediment in the coastal systems (Ducrotoy et al., 2010). At the same time, members of the polychaete family Oweniidae are represented as negative indicator species because their abundance is reduced with the increasing

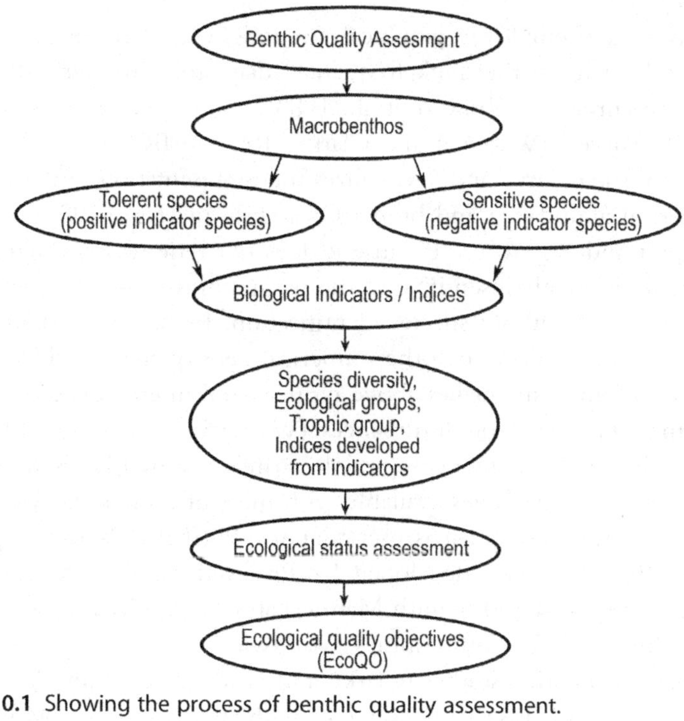

Figure 10.1 Showing the process of benthic quality assessment.

organic load. In general, the benthic indicators and indices can be classified into four categories based on diversity, ecological groups, trophic groups, and indices developed from indicators. Benthic communities respond to environmental disturbance through different adaptive strategies and towards pollution effects that can reflect both disturbance and sensitivity to stress. Among the univariate indices, benthic descriptors are the most common type of indices used for assessing the quality of coastal marine environments around the globe such as species number, abundance, and biomass. Most of the previous studies used different kinds of indices of diversity; it is a mathematical measurement of species diversity in an ecosystem. The most commonly used diversity indices are Shannon–Wiener diversity index (H′) (Shannon and Weaver, 1949), Simpson's indices of dominance, diversity, and evenness (Simpson, 1949), Brillouin indices of diversity and evenness (Brillouin, 1956), Pielou evenness index (J′) (Pielou, 1966), Margalef's index (Margalef, 1968), Hurlbert index (Margalef, 1968), Hill's diversity numbers and evenness measures (Hill, 1973), benthic pollution index (Leppäkoski, 1975), taxonomic diversity index, and taxonomic distinctness (Warwick and Clarke, 1995) (Table 10.1).

Many studies employed graphical methods which are much simple to understand such as the rank-frequency diagram (Frontier, 1977), K-dominance curves (Lambshead et al., 1983), abundance biomass comparison (ABC curves) (Warwick and Clarke, 1994). ABC curve is based on the k-dominance curve and it visualizes the representation of the evenness of numerical abundance and biomass (Lambshead et al., 1983). It is easy to interpret and easy to use because it does not require any reference site comparison, is mostly insensitive to the natural disturbance in populations, and can be used without species identification. In an undisturbed ecosystem biomass curve lies above the numerical density curve and in a highly disturbed system, number lies above biomass and intermediate levels indicate a moderate level of disturbances (Warwick, 1986). The Plymouth Routines in Multivariate Ecological Research (PRIMER) is one of the best-paid software packages available with most of these univariate indices or graphical methods and it is user-friendly. PRIMER Software package of PRIMER-E Limited developed by Professor K. Robert Clarke and Ray N. Gorley in the Plymouth Marine Laboratory, UK is now well recognized by scientific communities worldwide. This package is mainly developed for benthic studies (Clarke and Gorley, 2006) and later in the higher version of PRIMER offered with Permutational Multivariate

Table 10.1 Commonly used benthic indices in coastal and marine ecosystems.

Type of indices	Reference
Univariate indices	
Descriptors: Number of species (species richness), abundance (A), biomass (B)	
Indices of diversity	
Shannon–Wiener diversity index (H′)	Shannon and Weaver (1949)
Simpson's indices of dominance, diversity, and evenness	Simpson (1949)
Brillouin indices of diversity and evenness	Brillouin (1956)
Pielou evenness index (J′)	Pielou (1966)
Margalef's index	Margalef (1968)
Hurlbert index	Hurlbert (1971)
Hill's diversity numbers and evenness measures	Hill (1973)
Benthic pollution index (BPI)	Leppäkoski (1975)
Taxonomic diversity index and taxonomic distinctness	Warwick and Clarke (1995)
Graphical methods	
Rank-frequency diagram (RFD)	Frontier (1977)
K-dominance curves	Lambshead et al. (1983)
ABC curves	Warwick and Clarke (1994)
Ecological groups	
Indice Annélidien de Pollution	Bellan (1980)
Biotic index	Hily (1984)
Macrofauna monitoring index (MMI)	Roberts et al. (1998)
AZTI's marine biotic index (AMBI)	Borja et al. (2000)
BENTIX	Simboura and Zenetos (2002)
Indicator species index (ISI)	Rygg (2002)
Indice Biotique et Indice d'Evaluation de l'Endofaune Côtière (IE2C)	Grall and Glémarec (2003)
Benthic opportunistic polychaetes/amphipods ratio (BOPA)	Dauvin and Ruellet (2007)
Benthic index based on taxonomic sufficiency (BITS)	Mistri and Munari (2008)
Benthic opportunistic annelids/amphipods ratio (BO2A)	Dauvin and Ruellet (2009)

(*Continued*)

Table 10.1 (Continued)

Type of indices	Reference
p/aAMBI	Warwick et al. (2010)
NAMBI	Warwick et al. (2010)
BAMBI	Warwick et al. (2010)
PAMBI	Warwick et al. (2010)
gAMBI	Aylagas et al. (2018)
p/agAMBI	Aylagas et al., (2018)
Foram-AMBI	Alve et al. (2016)
microgAMBI	Aylagas et al. (2017)
Functional indices	
Infaunal trophic index (ITI)	Word (1979)
Ecofunctional quality index (EQI)	Fano et al. (2003)
Multimetric indices	
Pollution coefficient	Satsmadjis (1982)
Biological quality index (BiQI)	Jeffrey et al. (1985)
Organism sediment index	Rhoads and Germano (1986)
Infauna ratio-to-reference of sediment quality triad (RTR)	Chapman et al. (1987)
Benthic index of estuarine condition (BIEC)	Weisberg et al. (1993)
Benthic index of biotic integrity (B-IBI)	Ranasinghe et al. (1994)
Benthic condition index (BCI)	Engle et al. (1994)
Benthic habitat quality (BHQ)	Nilsson and Rosenberg (1997)
Virginia province benthic index (VPBI)	Paul et al. (2001)
Norwegian quality index (NQI)	Rygg (2002)
Index of environmental integrity (IEI)	Paul (2003)
Benthic quality index (BQI)	Rosenberg et al. (2004)
Infaunal quality index (IQI)	Prior et al. (2004)
Fuzzy index of environmental integrity for transitional environments (INES)	Mistri et al. (2005)
Marine biotic index tool (MarBIT)	Meyer et al. (2006)
Danske kvalitet indeks (DKI)	Borja et al. (2007)
Benthic ecosystem quality index (BEQI)	Van Hoey et al. (2007)
Brackish water benthic index (BBI)	Perus et al. (2007)
DAPHNE	Forni and Occhipinti-Ambrogi (2007)
Fuzzy index of ecosystem integrity (FINE)	Munari and Mistri (2008)
Macrobenthic index for semisheltered systems (MISS)	Lavesque et al. (2009)

(Continued)

Table 10.1 (Continued)

Type of indices	Reference
Benthic ecosystem quality index 2 (BEQI2)	van Loon et al. (2015)
ALien Biotic IndEX (ALEX)	Cinar and Bakir (2014)
Multivariate approaches, packages, and models	
Pollution load index (PLI)	Jeffrey et al. (1985)
Benthic response index (BRI)	Smith et al. (2001)
Multivariate-AMBI (M-AMBI)	Borja et al. (2004); Muxika et al. (2007)
Principal response curves (PRCs)	Pardal et al. (2004)
Typology and reference conditions for Portuguese and coastal waters (TICOR)	Bettencourt et al. (2004)
Combination of indices: B-IBI and TICOR	Chainho et al. (2008)
Acadian province benthic index (APBI)	Hale and Heltshe (2008)
Portuguese-benthic assessment tool (P-BAT)	Pinto et al. (2009)
TUrkish benthic index (TUBI)	Çinar et al. (2015)
M-BENTIX	Simboura et al. (2015)
M-BAMBI	Cai et al. (2014)
Genomic M-AMBI (M-gAMBI)	Aylagas et al. (2018)

Source: Modified from Borja, A., Franco, J., & Pérez, V., 2000. A marine biotic index to establish the ecological quality of soft-bottom benthos within European estuarine and coastal environments. Marine Pollution Bulletin 40, 1100–1114.

Analysis of Variance (PERMANOVA) package (Anderson, 2017). Most of these statistical analyses and graphics are now available with a variety of open-source packages developed by the R development core team (http://www.r-project.org/). The majority of the above-mentioned indices have its limitation to assess the quality of nature and to overcome these limitations, scientists developed indices which are based on an ecological group such as Annélidien de Pollution (Bellan, 1980), biotic index (Hily, 1984), macrofauna monitoring index (Roberts et al., 1998), AZTI's marine biotic index (AMBI) (Borja et al., 2000), BENTIX (Simboura and Zenetos, 2002), indicator species index (Rygg, 2002), Indice Biotique et Indice d'Evaluation de l'Endofaune Côtière (Grall and Glémarec, 2003), benthic opportunistic polychaetes/amphipods ratio (BOPA) (Dauvin and Ruellet, 2007), benthic index based on taxonomic sufficiency (Mistri and Munari, 2008), and benthic opportunistic annelids/amphipods ratio (Dauvin and Ruellet, 2009). Functional traits analysis can be used to assess the performance of organisms based on their morphological, physiological, phonological, and behavioral features or interaction between biological, physical, or chemical environments (Bremner et al., 2006a; Violle et al., 2014). Trait analysis along with the traditional taxonomic approach can be used for ecosystem monitoring (Rand et al., 2018). Biological trait analysis is based on this information on species life history, morphological and behavioral characteristics and it helps to understand the ecological responses (Bremner et al., 2006a). Different types of functional indices were employed in many coastal studies like the infaunal trophic index (ITI) (Word, 1979) and the ecofunctional quality index (Fano et al., 2003). Many studies considered multimetric indices which are much advanced indices used for the coastal marine benthic quality assessment such as pollution coefficient (Satsmadjis, 1982), biological quality index (BiQI) (Jeffrey et al., 1985), organism sediment index (Rhoads and Germano, 1986), infauna ratio-to-reference of sediment quality triad (Chapman et al., 1987), benthic index of estuarine condition (Weisberg et al., 1993), benthic index of biotic integrity (B-IBI) (Ranasinghe et al., 1994), benthic condition index (Engle et al., 1994), benthic habitat quality (Nilsson and Rosenberg, 1997), Virginia province benthic index (Paul et al., 2001), Norwegian quality index (Rygg, 2002), index of environmental integrity (Paul, 2003), benthic quality index (BQI) (Rosenberg et al., 2004), infaunal quality index (Prior et al., 2004), fuzzy index of environmental integrity for transitional environments (Mistri et al., 2005), marine biotic index tool (Meyer et al., 2006), Danske kvalitet indeks (Borja et al., 2007), benthic ecosystem quality index (Van Hoey et al., 2007), brackish water benthic index (Perus et al., 2007), DAPHNE (Forni and Occhipinti-Ambrogi, 2007), fuzzy index of ecosystem integrity (Munari and Mistri, 2008), and macrobenthic index for semisheltered systems (Lavesque et al., 2009). In addition to the multivariate approaches, packages and models are used by scientists. These indices incorporate multiple factors to get better results some time they compare the sites with different reference conditions such as multivariate-AMBI (M-AMBI) (Borja et al., 2004; Muxika et al., 2007), pollution load index (PLI) (Jeffrey et al., 1985), benthic response index (Smith et al., 2001), principal response curves (Pardal et al., 2004), typology and reference conditions for Portuguese transitional and coastal waters (TICOR) (Bettencourt et al., 2004), the combination of indices (B-IBI and TICOR) (Chainho et al., 2008), Acadian province benthic index (Hale and Heltshe, 2008), and Portuguese-benthic assessment tool (Pinto et al., 2009).

AMBI developed by Borja et al. (2000) laid the foundation for the development of many of the recent benthic biotic indices such as BENTIX (Simboura and Zenetos, 2002), M-AMBI (Borja et al., 2004; Muxika et al., 2007). Similarly, M-BENTIX was developed by following the concept of M-AMBI (Simboura et al., 2015). The basis of AMBI is the sensitivity of an ecological group to various stressors where these ecological groups are classified based on life span, growth rate, sexual maturation and larvae development strategy, the biomass production rate of species (Gray, 1979). Based on this, the benthic macroinvertebrates can be classified into five ecological groups (EGs) (Hily, 1984; Glemarec, 1986) such as species sensitive to disturbance (EG_I), indifferent to disturbance (EG_{II}), tolerant to disturbance (EG_{III}), second-order opportunistic species to disturbance (EG_{IV}), and first-order opportunistic species to disturbance (EG_V) (Borja et al., 2000, 2008). AMBI can be calculated with the free software package of AZTI-Tecnalia developed by Prof. Angel Borja and it can be downloaded from their website (https://ambi.azti.es/descarga-de-ambi/) (Borja et al., 2000).

$$AMBI = 0EG_I + 1.5EG_{II} + 3EG_{III} + 4.5EG_{IV} + 6EG_V$$

For the calculation of BENTIX index, the same ecological groups are classified into G_I and G_{II}. EG_1 and EG_{II} were considered as G_I, and EG_{III}, EG_{IV}, and EG_V were grouped into G_{II} (Simboura and Zenetos, 2002).

$$BENTIX = 6G_I + 2G_{II}$$

The concept of AMBI has also formed the basis for the Mediterranean occidental index (Pinedo and Jordana, 2007). Warwick et al. (2010) developed a different approach based on AMBI concepts such as p/aAMBI based on presence/absence data, NAMBI based on transformed abundance data, BAMBI based on biomass data, and PAMBI based on production data. Further development results in genomic AMBI (gAMBI), and p/agAMBI (Aylagas et al., 2014) with the support of recently developed metabarcoding tools. The AMBI concept also formed the basis for ALien Biotic IndEX (ALEX) (Çinar and Bakir, 2014) which is used for alien species assessment. In ALEX four groups of such as native species, casual alien species, established alien species, and invasive species are used instead of conventional ecological groups. The factorial analysis M-AMBI further developed into benthic assessment tool (Teixeira et al., 2009), M-BAMBI (Cai et al., 2014), the benthic ecosystem quality index 2 (van Loon et al., 2015), Turkish benthic index (Çinar et al., 2015), and genomic M-AMBI (Aylagas et al., 2018). In addition to these indices, AMBI is also applied to meiofaunal foraminifera (Alve et al., 2016), and microfaunal groups (Aylagas et al., 2017).

Most of these indices depend on the ecological response of species in particular communities. The diversity and abundance of most of these species vary with organic matter content in the sediment they live (Borja et al., 2000). Benthic opportunistic species are proliferating in organic matter enriched sediment. However, they respond to both natural and anthropogenic stress, so it is difficult to identify what is the major cause for such variation. Many studies are based on the combination of multiple measures like primary indices and derived community indices along with environmental data and other supporting data from the study site and reference conditions (Weisberg et al., 1997). Such multimeasure indices are most successful as compared to conventional indices but still, these indices need to refine from their present form (Borja et al., 2009; Lavesque et al., 2009). Among these, AMBI, M-AMBI, BQI, B-IBI, and ITI are among the most commonly used indices along the coastal and transitional zones around the world (ICES, 2008). Nevertheless, research on these indices is in progress to re-examine and calibrate to different conditions in the coastal marine environments that include transitional waters (Borja et al., 2009).

10.4 Commonly used marine biotic indices in Indian waters

Human activities have significantly impacted the coastal and transitional waters of India (Equbal et al., 2018; Feebarani et al., 2016; Jayachandran and Bijoy Nandan, 2012). This brings the attention of scientists to study the anthropogenic impact on the coastal marine ecosystems of India. Most of the studies are confined to chemical aspects of pollution and fewer studies are evaluated on biological components (Qasim et al., 1988). Recently, many studies have been published based on ecological quality assessment in the coastal systems of India (Sivadas et al., 2010). The benthic macroinvertebrate based simple benthic quality descriptors are conventionally used in the coastal monitoring programs of Indian waters, it is considered a powerful biological indicators of pollution (Ajmal Khan et al., 2004, 2014; Jayachandran et al., 2020; Mulik et al., 2017; Naidu et al., 2018; Pinto et al., 2009; Sigamani et al., 2015; Vijapure & Sukumaran, 2019; Zhang et al., 2013; Asha, 2017; Bijoy Nandan, 1991). The most commonly used benthic descriptors in the coastal waters o

India include species richness, abundance, and biomass. Many studies have used the possibility of diversity indices such as Margalef species richness (d), Shannon diversity index (H′), Pielou's evenness index (J′), Simpson dominance $(1-\lambda')$, and Hurlbert index (Hurlbert, 1971) (Ajmal Khan et al., 2004, 2014, Equbal et al., 2017, 2018, Jayachandran et al., 2019c, 2020; Mulik et al., 2017; Naidu et al., 2018; Rehitha et al., 2017; Sivadas et al., 2016; Srinivas et al., 2019; Vijapure & Sukumaran, 2019). The graphical methods were also widely used to understand the variability between impacted and control sites (Ajmal Khan et al., 2004; Equbal et al., 2017, 2018; Jayachandran et al., 2019c; Sigamani et al., 2019; Sivadas et al., 2016; Srinivas et al., 2019). However, recently more complex methods that are successfully implemented in the temperate regions were tested in the Indian waters to assess the ecological quality of coastal and estuarine waters. Mainly ecological sensitivity based indices such as AMBI, M-AMBI, BENTIX, BOPA, and BO2A were continuously tested in the coastal and transitional waters of India (Ajmal Khan et al., 2014; Dias et al., 2018; Equbal et al., 2018; Feebarani et al., 2016; Jayachandran et al., 2020; Mulik et al., 2020a; Rehitha et al., 2017; Sigamani et al., 2015, 2019; Sivadas et al., 2016). A study by Mulik et al. (2017) compared the most commonly used benthic indices in India and point out the major limitations of indices used in the coastal systems of India. Based on their study, AMBI is the most efficient indices in Indian waters. Some studies highlighted the effect of seasonality in results and tested different transformations of data for getting better results (Mulik et al., 2020b). Their result indicated that during monsoon season AMBI failed in the Ulhas estuary, Maharashtra for the correct assessment of ecological state. Sivaraj et al. (2014) studied the effluent discharge sites in the Nandgaon coastal waters of Maharashtra state, where M-AMBI value indicated moderate to high ecological status. Based on the AMBI and M-AMBI, the ecological quality of the Vellar estuary, Tamil Nadu was assessed as high ecological condition while in the Ennore estuary of same state, it was recorded as a poor state (Ajmal Khan et al., 2014; Selvaraj et al., 2019). An AMBI-based study on Coleroon estuaries of Tamil Nadu indicated a moderate level of disturbance (Sigamani et al., 2015). Similarly in the southern part of Vembanad lake, Kerala a good ecological condition was recorded in M-AMBI in contrast to moderately disturbed conditions recorded in the AMBI (Asha, 2017). Such contrasting results were also recorded in the other estuaries of India like the Vellar–Coleroon estuarine system (Ajmal Khan et al., 2014). While in the middle region of

Vembanad Lake, in the Cochin estuary, Kerala a moderate to poor ecological status were recorded (Feebarani et al., 2016). Based on the result of BO2A index, poor ecological conditions were recorded in the dredging zones of the Cochin estuary by Rehitha et al. (2017). At the same time, good to moderate levels recorded in AMBI, good to poor levels in M-AMBI, high to moderate levels in BENTIX, and high to poor levels in BOPA were recorded in the northern part of Vembanad Lake, Kodungallur-Azhikode estuary (Jayachandran et al., 2020). Based on the comprehensive study by Sivadas et al. (2016) suggested that a multibenthic indices approach is better to assess the coastal waters of India and need standardization for specific locations. Therefore future studies should be focused on the intercalibration of indices on a spatiotemporal scale to improve its efficiency in the Indian waters (Tables 10.2 and 10.3).

10.5 Major limitations of currently used benthic biotic Indices

Ecological quality assessments of coastal marine environments are increasingly studied by scientists around the world to develop simple and reliable tools for the general public, nonspecialists and ecosystem managers. However, the increasing number of indices indicates that these indices are not accepted by ecosystem managers or scientists due to their inefficiency for a successful evaluation. Scientists are seeking an index that integrates multivariate data set to a simple value that can simply indicate the state of the ecosystem. It should be indicative of ecological integrity, response to anthropogenic disturbance, capable to indicate spatial changes, and indicative of causes of changes. Most of the currently used indices need to test their efficiency to an individual or multiple human pressures in different localities. Still, studies are trying to intercalibrate different approaches for a long period to get better consistency in the results. Many researchers suggest that need much more specific indices to different stressors, communities, or locations. Studies also recommend developing better indices that are susceptible to natural environmental gradients such as salinity, sediment type, and locality in addition to seasonality variations. A major limitation is that most of these benthic species are not uniformly sensitive to all kinds of pollutants. They represent a static expression of an ecological condition and are not explicitly linked to changes in ecological function, not be specific concerning different kinds of stressors, they subject to underlying taxonomic changes across estuarine gradients, their use can be

Major studies on benthic biotic indices in Indian waters (+ present, − absent).

Sl. no.	Location	Ecosystem type	AMBI	BENTIX	BOPA	BO2A	ITI	NQI	IQI	DKI	BEQI	M-AMBI	References
1	Ulhas, Amba, Kundalika, Savitri, Vellar, Uppanar, and Vellar estuary	Estuarine	+	−	−	−	−	−	−	−	−	−	Srinivas et al. (2019)
2	Uppanar and Vellar estuary	Estuarine	+	−	−	−	−	−	−	−	−	+	Ajmal Khan et al. (2014)
3	Uppanar estuary	Estuarine	−	−	−	−	−	−	−	−	−	−	Ajmal Khan et al. (2004)
4	Cochin estuary	Estuarine	+	−	−	−	−	−	−	−	−	+	Feebarani et al. (2016)
5	Kodungallur-Azhikode estuary	Estuarine	+	+	+	−	−	−	−	−	−	+	Jayachandran et al. (2020)
6	Ulhas estuary	Estuarine	+	+	+	+	−	−	−	−	−	+	Mulik et al. (2017)
7	Ulhas, Amba, and Savitri estuaries	Estuarine	+	+	+	+	−	−	−	−	−	+	Mulik et al. (2020a)
8	Vembanad Lake	Estuarine	+	−	−	−	−	−	−	−	−	+	Asha et al. (2016)
9	Cochin estuary	Estuarine	−	−	−	+	−	−	−	−	−	−	Rehitha et al. (2017)

(*Continued*)

Table 10.2 (Continued)

Sl. no.	Location	Ecosystem type	AMBI	BENTIX	BOPA	BO2A	ITI	NQI	IQI	DKI	BEQI	M-AMBI	References
10	Vellar–Coleroon	Estuarine	+	–	–	–	–	–	–	–	–	+	Sigamani et al. (2015)
11	Cuddalore	Coastal	–	–	–	–	–	–	–	–	–	–	Sigamani et al. (2019)
13	Nandgaon	Coastal	+	–	–	–	–	–	–	–	–	+	Sivaraj et al. (2014)
14	Kochi	Coastal	–	+	–	–	–	–	–	–	–	–	Naidu et al. (2018)
15	Port Blair Bay	Bay	+	+	+	–	–	–	–	–	–	+	Equbal et al. (2017)
16	South Andaman	Intertidal	+	+	+	–	–	–	–	–	–	+	Equbal et al. (2018)
17	Mumbai to Palk Bay	Estuaries, bays, and coastal	+	–	+	–	+	+	+	+	+	+	Sivadas et al. (2016)
18	Vellar and Ennore estuaries	Estuarine	+	–	–	–	–	–	–	–	–	+	Selvaraj et al. (2019)
19	Ulhas estuary	Estuarine	+	–	–	–	–	–	–	–	–	–	Mulik et al. (2020b)

Table 10.3 Ecological quality (EcoQ) status classes of commonly used benthic indices to classify coastal waters of India.

	EcoQ status					
	High	Good	Moderate	Poor	Bad	
H′log2	> 4	4–3	3–2	2–1	< 1	Blanchet et al. (2008)
BOPA	0.00–0.045	0.045–0.139	0.139–0.193	0.193–0.267	0.267–0.301	Dauvin and Ruellet (2007)
AMBI	0.0–1.2	1.2–3.3	3.3–4.3	4.3–5.5	> 5.5	Borja et al. (2000)
M-AMBI	> 0.77	0.77–0.53	0.53–0.38	0.38–0.20	< 0.20	Muxika et al. (2007)
BENTIX	6–4.5	4.5–3.5	3.5–2.5	2.5–2.0	0	Simboura and Zenetos (2002)
BO2A	< 0.025	0.025–0.130	0.130–0.199	0.199–0.255	> 0.255	Dauvin and Ruellet (2009)
BPOFA	0.031	0.031–0.126	0.126–187	0.187–0.237	> 0.237	Dauvin et al. (2016)
BPA	0.135	0.135–0.211	0.211–0.260	0.260–0.300	< 0.300	Dauvin et al. (2016)

labor intensive, and it cannot be applied always across biogeographic regions. One of the major issues for the currently used conventional marine benthic biotic indices is with time-consuming conventional taxonomic identification and lack of taxonomic experts, and it has low upscaling potential to resolve change. It is very difficult to identify and quantify the rare or cryptic species and juvenile stages. The incorporation of modern methods like microarray, Real-Time quantitative PCR (qPCR), metabarcoding tools, remote sensing, acoustic methods, in situ monitoring instruments may help to resolve some of these difficulties and it is much cheaper and faster than traditional methods. The major issue with the genomic method is that quantitative evaluation is much difficult and it requires careful standardization before implementation.

Acknowledgments

The authors are also thankful to the Head, Department of Marine Biology, Microbiology and Biochemistry, Cochin University of Science and Technology, for providing necessary facilities. The first author is thankful to Ángel Borja, AZTI, Marine Research Division, Spain, for his pieces of advice and help for work.

References

Ajmal Khan, S., Manokaran, S., & Lyla, P. S. (2014). Assessment of ecological quality of Vellar and Uppanar estuaries, southeast coast of India, using benthos. *Mitigation and Adaptation Strategies for Global Change*, *43*, 1989–1995.

Ajmal Khan, S., Murugesan, P., Lyla, P. S., & Jaganathan, S. (2004). A new indicator macroinvertebrate of pollution and utility of graphical tool and diversity indices in pollution monitoring studies. *Current Science*, *87*, 1508–1510.

Alve, E., Korsun, S., Schonfeld, J., Dijkstra, N., Golikova, E., Hess, S., Husum, K., & Panieri, G. (2016). Foram-AMBI: A sensitivity index based on benthic foraminiferal faunas from North-East Atlantic and Arctic fjords, continental shelves and slopes. *Marine Micropaleontology*, *122*, 1–12.

Anderson, M. J. (2017). Permutational multivariate analysis of variance (PERMANOVA). In N. Balakrishnan, T. C. B. Everitt, W. Piegorsch, F. Ruggeri, & J. L. Teugels (Eds.), *Wiley StatsRef: Statistics reference online* (pp. 1–15). Wiley.

Anu, P. R., Bijoy Nandan, S., Jayachandran, P. R., & Don Xavier, N. D. (2016). Toxicity effects of copper on the marine diatom, *Chaetoceros calcitrans*. *Regional Studies in Marine Science*, *8*, 498–504.

Asha, C. V. (2017). Assessing the structural and functional ecology of benthic fauna in Vembanad estuarine system, India. Cochin University of Science and Technology, Kochi, Kerala, India.

Asha, C. V., Retina, I. C., Suson, P. S., & Bijoy Nandan, S. (2016). Ecosystem analysis of the degrading Vembanad wetland ecosystem, the largest Ramsar site on the south west coast of India—Measures for its sustainable management. *Regional Studies in Marine Science*, *8*, 408–421.

Aylagas, E., Borja, Á., Muxika, I., & Rodríguez-Ezpeleta, N. (2018). Adapting metabarcoding-based benthic biomonitoring into routine marine ecological status assessment networks. *Ecological Indicators, 95*(1), 194–202.

Aylagas, E., Borja, A., & Rodriguez-Ezpeleta, N. (2014). Environmental status assessment using DNA metabarcoding: Towards a genetics-based marine biotic index (gAMBI). *PLoS One, 9*, e90529.

Aylagas, E., Borja, A., Tangherlini, M., Dell'Anno, A., Corinaldesi, C., Michell, C. T., Irigoien, X., Danovaro, R., & Rodríguez-Ezpeleta, N. (2017). A bacterial community-based index to assess the ecological status of estuarine and coastal environments. *Marine Pollution Bulletin, 114*, 679–688.

Bald, J., Borja, A., Muxika, I., Franco, J., & Valencia, V. (2005). Assessing reference conditions and physicochemical status according to the European Water Framework Directive: A case-study from the Basque Country (Northern Spain). *Marine Pollution Bulletin, 50*, 1508–1522.

Bellan, G. (1980). Relationships of pollution to rocky substratum polychaetes on the French Mediterranean coast. *Marine Pollution Bulletin, 11*, 318–321.

Bettencourt, A. M., Bricker, S. B., Ferreira, J. G., Franco, A., Marques, J. C., Melo, J. J., Nobre, A., Ramos, L., Reis, C. S., Salas, F., Silva, M. C., Simas, T., & Wolff, W. J. (2004). *Typology and reference conditions for Portuguese transitional and coastal waters* (p. 98) Instituto da Agua (INAG) and Institute of Marine Research (IMAR).

Bijoy Nandan, S. (1991). Effect of coconut husk retting on the water quality and biota of an aquatic biotope in Kerala. University of Kerala.

Bijoy Nandan, S. (2008). Current status and biodiversity modification in the coastal wetland ecosystems of India with objectives for its sustainable management, In Proceedings of the conserv-vision conference, The University of Waikato, http://www.waikato.ac.n/wtass/conserve-vision.

Blanchet, H., Lavesque, N., Ruellet, T., Dauvin, J. C., Sauriau, P. G., Desroy, N., Desclaux, C., Leconte, M., Bachelet, G., Janson, A. L., Bessineton, C., Duhamel, S., Jourde, J., Mayot, S., Simon, S., & de Montaudouin, X. (2008). Use of biotic indices in semi-enclosed coastal ecosystems and transitional waters habitats—Implications for the implementation of the European Water Framework Directive. *Ecological Indicators, 8*, 360–372.

Borja, A., Aguirrezabalaga, F., Martínez, J., Sola, J. C., García-Arberas, L., & Gorostiaga, J. M. (2004). Benthic communities, biogeography and resources management. In A. Borja, & M. Collins (Eds.), *Oceanography and marine environment of the Basque Country. Elsevier oceanography series* (pp. 455–492). Elsevier.

Borja, A., Basset, A., Bricker, S., Dauvin, J., Elliott, M., Harrison, T., Marques, J., Weisberg, S., & West, R. J. (2011). *Classifying ecological quality and integrity of estuaries. Treatise on estuarine and coastal science, volume 1: Classification of estuarine and nearshore coastal ecosystems* (pp. 125–162). Elsevier.

Borja, A., Dauer, D. M., Díaz, R., Llansó, R. J., Muxika, I., Rodríguez, J. G., & Schaffner, L. (2008). Assessing estuarine benthic quality conditions in the Chesapeake Bay: A comparison of three indices. *Ecological Indicators, 8*, 395–403.

Borja, A., Franco, J., & Pérez, V. (2000). A marine biotic index to establish the ecological quality of soft-bottom benthos within European estuarine and coastal environments. *Marine Pollution Bulletin, 40*, 1100–1114.

Borja, A., Josefson, A. B., Miles, A., Muxika, I., Olsgard, F., Phillips, G., Rodríguez, J. G., & Rygg, B. (2007). An approach to the intercalibration of benthic ecological status assessment in the North Atlantic ecoregion, according to the European Water Framework Directive. *Marine Pollution Bulletin, 55*, 42–52.

Borja, A., Ranasinghe, A., & Weisberg, S. B. (2009). Assessing ecological integrity in marine waters, using multiple indices and ecosystem components: Challenges for the future. *Marine Pollution Bulletin, 59*, 1–4.

Bremner, J., Rogers, S. I., & Frid, C. L. J. (2006a). Methods for describing ecological functioning of marine benthic assemblages using biological traits analysis (BTA). *Ecological Indicators, 6*, 609–622.

Brillouin, L. (1956). *Science and information theory*. Academic Press.

Cai, W., Borja, A., Liu, L., Meng, W., Muxika, I., & Rodnguez, J. G. (2014). Assessing benthic health under multiple human pressures in Bohai Bay (China), using density and biomass in calculating AMBI and M-AMBI. *Marine Ecology, 35*, 180–192.

Chainho, P., Chaves, M. L., Costa, J. L., Costa, M. J., & Dauer, D. M. (2008). Use of multimetric indices to classify estuaries with different hydromorphological characteristics and different levels of human pressure. *Marine Pollution Bulletin, 56*, 1128–1137.

Chapman, P. M., Dexter, R. N., & Long, E. R. (1987). Synoptic measures of sediment contamination, toxicity and infauna community composition (the Sediment Quality Triad) in San Francisco Bay. *Marine Ecology Progress Series, 37*, 75–96.

Çinar, M. E., & Bakir, K. (2014). ALien Biotic IndEX (ALEX)—A new index for assessing impacts of alien species on benthic communities. *Marine Pollution Bulletin, 87*, 171–179.

Çinar, M. E., Bakır, K., Öztürk, B., Katağan, T., Dağlı, E., Açık, S., Doğan, A., & Bakır, B. B. (2015). TUBI (TUrkish benthic index): A new biotic index for assessing impacts of organic pollution on benthic communities. *Journal of the Black Sea/Mediterranean Environment, 21*, 135–168.

Clarke, K. R., & Gorley, R. N. (2006). PRIMER v6: User manual/tutorial (Plymouth routines in multivariate ecological research). PRIMER-E, Plymouth.

Coull, B. C. (1977). *Ecology of marine benthos* (p. 467) University of South Carolina Press.

Danovaro, R., Carugati, L., Berzano, M., Cahill, A. E., Carvalho, S., Chenuil, A., Corinaldesi, C., Cristina, S., David, R., Dell'Anno, A., Dzhembekova, N., Garcés, E., Gasol, J. M., Goela, P., Féral, J.-P., Ferrera, I., Forster, R. M., Kurekin, A. A., Rastelli, E., ... Borja, A. (2016). Implementing and innovating marine monitoring approaches for assessing marine environmental status. *Frontiers in Marine Science, 3*, 213.

Dauvin, J. C. (2007). Paradox of estuarine quality: Benthic indicators and indices, consensus or debate for the future. *Marine Pollution Bulletin, 55*, 271–281.

Dauvin, J. C., Andrade, H., de-la-Ossa-Carretero, J. A., Del-Pilar-Ruso, Y., & Riera, R. (2016). Polychaete/amphipod ratios: An approach to validating simple benthic indicators. *Ecological Indicators, 63*(89-99).

Dauvin, J. C., & Ruellet, T. (2007). Polychaete/amphipod ratio revisited. *Marine Pollution Bulletin, 55*, 215–224.

Dauvin, J. C., & Ruellet, T. (2009). The estuarine quality paradox: Is it possible to define an ecological quality status for specific modified and naturally stressed estuarine ecosystems? *Marine Pollution Bulletin, 59*, 38–47.

Dias, H. Q., Sukumaran, S., Srinivas, T., & Mulik, J. (2018). Ecological quality status evaluation of a monsoonal tropical estuary using benthic indices: Comparison via a seasonal approach. *Environmental Science and Pollution Research, 25*, 22672–22688.

Don Xavier, N. D., Bijoy Nandan, S., Jayachandran, P. R., Anu, P. R., Midhun, A. M., & Mohan, D. (2019). Chronic effects of copper and zinc on the fish, *Etroplus suratensis* (Bloch, 1790) by continuous flow-through (CFT) bioassay. *Marine Environmental Research, 143*, 141–157.

Ducrotoy, J. P., Mazik, K., & Elliott, M. (2010). *Bio-sedimentary indicators for estuaries: A critical review* (pp. 1–77). Océanis Institut Océanographique.

Engle, V. D., Summers, J. K., & Gaston, G. R. (1994). A benthic index of the environmental condition of Gulf of Mexico estuaries. *Estuaries, 17*, 372–384.

Equbal, J., Lakra, R. K., Savurirajan, M., Satyam, K., & Thiruchitrambalam, G. (2018). Testing performances of marine benthic biotic indices under the strong seasonality in the tropical intertidal habitats, South Andaman, India. *Marine Pollution Bulletin, 135*, 266–282.

Equbal, J., Thiruchitrambalam, G., Lakra, R. K., Savurirajan, M., & Satyam, K. (2017). Assessing the ecological quality of the Port Blair coast (South Andaman, India) using different suites of benthic biotic indices. *Journal of the Marine Biological Association of the United Kingdom, 97*, 1007–1021.

Fano, E. A., Mistri, M., & Rossi, R. (2003). The ecofunctional quality index (EQI): A new tool for assessing lagoonal ecosystem impairment. *Estuarine, Coastal and Shelf Science, 56*, 709–716.

Farrington, J. W., Tripp, B. W., Tanabe, S., Subramanian, A., Sericano, J. L., Wade, T. L., & Knap, A. H. (2016). Edward D. Goldberg's proposal of "the Mussel Watch": Reflections after 40 years. *Marine Pollution Bulletin, 110*, 501–510.

Feebarani, J., Joydas, T. V., Damodaran, R., & Borja, A. (2016). Benthic quality assessment in a naturally- and human-stressed tropical estuary. *Ecological Indicators, 67*, 380–390.

Forni, G., & Occhipinti-Ambrogi, A. (2007). *Daphne: A new multimetric benthic index for the quality assessment of marine coastal environment in the Northern Adriatic Sea. Chemistry and Ecology* (23, pp. 427–442).

Frontier, S. (1977). Utilisation des diagrammes rang-fréquence dans l'analyse des systèmes. *Journal de Recherches Océanographiques, 1*, 33–58.

Glemarec, M. (1986). Ecological impact of an oil-spill: Utilisation of biological indicators. *IAWPRC Journal, 18*, 203–211.

Grall, J., & Glémarec, M. (2003). L'indice d'évaluation de l'endofaune côtière. In C. Alzieu (Ed.), *Bioévaluation de la qualité environnementale des sédiments portuaires et des zones d'immersion, Ifremer* (pp. 51–85). Ifremer.

Gray, J. S. (1979). Pollution-induced changes in populations. *Philosophical Transactions of the Royal Society of London. Series B, Biological Sciences, 286*, 545–561.

Hale, S. S., & Heltshe, J. F. (2008). Signals from the benthos: Development and evaluation of a benthic index for the nearshore Gulf of Maine. *Ecological Indicators, 8*, 338–350.

Hawkins, A. J. S., Bayne, B. L., Menon, N. R., & Damodaran, R. (1986). The mussels *Perna viridis* and Perna indica as transplantable indicators of pollution: Comparison of their metabolic response to reductions of both oxygen tension and salinity. In N. R. Menon (Ed.), *National seminar on Mussel Watch* (pp. 51–62). Cochin University of Science and Technology.

Hayes, K. R., Lynne, V., Dambacher, J. M., Sharples, R., & Smith, R. (2008). Ecological Indicators for the Exclusive Economic Zone Waters of the South West Marine Region Final Report (08/82) for the Australian Government Department of the Environment and Heritage, p. 100.

Hill, M. O. (1973). Diversity and evenness: A unifying notation and its consequences. *Ecology, 54*, 427–432.

Hily, C. (1984). Variabilité de la macrofaune benthique dans les milieux hypertrophiques de la Rade de Brest. Thèse de Doctorat d'Etat, Université de Bretagne Occidentale Vol. 2, 337 pp.

ICES. (2008). Report of the workshop on benthos related environment metrics (WKBEMET). Oostende, Belgium, 11–14 February 2008, ICES CM2008/MHC:01.

Jayachandran, P. R., Aneesh, B. P., Oliver, P. G., Philomina, J., Jima, M., Harikrishnan, K., & Bijoy Nandan, S. (2019a). First record of the alien invasive biofouling mussel *Mytella strigata* (Hanley, 1843) (Mollusca: Mytilidae) from Indian waters. *BioInvasions Records, 8*, 828–837.

Jayachandran, P. R., & Bijoy Nandan, S. (2012). Assessment of trophic change and its probable impact on tropical estuarine environment (the Kodungallur-Azhikode estuary, India). *Mitigation and Adaptation Strategies for Global Change, 17*, 837–847.

Jayachandran, P. R., Bijoy Nandan, S., Jima, M., Joseph, P., Don Xavier, N. D., Sreedevi, O. K., Prabhakaran, M. P., & Joseph, K. J. (2019b). Macrobenthic functional feeding groups in a microtidal monsoonal estuary (Kodungallur–Azhikode estuary, India). *Regional Studies in Marine Science, 25*, 100444.

Jayachandran, P. R., Bijoy Nandan, S., Jima, M., Sreedevi, O. K., Philomina, J., & Prabhakaran, M. P. (2019c). Bioecology of macrobenthic communities in the microtidal monsoonal Kodungallur–Azhikode Estuary, southwest coast of India. *Lakes & Reservoirs: Science, Policy and Management for Sustainable Use, 24*, 372–390.

Jayachandran, P. R., Bijoy Nandan, S., Sanu, V. F., Anu, P. R., Jima, M., Midhun, A. M., Don Xavier, N. D., Philomina, J., & Asha, C. V. (2018a). Authentication of Nassodonta insignis H. Adams, 1867 (Gastropoda: Nassariidae) from the Kodungallur—Azhikode backwater, southwest coast of India using mitochondrial COI marker. *Indian Journal of Geo-Marine Sciences, 47*, 623–628.

Jayachandran, P. R., Bijoy Nandan, S., & Sreedevi, O. K. (2012). Water quality variations and nutrient characteristics of Kodungallur-Azhikode Estuary, Kerala, India. *Indian Journal of Geo-Marine Sciences, 41*, 180–187.

Jayachandran, P. R., Jima, M., Philomina, J., & Bijoy Nandan, S. (2020). Assessment of benthic macroinvertebrate response to anthropogenic and natural disturbances in the Kodungallur-Azhikode estuary, southwest coast of India. *Environmental Monitoring and Assessment, 192*, 626.

Jayachandran, P. R., Jima, M., Philomina, J., Sanu, V. F., & Bijoy Nandan, S. (2018b). Invasion of biofouling mussel Mytilopsis Conrad, 1857 (Bivalvia: Dreissenacea) in the Cochin backwaters, southwest coast of India. *Current Science, 115*, 2198–2200.

Jeffrey, D. W., Wilson, J. G., Harris, C. R., & Tomlinson, D. L. (1985). The application of two simple indices to Irish estuary pollution status. In J. G. Wilson (Ed.), *Estuarine management and quality assessment* (pp. 147–165). Plenum.

Jørgensen, S. E., Xu, F. L., & Costanza, R. (2010). *Handbook of ecological indicators for assessment of ecosystem health*. CRC Press/Taylor and Francis.

Krishnakumar, P. K., Asokan, P. K., & Pillai, V. K. (1990). Physiological and cellular responses to copper and mercury in the green mussel Perna viridis (Linnaeus). *Aquatic Toxicology, 18*, 163–173.

Lalli, C. M., & Parsons, T. R. (1997). Benthos. In C. M. Lalli, & T. R. Parsons (Eds.), *Biological oceanography: An introduction* (2nd ed., pp. 177–195). Butterworth-Heinemann.

Lambshead, P. J. D., Platt, H. M., & Shaw, K. W. (1983). The detection of differences among assemblages of marine benthic species based on an assessment of dominance and diversity. *Journal of Natural History, 17*, 859–874.

Lavesque, N., Blanchet, H., & de Montaudouin, X. (2009). Development of a multimetric approach to assess perturbation of benthic macrofauna in *Zostera noltii*. *Journal of Experimental Marine Biology and Ecology, 368*, 101–112.

Leppäkoski, E. (1975). *Assessment of degree of pollution on the basis of macrozoobenthos in marine and brackish-water environments*. Acta Academiae.

Levin, L. A. (2000). Polychaetes as environmental indicators: Response to low oxygen and organic enrichment. *Bulletin of Marine Science, 67*, 668–678.

Mare, M. F. (1942). A study of a marine benthic community with special reference to the micro organisms. *Journal of the Marine Biological Association of the United Kingdom, 25*, 517–554.

Margalef, R. (1968). *Perspectives in ecological theory*. University Chicago Press.

McLusky, D. S., & Elliott, M. (2004). *The estuarine ecosystem: Ecology, threats, and management* (3rd ed.). Oxford University Press.

Meyer, T., Reincke, T., & Fürhaupter, K. (2006). *Ostsee makrozoobenthos klassifizie rungssystem für die Wasserrahmenrichtlinie*. University of Rostock.

Mistri, M., & Munari, C. (2008). BITS: A SMART indicator for soft-bottom, non-tidal lagoons. *Marine Pollution Bulletin, 56*, 587–599.

Mistri, M., Munari, C., & Marchini, A. (2005). INES: A new fuzzy index of environmental integrity for transitional environments. In *15th Meeting of the Italian Society of Ecology*. http://www.xvcongresso.societaitalianaecologia.org/articles/ (accessed July 2010).

Mulik, J., Sukumaran, S., & Dias, H. Q. (2020a). Can the ecological status of three differentially impacted monsoonal tropical estuaries in NW India, be adequately assessed by a common estuarine benthic index? *Ecological Indicators, 119*, 106807.

Mulik, J., Sukumaran, S., & Dias, H. Q. (2020b). Is the benthic index AMBI impervious to seasonality and data transformations while evaluating the ecological status of an anthropized monsoonal estuary? *Ocean & Coastal Management, 186*, 105080.

Mulik, J., Sukumaran, S., Srinivas, T., & Vijapure, T. (2017). Comparative efficacy of benthic biotic indices in assessing the Ecological Quality Status (EcoQS) of the stressed Ulhas estuary, India. *Marine Pollution Bulletin, 120*, 192–202.

Munari, C., & Mistri, M. (2008). The performance of benthic indicators of ecological change in Adriatic coastal lagoons: Throwing the baby with the water. *Marine Pollution Bulletin, 56*, 95–105.

Muxika, I., Borja, Á., & Bald, J. (2007). Using historical data, expert judgement and multivariate analysis in assessing reference conditions and benthic ecological status, according to the European Water Framework Directive. *Marine Pollution Bulletin, 55*, 16–29.

Naidu, S. A., Rao, G. D., Vardhan, K. V., Kumaraswami, M., Rao, V. R., & Ramu, K. (2018). Impact of natural and anthropogenic disturbances on the benthic ecological quality of the coastal waters of Kochi, southwest coast of India. *International Journal of Environmental Research, 12*, 861–875.

Nilsson, H. C., & Rosenberg, R. (1997). Benthic habitat quality assessment of an oxygen stressed fjord by surface and sediment profile images. *Journal of Marine Systems, 11*, 249–264.

Pardal, M. A., Cardoso, P. G., Sousa, J. P., Marques, J. C., & Raffaelli, D. (2004). Assessing environmental quality: A novel approach. *Marine Ecology Progress Series, 267*, 1–8.

Paul, J. F. (2003). Developing and applying an index of environmental integrity for the US Mid-Atlantic region. *Journal of Environmental Management, 67*, 175–185.

Paul, J. F., Scott, K. J., Campbell, D. E., Gentile, J. H., Strobel, C. S., Valente, R. M., Weisberg, S. B., Holland, A. F., & Ranasinghe, J. A. (2001). Developing and applying a benthic index of estuarine condition for the Virginian biogeographic province. *Ecological Indicators, 1*, 83–99.

Pearson, T., & Rosenberg, R. (1978a). Macrobenthic succession in relation to organic enrichment and pollution of the marine environment. *Oceanography and marine biology—An annual review, 16*, 229–311.

Pearson, T. H., & Rosenberg, R. (1978b). Macrobenthic succession in relation to organic enrichment and pollution of the marine environment. In H. Barnes (Ed.), *Oceanographic marine biology annual review* (pp. 229–311). Aberdeen University Press.

Perus, J., Bonsdorff, E., Bäck, S., Lax, H. G., Villnäs, A., & Westberg, V. (2007). Zoobenthos as indicators of ecological status in coastal brackish waters: A comparative study from the Baltic Sea. *Ambio, 36*, 250–256.

Pielou, E. C. (1966). Species diversity and pattern diversity in the study of ecological succession. *Journal of Theoretical Biology, 10*, 372–383.

Pinedo, S., & Jordana, E. (2007). Spain (Catalonia and Balearic Islands). In A. Carletti, & A.-S. Heiskanen (Eds.), *Water Framework Directive Intercalibration Technical Report Part 3: Coastal and Transitional Waters. JRC, IES* (pp. 62–70). OPOCE.

Pinto, R., Patricio, J., Baeta, A., Fath, B. D., Neto, J. M., & Marques, J. C. (2009). Review and evaluation of estuarine biotic indices to assess benthic condition. *Ecological Indicators, 9*, 1–25.

Prior, A., Miles, A. C., Sparrow, A. J., & Price, N. (2004). Development of a classification scheme for the marine benthic invertebrate component, Water Framework Directive, Phase I & II—Transitional and coastal waters. Environment Agency (UK), R&D Interim Technical Report, E1-116, E1-132, 103 pp. (+appendix).

Qasim, S. Z., Gupta, R. S., & Kureishy, T. W. (1988). Pollution of the seas around India. *Proceedings: Animal Sciences, 97*, 117–131.

Rakocinski, C. F., & Zapfe, G. A. (2005). Macrobenthic process indicators of estuarine condition. In S. A. Bortone (Ed.), *Estuarine indicators. CRC marine science series* (pp. 316–331). CRC Press.

Ranasinghe, J. A., Weisberg, S. B., Dauer, D. M., Schaffner, L. C., Diaz, R. J., & Frithsen, J. B. (1994). Chesapeake Bay benthic community restoration goals. CBP/TRS 107/94, Chesapeake Bay Program Office. USEPA, Annapolis, MD, USA, p. 49.

Rand, K., Logerwell, E., Bluhm, B., Chenelot, H., Danielson, S., Iken, K., & Sousa, L. (2018). Using biological traits and environmental variables to characterize two Arctic epibenthic invertebrate communities in and adjacent to Barrow Canyon Deep. *Deep Sea Research Part II Topical Studies in Oceanography, 152*, 154–169.

Rehitha, T. V., Ullas, N., Vineetha, G., Benny, P. Y., Madhu, N. V., & Revichandran, C. (2017). Impact of maintenance dredging on macrobenthic community structure of a tropical estuary. *Ocean & Coastal Management, 144*, 71–82.

Reiss, H., & Kröncke, I. (2005). Seasonal variability of benthic indices: An approach to test the applicability of different indices for ecosystem quality assessment. *Marine Pollution Bulletin, 50*, 1490–1499.

Rhoads, D. C., & Germano, J. D. (1986). Interpreting long-term changes in benthic community structure: A new protocol. *Hydrobiologia, 142*, 291–308.

Roberts, D., Gregory, R., & Foster, A. (1998). Developing an efficient macrofauna monitoring index from an impact study a dredge spoil example. *Marine Pollution Bulletin, 36*, 231–235.

Rosenberg, R., Blomqvist, M., Nilsson, H. C., Cederwall, H., & Dimming, A. (2004). Marine quality assessment by use of benthic species-abundance distributions: A proposed new protocol within the European Union Water Framework Directive. *Marine Pollution Bulletin, 49*, 728–739.

Rygg, B. (1985). Distribution of species along pollution-induced diversity gradients in benthic communities in Norwegian fjords. *Marine Pollution Bulletin, 16*, 469–474.

Rygg, B. (2002). *Indicator species index for assessing benthic ecological quality in marine waters of Norway* (pp. 1–32). Norwegian Institute for Water Research.

Satsmadjis, J. (1982). Analysis of benthic data and measurement of pollution. *Revue internationale d'Océanographie Medicale, 66–67*, 103–107.

Selvaraj, P., Murugesan, P., Punniyamoorthy, R., Parthasarathy, P., & Marigoudar, S. R. (2019). Assessment of the ecological health of Vellar and Ennore estuarine ecosystems using health indices. *Indian Journal of Geo-Marine Sciences, 48*, 1580–1592.

Shannon, C. E., & Weaver, W. (1949). *The mathematical theory of communication* (p. 115) The University of Illinois Press.

Sigamani, S., Perumal, M., Arumugam, S., Preetha Mini Jose, H. M., & Veeraiyan, B. (2015). AMBI indices and multivariate approach to assess the ecological health of

Vellar–Coleroon estuarine system undergoing various human activities. *Marine Pollution Bulletin, 100,* 334–343.

Sigamani, S., Samikannu, M., & Alagiri, T. G. (2019). Assessment of effluent stressed ecosystem of Cuddalore Coastal Waters@ a bio-indicator approach. *Thalassas: An International Journal of Marine Sciences, 35,* 437–449.

Simboura, N., Tsapakis, M., Pavlidou, A., Assimakopoulou, G., Pagou, K., Kontoyannis, H., Zeri, C., Krasakopoulou, E., Rousselaki, E., Katsiaras, N., Diliberto, S., Naletaki, M., Tsiamis, K., Gerakaris, V., Drakopoulou, P., & Panayotidis, P. (2015). Assessment of the environmental status in Hellenic coastal waters (Eastern Mediterranean): From the Water Framework Directive to the Marine Strategy Framework Directive. *Mediterranean Marine Science, 16,* 46–64.

Simboura, N., & Zenetos, A. (2002). Benthic indicators to use in ecological quality classification of Mediterranean soft bottom marine ecosystems, including a new biotic index. *Mediterranean Marine Science, 3,* 35.

Simpson, E. H. (1949). *Measurement of diversity. Nature* (163, p. 688).

Sivadas, S., Ingole, B. S., & Nanajkar, M. (2010). Benthic polychaetes as good indicators of anthropogenic impact. *Indian Journal of Marine Sciences, 39,* 201–211.

Sivadas, S. K., Nagesh, R., Gupta, G. V. M., Gaonkar, U., Mukherjee, I., Ramteke, D., & Ingole, B. S. (2016). Testing the efficiency of temperate benthic biotic indices in assessing the ecological status of a tropical ecosystem. *Marine Pollution Bulletin, 106,* 62–76.

Sivaraj, S., Murugesan, P., Muthuvelu, S., Vivekanandan, K. E., & Vijayalakshmi, S. (2014). AMBI and M-AMBI indices as a robust tool for assessing the effluent stressed ecosystem in Nandgaon Coastal waters, Maharashtra, India. *Estuarine, Coastal and Shelf Science, 146,* 60–67.

Smith, R. W., Bergen, M., Weisberg, S. B., Cadien, D., Dalkey, A., Montagne, D., Stull, J. K., & Velarde, R. G. (2001). Benthic response index for assessing infaunal communities on the southern California mainland shelf. *Ecological Applications, 11,* 1073–1087.

Srinivas, T., Sukumaran, S., Mulik, J., & Dias, H. Q. (2019). Community structure of benthic amphipods in four estuaries of northwest India. *Regional Studies in Marine Science, 27,* 100532.

Teixeira, H., Neto, J. M., Patrıcio, J., Verıssimo, H., Pinto, R., Salas, F., & Marques, J. C. (2009). Quality assessment of benthic macroinvertebrates under the scope of WFD using BAT, the benthic assessment tool. *Marine Pollution Bulletin, 58,* 1477–1486.

Van Hoey, G., Drent, J., Ysebaert, T., & Herman, P. (2007). The benthic ecosystem quality index (BEQI), intercalibration and assessment of Dutch coastal and transitional waters for the Water Framework Directive—Final report.

van Loon, W. M. G. M., Boon, A. R., Gittenberger, A., Walvoort, D. J. J., Lavaleye, M., Duineveld, G. C. A., & Verschoor, A. J. (2015). Application of the benthic ecosystem quality index 2 to benthos in Dutch transitional and coastal waters. *Journal of Sea Research, 103,* 1–13.

Vijapure, T., & Sukumaran, S. (2019). Optimization of the taxonomic resolution of an indicator taxon for cost-effective ecological monitoring: Perspectives from a heterogeneous tropical coastline. *Journal of Environmental Management, 247,* 474–483.

Violle, C., Reich, P. B., Pacala, S. W., Enquist, B. J., & Kattge, J. (2014). The emergence and promise of functional biogeography. *Proceedings of the National Academy of Sciences of the United States of America, 111,* 13690–13696.

Vitousek, P. M., Mooney, H. A., Lubchenco, J., & Melillo, J. M. (1997). Human domination of Earth's ecosystems. *Science (New York, N.Y.), 27,* 494–499.

Warwick, R., & Clarke, K. R. (1994). Relating the ABC: Taxonomic changes and abundance/biomass relationship in disturbed benthic communities. *Marine Biology, 118*, 739–744.

Warwick, R. M. (1986). A new method for detecting pollution effects on marine macrobenthic communities. *Marine Biology, 92*, 557–562.

Warwick, R. M., & Clarke, K. R. (1995). New "biodiversity" measures reveal a decrease in taxonomic distinctness with increasing stress. *Marine Ecology Progress Series, 129*, 301–305.

Warwick, R. M., Robert Clarke, K., & Somerfield, P. J. (2010). Exploring the marine biotic index (AMBI): Variations on a theme by Angel Borja. *Marine Pollution Bulletin, 60*, 554–559.

Weisberg, S. B., Frithsen, J. B., Holland, A. F., Paul, J. F., Scott, K. J., Summers, J. K., Wilson, H. T., Heimbuch, D. G., Gerritsen, J., Schimmel, S. C., & Latimer, R. W. (1993). *Virginian Province demonstration project report, EMAP-Estuaries, 1990. EPA/ 620/ R-93/006, Office of Research and Development*. USEPA.

Weisberg, S. B., Ranasinghe, J. A., Dauer, D. M., Schaffner, L. C., Diaz, R. J., & Frithsen, J. B. (1997). An estuarine benthic index of biotic inegrity (I-IBI) for Chesapeake Bay. *Estuaries, 20*, 149–158.

Word, J. Q. (1979). *The infaunal trophic index. California coastal water research project annual report* (pp. 19–39). El Segundo.

Zhang, Y., Lv, Z., Guan, B., Liu, Y., Li, F., Li, S., Ma, Y., Yu, J., & Li, Y. (2013). Status of macrobenthic community and its relationships to trace metals and natural sediment characteristics. *CLEAN—Soil, Air, Water, 41*, 1027–1034.

Index

Note: Page numbers followed by "*f*" and "*t*" refer to figures and tables, respectively.

C

D

E

N

O

P

U

V

W

Made in the USA
Monee, IL
13 September 2026

63138398R20214